Computational Methods and Mathematical Modeling in Cyberphysics and Engineering Applications 2

SCIENCES

Mathematics, Field Director – Nikolaos Limnios

Mathematics in Engineering, Subject Head – Dmitri Koroliouk

Computational Methods and Mathematical Modeling in Cyberphysics and Engineering Applications 2

Coordinated by

Dmitri Koroliouk

Sergiy Lyashko

Nikolaos Limnios

iSTE

WILEY

First published 2026 in Great Britain and the United States by ISTE Ltd and John Wiley & Sons, Inc.

ISTE Ltd
27-37 St George's Road
London SW19 4EU
UK

www.iste.co.uk

John Wiley & Sons, Inc.
111 River Street
Hoboken, NJ 07030
USA

www.wiley.com

Library of Congress Control Number: 2026934957

British Library Cataloguing-in-Publication Data
A CIP record for this book is available from the British Library
ISBN 978-1-78945-242-6

ERC code:
PE1 Mathematics
PE1_21 Application of mathematics in sciences
PE1_22 Application of mathematics in industry and society
PE6 Computer Science and Informatics
PE6_2 Distributed systems, parallel computing, sensor networks, cyber-physical systems

The manufacturer's authorized representative according to the EU General Product Safety Regulation is Wiley-VCH GmbH, Boschstr. 12, 69469 Weinheim, Germany, e-mail: Product_Safety@wiley.com.

Contents

Chapter 1. Solution of Differential Equations Systems that Arise During the Analysis of Complex Multicomponent Environments 1
V. Bohaienko, O. Marchenko and T. Samoilenko

1.1. Construction of an approximate solution of the axisymmetric parabolic problem. 1
1.2. The analysis of numerical modeling of soil mass dynamics in the presence of unsteady pressure filtration 14
1.3. The analysis of numerical simulation of non-isothermal processes in soil. 26
1.4. The study of soil massif state in the foundations of hydraulic structures. 34
1.5. Parallel algorithm for AMLI preconditioner and its application to model soil massif state. 46
1.6. References 68

Chapter 2. Computer Simulation of Transdermal Drug Delivery Using Soluble Microneedles 71
D.A. Klyushin, S. Lyashko, V.V. Onotskyi and O.S. Bondar

2.1. Introduction 71
2.2. Mathematical model. 72
2.3. Mathematical methods 78
2.4. Numerical experiments 80
2.5. Results and discussion. 81
2.6. Conclusion 91
2.7. References 91

Chapter 3. Homogenization and Modeling of Processes in Composites Similar to Photonic Crystals 95
G.V. SANDRAKOV

3.1. Introduction 95
3.2. Composite media with periodic structures and contrast properties 98
3.3. Regular homogenized asymptotic expansions of solutions. 100
3.4. Singular homogenized asymptotic expansions of solutions 107
3.5. Computational aspects of modeling by homogenization 117
3.6. Spectral aspects of modeling by homogenization. 119
3.7. Conclusion 129
3.8. References 129

Chapter 4. Polynomial Operator Interpolation and its Applications 133
V.L. MAKAROV and O.F. KASHPUR

4.1. Introduction 133
4.2. Formulation of Lagrange's operator interpolation problem 135
4.3. Solution of the Lagrange's operator interpolation problem 136
4.4. Solution operator equations by the interpolation method. 139
4.5. Interpolation in Euclidean spaces 143
4.6. Construction of surfaces. 146
4.6.1. Example 146
4.6.2. Example 147
4.6.3. Example 150
4.7. Conclusion 151
4.8. References 151

Chapter 5. New Fractional Differential Analogues of the Biparabolic Evolution Equation and Some Boundary Value Problems. 155
V.M. BULAVATSKY and S. LYASHKO

5.1. Introduction 155
5.2. Some boundary value problems for the fractional–differential analogue of the biparabolic equation with non-locality in time and space 159
5.3. Generalization of the model equation based on Hilfer-type fractional derivatives 169
5.4. Fractional–differential analogue of the biparabolic evolution equation with Caputo and Caputo–Fabrizio derivatives. 173
5.5. Conclusions 178
5.6. References 178

Chapter 6. Optimal Control for Integro-differential Systems of Hyperbolic Type 181
A.V. ANIKUSHYN, Kh.M. HRANISHAK, V.S. LYASHKO and O.S. BONDAR

6.1. Introduction 181
6.2. Main notations and spaces 189
6.3. Generalized control problem 191
6.4. A priori inequalities for the differential part of the operator 197
6.5. A priori inequalities for the integro-differential operator. 205
6.6. Example of an optimal control problem 212
6.7. Conclusion 219
6.8. References 219

Chapter 7. Self-adaptive Operator Extrapolation Method for Operator Inclusions in Banach Space 223
V. SEMENOV and S. DENYSOV

7.1. Introduction 223
7.2. Preliminaries 227
7.3. Algorithm. 231
7.4. Convergence 233
7.5. Variants. 239
7.6. Application to variational inequalities 241
7.7. Conclusion 243
7.8. Acknowledgments 243
7.9. References 243

Chapter 8. Forecasting Algorithms Based on Intellectual Analysis of Polynomial Extrapolation and Divided Differences 247
Y. TURBAL, M. TURBAL and A. BOMBA

8.1. Introduction – Problem of the time series forecasting 247
8.2. Method of finding the predictive value based on a polynomial of any degree without finding the coefficients of the polynomial 249
8.3. The optimal polynomial extrapolation problem. 254
8.4. Condition of forecast efficiency based on the arithmetic mean of polynomial forecasts 260
8.5. Improved algorithm for optimal polynomial forecasting. 262
8.6. Numerical results of the polynomial forecast 266
8.7. Pyramidal method of extrapolation. 271
8.8. Numerical results for the pyramidal methods 283
8.9. Conclusions 285
8.10. References 286

Chapter 9. Transformer with BPE Tokenization for Analysis of Interactions of Chemical Substances and Proteins 289
M. ZOZIUK, P. KRYSENKO, S. DOVGIY, V. MAKAROV, Y. YAKIMENKO and D. KOROLIOUK

9.1. Introduction 290
9.2. Methods and data 291
9.3. The model's architecture 293
9.4. The model's training. 295
9.5. Conclusion 298
9.6. References 298

List of Authors 301

Index. 305

1

Solution of Differential Equations Systems that Arise During the Analysis of Complex Multicomponent Environments

V. BOHAIENKO[1], O. MARCHENKO[1] and T. SAMOILENKO[2]

[1] *V.M. Glushkov Institute of Cybernetics of the NAS of Ukraine, Kyiv, Ukraine*

[2] *National Technical University of Ukraine "Igor Sikorsky Kyiv Polytechnic Institute", Kyiv, Ukraine*

1.1. Construction of an approximate solution of the axisymmetric parabolic problem

In this section, we propose an algorithm for calculating the dynamics of the non-isothermal process of moisture transfer in an axisymmetric setting, which is essential when studying the state of soil moisture around, for example, vertical drains, wells, piles, etc. We formulate the initial-boundary value problem for a system of non-stationary moisture and heat transfer equations for an isotropic environment in a cylindrical coordinate system with heterogeneous mixed boundary conditions, including the conditions of heat exchange with the external environment (Marchenko and Samoilenko 2020).

The obtained results are also important for further research in cylindrical coordinates of problems simulating the migration of moisture in the process of seasonal freezing of the soil, as a continuation of the approach presented in Pavlov and Permyakov (1983), Marchenko and Lezhnina (2002) and Marchenko and

Computational Methods and Mathematical Modeling in Cyberphysics and Engineering Applications 2,
coordinated by Dmitri KOROLIOUK, Sergiy LYASHKO and Nikolaos LIMNIOS.

Samoilenko (2012), in which phase transitions from unfrozen water to ice are taken into account in the entire volume of soil massif without highlighting the crystallization front. In this case, moisture exchange and heat exchange characteristics are functions of the total moisture content, and the moisture transfer equation is written relative to the "fictitious" moisture content. Taking into account the directions of moisture migration relative to the freezing/thawing front, we consider the convective heat transfer along the vertical axis to be essential, which allows us to achieve sufficient agreement with experimental data (Kislitsyn et al. 2018).

The system of equations has the following form:

$$r\frac{\partial W}{\partial t}=\frac{\partial}{\partial r}\left(rk(W,T)\frac{\partial W}{\partial r}\right)+r\frac{\partial}{\partial z}\left(k(W,T)\frac{\partial W}{\partial z}\right),$$

$$c_T r\frac{\partial T}{\partial t}=\frac{\partial}{\partial r}\left(r\tilde{\lambda}\frac{\partial T}{\partial r}\right)+r\frac{\partial}{\partial z}\left(\tilde{\lambda}\frac{\partial T}{\partial z}-c_W v(W)T\right), \quad [1.1]$$

where $(r,z,t)\in\Omega_{\tilde{T}}$, $\Omega_{\tilde{T}}=\Omega\times(0,\tilde{T}]$, $0<r_0\le r\le r_1<\infty$, $0\le z\le z_1<\infty$, $W(r,z,t)$ is the volumetric moisture content, $T(r,z,t)$ is the temperature, $k(W,T)$ is the hydraulic conductivity, $\tilde{\lambda}$ is the thermal conductivity coefficient, c_T, c_W are the volume heat capacities of soil and liquid in pores and $v(W)$ is the moisture transfer-filtration rate along the z axis.

Let us set the boundary conditions in the form

$$k(W,T)\frac{\partial W}{\partial r}=0,\ \tilde{\lambda}\frac{\partial T}{\partial r}=0,\ (r,z,t)\in(\Gamma_1\cup\Gamma_3)\times(0,\tilde{T}], \quad [1.2]$$

$$k(W,T)\frac{\partial W}{\partial z}=0,\ (r,z,t)\in(\Gamma_2\cup\Gamma_4)\times(0,\tilde{T}], \quad [1.3]$$

$$\tilde{\lambda}\frac{\partial T}{\partial z}=-\alpha(T(r,z,t)-T_a),\ (r,z,t)\in\Gamma_2\times(0,\tilde{T}], \quad [1.4]$$

$$\tilde{\lambda}\frac{\partial T}{\partial z}=0,\ (r,z,t)\in\Gamma_4\times(0,\tilde{T}], \quad [1.5]$$

$$\Gamma_1\cup\Gamma_2\cup\Gamma_3\cup\Gamma_4=\partial\Omega,\ \Gamma_1=\{(r,z):\ r=r_0,\ 0\le z\le z_1\},$$

$$\Gamma_2=\{(r,z):\ r_0\le r\le r_1,\ z=z_1\},\ \Gamma_3=\{(r,z): r=r_1, 0\le z\le z_1\},$$

$$\Gamma_4=\{(r,z):\ r_0\le r\le r_1,\ z=0\};$$

T_a is the ambient temperature.

The initial conditions are as follows:

$$W(r,z,0) = W_0(r,z), T(r,z,0) = T_0(r,z), (r,z) \in \bar{\Omega}. \qquad [1.6]$$

We assume that $\alpha = const > 0$ and the given functions $k(W,T)$, $\upsilon(W)$, $W_0(r,z)$, $T_0(r,z)$ are sufficiently smooth on $\Omega_{\tilde{T}}$, are bounded and satisfy the Lipschitz condition, i.e.:

$$0 < \bar{k} \le k(v_1,u_1) \le \bar{\bar{k}} < \infty, |k(v_1,u_1) - k(v_2,u_2)| \le K_1(|v_1 - v_2| + |u_1 - u_2|),$$

$$0 < \bar{\upsilon} \le \upsilon(v_1) \le \bar{\bar{\upsilon}} < \infty, |\upsilon(v_1) - \upsilon(v_2)| \le K_2|v_1 - v_2| \qquad [1.7]$$

$$\forall v_i, u_j \in R, i,j = 1,2; \bar{k},\ \bar{\bar{k}},\ \bar{\upsilon},\ \bar{\bar{\upsilon}},\ K_1,\ K_2 = const.$$

The classical solution of the initial-boundary value problem [1.1]–[1.6] is a vector function $h(r,z,t) = (W(r,z,t), T(r,z,t))^T = (h_1(r,z,t), h_2(r,z,t))^T$, whose components together with their partial derivatives $\frac{\partial h_i}{\partial r}, \frac{\partial h_i}{\partial z}$, $i = 1,2$, are continuous in the domain $\bar{\Omega}_{\tilde{T}}$, have bounded continuous partial derivatives $\frac{\partial h_i}{\partial t}, \frac{\partial^2 h_i}{\partial r^2}, \frac{\partial^2 h_i}{\partial z^2}$, $i = 1,2$ in $\Omega_{\tilde{T}}$, satisfy the equation [1.1] and the conditions [1.2]–[1.6].

We denote by Z the set of vector functions $h(r,z,t) = (W(r,z,t), T(r,z,t))^T = (h_1(r,z,t), h_2(r,z,t))^T$, whose components $\forall t \in (0,\tilde{T}]$ belong to the Sobolev space $W_2^1(\Omega)$ and $\frac{\partial h_i}{\partial t}(r,z,t)$ $\forall t \in (0,\tilde{T}]$, $h_i(r,z,0), i = 1,2$, belong to $L_2(\Omega)$. Accordingly, we denote by Z_0 the set of vector functions $v(r,z) = (v_1(r,z), v_2(r,z))^T$, whose components belong to the space $W_2^1(\Omega)$.

The generalized Galerkin solution of the initial-boundary value problem for the system [1.1]–[1.6] is the vector function $h(r,z,t) \in Z$, which, for an arbitrary vector function $v(r,z) \in Z_0$, satisfies the integral relations

$$m\left(\frac{\partial h}{\partial t}, v\right) + A(h;h,v) = (F,v)\ \forall t \in 0,\tilde{T}\ \forall v \in Z_0, \qquad [1.8]$$

$$(h(\cdot,\cdot,0), v) = (h_0, v)\ \forall v \in Z_0 \qquad [1.9]$$

where

$$m\left(\frac{\partial h}{\partial t}, v\right) = \iint_\Omega r\left(\frac{\partial W}{\partial t}v_1 + c_T\frac{\partial T}{\partial t}v_2\right)d\Omega,$$

$$A(h;h,v) = a(h;h,v) - b_1(h;h,v) + b_2(h,v),$$

$$a(h;h,v)=\iint_{\Omega} r\left(k(W,T)\left(\frac{\partial W}{\partial r}\frac{\partial v_1}{\partial r}+\frac{\partial W}{\partial z}\frac{\partial v_1}{\partial z}\right)+\tilde{\lambda}\left(\frac{\partial T}{\partial r}\frac{\partial v_2}{\partial r}+\frac{\partial T}{\partial z}\frac{\partial v_2}{\partial z}\right)\right)d\Omega,$$

$$b_1(h;h,v)=\iint_{\Omega} r c_w v(W) T\frac{\partial v_2}{\partial z}d\Omega,$$

$$b_2(h,v)=\int_{\Gamma_2} r\alpha T\ v_2 d\Gamma_2,\ (F,v)=\int_{\Gamma_2} r\alpha T_a v_2 d\Gamma_2, \qquad [1.10]$$

(,) is the scalar product in $L_2(\Omega)$, $h_0=(W_0,T_0)^T$.

We will search the approximate generalized solution of this Cauchy problem in a finite-dimensional subspace $Z^N\subset Z$ using the finite element method (FEM) in the following form:

$$h^N(r,z,t)=\sum_{i=1}^{2N}\alpha_i(t)\Phi_i(r,z) \qquad [1.11]$$

where $\alpha_i(t)$, $i=\overline{1,2N}$ are the functions integrable together with their second derivatives on $[0,\tilde{T}]$

$$\Phi_i=\begin{pmatrix}\varphi_i\\0\end{pmatrix},\Phi_{N+i}=\begin{pmatrix}0\\\varphi_i\end{pmatrix}, i=\overline{1,N} \qquad [1.12]$$

is the basis of the space Z_t^N obtained from Z^N by fixing $\forall t\in[0,\tilde{T}]$; $\{\varphi_i(r,z)\}$, $i=\overline{1,N}$ is the set of linearly independent functions corresponding to the nodal points of the FEM, which are built on complete polynomials of degree k ($k=1,2,3$) and have a limited support in $\bar{\Omega}$.

The basis of the space $Z_0^N\subset Z_0$ similarly consists of $2N$ vector functions $\Phi_i(r,z)$ so that an arbitrary function $v^N(r,z)\in Z_0^N$ can be presented in the form:

$$v^N(r,z)=\sum_{i=1}^{2N}\beta_i\Phi_i(r,z) \qquad [1.13]$$

where β_i are constants.

The approximate generalized solution $h^N(r,z,t)\in Z^N$ of the problem [1.8], [1.9] satisfies the integral relations

$$m\left(\frac{\partial h^N}{\partial t},v^N\right)+A\left(h^N;h^N,v^N\right)=(F,v^N)\ \forall t\in 0,\tilde{T}\ \forall v^N(r,z)\in Z_0^N, \qquad [1.14]$$

$$\left(h^N(\cdot,\cdot,0),v^N\right)=(h_0,v^N)\ \forall v^N(r,z)\in Z_0^N. \qquad [1.15]$$

Let us use the following notations:

$$\|u\|^2_{L_2\times L_2} = \int_0^{\tilde{T}} \|u\|^2_{L_2} dt,\ \|u\|^2_{H_0^1\times L_2} = \int_0^{\tilde{T}} \|u\|^2_{H_0^1} dt,\ \|u\|^2_{W_2^1\times L_2} = \int_0^{\tilde{T}} \|u\|^2_{W_2^1} dt,$$

$$\|u\|^2_{L_2\times L_\infty} = \max_{t\in[0,\tilde{T}]} \|u\|^2_{L_2},\ \|u\|^2_{L_2(\Gamma)\times L_2} = \int_0^{\tilde{T}} \|u\|^2_{L_2(\Gamma)} dt,\ \forall u = (u_1, u_2) \in Z. \quad [1.16]$$

The following theorem is true.

THEOREM 1.1.– *Let the classical solution of the initial-boundary value problem [1.1]–[1.6] $h(r,z,t) = (W(r,z,t), T(r,z,t))^T = (h_1(r,z,t), h_2(r,z,t))^T$ has bounded continuous partial derivatives $\frac{\partial^{k+2} h_i}{\partial r^\kappa \partial z^\upsilon \partial t}$ $(\kappa + \upsilon = k + 1)$ on $\bar{\Omega}_{\tilde{T}}$ $(i = 1,2)$.*

Then, for the approximate generalized solution $h^N(r,z,t) = (h_1^N(r,z,t), h_2^N(r,z,t)) \in Z^N$ of the problem [1.8], [1.9], there exists a constant $C > 0$ such that the following estimate holds:

$$\|h - h^N\|^2_{W_2^1\times L_2} + \|h_2 - h_2^N\|^2_{L_2(\Gamma_2)\times L_2} \le C\bar{h}^{2k},$$

where $\bar{h}$ is the maximum length of the sides of the triangles, k =1,2,3 is the degree polynomials of FEM.

PROOF.– Let h and h^N be the classical and the approximate generalized solution of the problem [1.1]–[1.6]; then, $\forall \tilde{h} \in Z^N$

$$m\left(\frac{\partial(h - h^N)}{\partial t}, \tilde{h} - h^N\right) + A\left(h; h, \tilde{h} - h^N\right) - A\left(h^N; h^N, \tilde{h} - h^N\right) = 0. \quad [1.17]$$

Using the relation [1.17], consider the expression (Deineka et al. 1995)

$$m\left(\frac{\partial(h - h^N)}{\partial t}, h - h^N\right) + A\left(h^N; h - h^N, h - h^N\right) =$$

$$= m\left(\frac{\partial(h - h^N)}{\partial t}, \tilde{h} - h^N\right) + A\left(h; h, \tilde{h} - h^N\right) - A\left(h^N; h^N, \tilde{h} - h^N\right) +$$

$$+m\left(\frac{\partial(h - h^N)}{\partial t}, h - \tilde{h}\right) - A\left(h; h, \tilde{h} - h^N\right) + A\left(h^N; h, h - h^N\right)$$
$$- A\left(h^N; h^N, h - \tilde{h}\right) =$$

$$= m\left(\frac{\partial\left(h-h^N\right)}{\partial t}, h-\tilde{h}\right) + A\left(h^N; h-h^N, h-\tilde{h}\right) +$$

$$+A\left(h^N; h, \tilde{h}-h^N\right) - A\left(h; h, \tilde{h}-h^N\right). \quad [1.18]$$

Let us estimate the left and right parts of the relation [1.18] taking into account [1.10], [1.7] as

$$m\left(\frac{\partial(h-h^N)}{\partial t}, h-h^N\right) \geq c_1 \frac{d}{dt}\|h-h^N\|_{L_2}^2, a(h^N; h-h^N, h-h^N) \geq c_2\|h-h^N\|_{H_0^1}^2,$$

$$b_2(h-h^N, h-h^N) \geq r_0\alpha\|h_2-h_2^N\|_{L_2(\Gamma_2)}^2,$$

$$|b_1(h^N; h-h^N, h-h^N)| \leq c_3\|h-h^N\|_{L_2}\|h-h^N\|_{H_0^1} \leq$$

$$\leq c_3\left(\varepsilon_1\|h-h^N\|_{L_2}^2 + \frac{1}{4\varepsilon_1}\|h-h^N\|_{H_0^1}^2\right),$$

$$\left|A\left(h^N; h-h^N, h-\tilde{h}\right)\right| \leq c_4\|h-h^N\|_{H_0^1}\left\|h-\tilde{h}\right\|_{H_0^1} + c_3\|h-h^N\|_{L_2}\left\|h-\tilde{h}\right\|_{H_0^1} +$$

$$+r_1\alpha\|h_2-h_2^N\|_{L_2(\Gamma_2)}\left\|h_2-\tilde{h}_2\right\|_{L_2(\Gamma_2)} \leq c_4\left(\varepsilon_2\|h-h^N\|_{H_0^1}^2 + \frac{1}{4\varepsilon_2}\left\|h-\tilde{h}\right\|_{H_0^1}^2\right) +$$

$$+c_3\left(\varepsilon_3\|h-h^N\|_{L_2}^2 + \frac{1}{4\varepsilon_3}\left\|h-\tilde{h}\right\|_{H_0^1}^2\right) +$$

$$+r_1\alpha\left(\varepsilon_4\|h_2-h_2^N\|_{L_2(\Gamma_2)}^2 + \frac{1}{4\varepsilon_4}\left\|h_2-\tilde{h}_2\right\|_{L_2(\Gamma_2)}^2\right),$$

$$\left|A\left(h^N; h, \tilde{h}-h^N\right) - A\left(h; h, \tilde{h}-h^N\right)\right| \leq c_5\|h-h^N\|_{L_2}\left\|\tilde{h}-h^N\right\|_{H_0^1} +$$

$$+c_6\|h-h^N\|_{L_2}\left\|\tilde{h}-h^N\right\|_{H_0^1} \leq (c_5+c_6)\left(\|h-h^N\|_{L_2}\left\|h-\tilde{h}\right\|_{H_0^1} +\right.$$

$$\left.+\|h-h^N\|_{L_2}\|h-h^N\|_{H_0^1}\right) \leq (c_5+c_6)\left((\varepsilon_5+\varepsilon_6)\|h-h^N\|_{L_2}^2 +\right.$$

$$\left.+\frac{1}{4\varepsilon_6}\|h-h^N\|_{H_0^1}^2 + \frac{1}{4\varepsilon_5}\left\|h-\tilde{h}\right\|_{H_0^1}^2\right) \quad [1.19]$$

where ε_i, $i = \overline{1,6}$ are some positive constants of ε-inequalities,

$c_1 = 0{,}5r_0\ \min\{1, c_T\}$, $c_2 = r_0 \min\{\tilde{\lambda},\ \bar{k}\}$, $c_3 = r_1 c_w\ \bar{\bar{v}}$, $c_4 = r_1\ \max\{\tilde{\lambda},\ \bar{\bar{k}}\}$,

$$c_5 = r_1\ \tilde{c}\ max\{\tilde{\lambda},\ K_1\}, c_6 = r_1 c_w\ \tilde{c}\ K_2, \tilde{c} = \max_{\bar{\Omega}_{\tilde{T}}}\left\{|h_i|, \left|\frac{\partial h_i}{\partial r}\right|, \left|\frac{\partial h_i}{\partial z}\right|, i = 1{,}2\right\}. \quad [1.20]$$

From [1.18], [1.19], we obtain the inequality in the form

$$\frac{d}{dt}\left\|h - h^N\right\|_{L_2}^2 + \delta_1\left\|h - h^N\right\|_{H_0^1}^2 + \delta_2\left\|h_2 - h_2^N\right\|_{L_2(\Gamma_2)}^2 \leq$$

$$\leq c\left(\left\|h - h^N\right\|_{L_2}^2 + \left\|h - \tilde{h}\right\|_{H_0^1}^2 + \left\|h_2 - \tilde{h}_2\right\|_{L_2(\Gamma_2)}^2\right) + \frac{1}{c_1} m\left(\frac{\partial(h-h^N)}{\partial t}, h - \tilde{h}\right) \quad [1.21]$$

where constants $\delta_1 = \delta_1(c_j, \varepsilon_i)$, $\delta_2 = \frac{\alpha(r_0 - r_1\varepsilon_4)}{c_1}$, $c = c(c_j, \varepsilon_i, \alpha, r_1)$ are positive due to appropriately selected ε_i, $i = \overline{1{,}6}$.

Multiplying both parts of the inequality [1.21] by e^{-ct} and integrating the result over the variable from 0 to s, $s \in (0, \tilde{T}]$, we obtain the inequality

$$\int_0^s \left(e^{-ct}\frac{d}{dt}\left\|h - h^N\right\|_{L_2}^2 - ce^{-ct}\left\|h - h^N\right\|_{L_2}^2\right) dt + \delta_1 \int_0^s e^{-ct}\left\|h - h^N\right\|_{H_0^1}^2 dt +$$

$$+\delta_2 \int_0^s e^{-ct}\left\|h_2 - h_2^N\right\|_{L_2(\Gamma_2)}^2 dt \leq c\left(\int_0^s e^{-ct}\left\|h - \tilde{h}\right\|_{H_0^1}^2 dt +\right.$$

$$\left. + \int_0^s e^{-ct}\left\|h_2 - \tilde{h}_2\right\|_{L_2(\Gamma_2)}^2 dt\right) + \frac{1}{c_1}\int_0^s e^{-ct} m\left(\frac{\partial(h-h^N)}{\partial t}, h - \tilde{h}\right) dt \quad [1.22]$$

Let us obtain an upper estimate to the last term of the relation [1.22]. Integrating by parts and taking into account the Cauchy–Buniakovsky inequality, ε-inequality and notation [1.10], we have

$$\left|\int_0^s e^{-ct} m\left(\frac{\partial(h - h^N)}{\partial t}, h - \tilde{h}\right) dt\right| \leq \left|\int_0^s \frac{\partial}{\partial t}\left(e^{-ct} m(h - h^N, h - \tilde{h})\right) dt\right| +$$

$$+\left|\int_0^s e^{-ct} m\left(h - h^N, \frac{\partial(h - \tilde{h})}{\partial t}\right) dt\right| + c\left|\int_0^s e^{-ct} m(h - h^N, h - \tilde{h})\, dt\right| \leq$$

$$\leq \tilde{c}_2\varepsilon_1\left\|h - h^N\right\|_{L_2}^2(s) + \frac{\tilde{c}_2}{4\varepsilon_1}\left\|h - \tilde{h}\right\|_{L_2}^2(s) + \tilde{c}_2\varepsilon_2\left\|h - h^N\right\|_{L_2}^2(0) +$$

$$+\frac{\tilde{c}_2}{4\varepsilon_2}\left\|h - \tilde{h}\right\|_{L_2}^2(0) + \tilde{c}_2\varepsilon_3\int_0^s \left\|h - h^N\right\|_{L_2}^2 dt + \frac{\tilde{c}_2}{4\varepsilon_3}\int_0^s \left\|\frac{\partial(h - \tilde{h})}{\partial t}\right\|_{L_2}^2 dt +$$

$$+c\tilde{c}_2\varepsilon_4\int_0^s\left\|h-h^N\right\|_{L_2}^2dt+\frac{c\tilde{c}_2}{4\varepsilon_4}\int_0^s\left\|h-\tilde{h}\right\|_{L_2}^2dt \qquad [1.23]$$

where $\tilde{c}_2=r_1\,max\{1,c_T\}$.

Taking into account the following estimate of the norm $\forall u\in W_2^1(\Omega)$ on the boundary $\partial\Omega$ of the given domain:

$$\|u\|_{L_2(\partial\Omega)}^2=\int_{\Gamma_1\cup\Gamma_2}u^2\cos(n,r)d\Gamma+\int_{\Gamma_2\cup\Gamma_4}u^2\cos(n,z)d\Gamma=$$

$$=\iint_\Omega\left(\frac{\partial}{\partial r}(u^2)+\frac{\partial}{\partial z}(u^2)\right)d\Omega=2\iint_\Omega u\left(\frac{\partial u}{\partial r}+\frac{\partial u}{\partial z}\right)d\Omega\le$$

$$\le C_1\left(\iint_\Omega u^2d\Omega\right)^{1/2}\left(\iint_\Omega\left(\left(\frac{\partial u}{\partial r}\right)^2+\left(\frac{\partial u}{\partial z}\right)^2\right)d\Omega\right)^{1/2}=C_1\|u\|_{L_2(\Omega)}\|u\|_{H_0^1(\Omega)}.$$

Friedrichs' inequality (Rectoris 1977), written in the general case, and ε-inequality, we have

$$\|u\|_{L_2(\Omega)}^2\le C_2\|u\|_{H_0^1(\Omega)}^2+C_3\|u\|_{L_2(\partial\Omega)}^2\le C_2\|u\|_{H_0^1(\Omega)}^2+C_3C_1\|u\|_{L_2(\Omega)}\|u\|_{H_0^1(\Omega)}\le$$

$$\le\left(C_2+\frac{C_3C_1}{4\varepsilon}\right)\|u\|_{H_0^1(\Omega)}^2+C_3C_1\varepsilon\|u\|_{L_2(\Omega)}^2\ .$$

In the last inequality, we choose ε such that $1-C_3C_1\varepsilon>0$. So, for the considered domain $\Omega\ \exists\ C_4=const>0$ such that $\forall u\in W_2^1(\Omega)$

$$\|u\|_{L_2(\Omega)}^2\le C_4\|u\|_{H_0^1(\Omega)}^2.$$

In view of the last estimate and the inequality [1.23], we rewrite the estimate [1.22] as follows:

$$\left(e^{-c\tilde{T}}-\frac{\tilde{c}_2}{c_1}\varepsilon_1\right)\left\|h-h^N\right\|_{L_2}^2(s)+\left(\delta_1e^{-c\tilde{T}}-C_4\frac{\tilde{c}_2}{c_1}(\varepsilon_3+c\varepsilon_4)\right)\int_0^s\left\|h-h^N\right\|_{H_0^1}^2dt+$$

$$+\delta_2e^{-c\tilde{T}}\int_0^s\left\|h_2-h_2^N\right\|_{L_2(\Gamma_2)}^2dt\le\left(1+\frac{\tilde{c}_2}{c_1}\varepsilon_2\right)\left\|h-h^N\right\|_{L_2}^2(0)+$$

$$\frac{\tilde{c}_2}{c_1}\frac{1}{4\varepsilon_2}\left\|h-\tilde{h}\right\|_{L_2}^2(0)+\frac{\tilde{c}_2}{c_1}\frac{1}{4\varepsilon_1}\left\|h-\tilde{h}\right\|_{L_2}^2(s)+c\left(1+C_4\frac{\tilde{c}_2}{c_1}\frac{1}{4\varepsilon_4}\right)\int_0^s\left\|h-\tilde{h}\right\|_{H_0^1}^2dt+$$

$$+\frac{\tilde{c}_2}{c_1}\frac{1}{4\varepsilon_3}\int_0^s\left\|\frac{\partial(h-\tilde{h})}{\partial t}\right\|_{L_2}^2 dt + c\int_0^s\left\|h_2-\tilde{h}_2\right\|_{L_2(\Gamma_2)}^2 dt. \qquad [1.24]$$

From the equalities [1.9], [1.15], we have

$$\|h-h^N\|_{L_2}(0) \le \left\|h-\tilde{h}\right\|_{L_2}(0) \le \left\|h-\tilde{h}\right\|_{L_2\times L_\infty} \ \forall\tilde{h}\in Z^N.$$

Taking into account the last relation, from the inequality [1.24] and the notations [1.16], we obtain

$$\left\|h-h^N\right\|_{L_2\times L_\infty}^2 + \delta\left\|h-h^N\right\|_{H_0^1\times L_2}^2 + \bar{\bar{\delta}}\left\|h_2-h_2^N\right\|_{L_2(\Gamma_2)\times L_2}^2 < \bar{C}\left(\left\|h-\tilde{h}\right\|_{L_2\times L_\infty}^2\right.$$

$$\left.+\left\|h-\tilde{h}\right\|_{H_0^1\times L_2}^2 + \left\|\frac{\partial(h-\tilde{h})}{\partial t}\right\|_{L_2\times L_2}^2 + \left\|h_2-\tilde{h}_2\right\|_{L_2(\Gamma_2)\times L_2}^2\right).$$

The statement of the theorem follows from the last inequality and well-known estimates of interpolation by FEM polynomials (Deineka et al. 1995). □

The problem [1.14], [1.15] taking into account [1.11]–[1.13] can be rewritten in the matrix form

$$\bar{M}\dot{\alpha}(t) + A\big(\alpha(t)\big)\alpha(t) = F(t)\ \forall t\in(0,\tilde{T}], \qquad [1.25]$$

$$M\alpha(0) - \bar{H}_0,\ t=0 \qquad [1.26]$$

where

$$\alpha(t) = (\alpha_1(t),\alpha_2(t),\dots,\alpha_{2N}(t))^T,\ \alpha_i(t)=\alpha_i^1(t),\ \alpha_{i+N}(t)=\alpha_i^2(t),\ i=\overline{1,N},$$

$$\bar{M}=\begin{pmatrix}M^1 & 0\\ 0 & c_T M^2\end{pmatrix}, M=\begin{pmatrix}M^1 & 0\\ 0 & M^2\end{pmatrix}, A=\begin{pmatrix}A^1 & 0\\ 0 & A^2\end{pmatrix}, F(t)=\begin{pmatrix}F^1(t)\\ F^2(t)\end{pmatrix}, \bar{H}_0=\begin{pmatrix}\bar{H}_0^1\\ \bar{H}_0^2\end{pmatrix},$$

elements of matrices $M^s=\{m_{ij}^s\}_{i,j=1}^N$, $A^s=\{a_{ij}^s\}_{i,j=1}^N$ and vectors $F^s(t)=\{F_i^s(t)\}_{i=1}^N$, $\bar{H}_0^s=\{\bar{H}_{0i}^s\}_{i=1}^N$, $s=1,2$, are calculated according to the formulas

$$m_{ij}^s = \iint_\Omega r\ \varphi_i(r,z)\varphi_j(r,z)d\Omega,$$

$$a_{ij}^1 = \iint_\Omega r\ k\Big(\sum_{i=1}^N\alpha_i^1\varphi_i,\sum_{i=1}^N\alpha_i^2\varphi_i\Big)\left(\frac{\partial\varphi_i}{\partial r}\frac{\partial\varphi_j}{\partial r}+\frac{\partial\varphi_i}{\partial z}\frac{\partial\varphi_j}{\partial z}\right)drdz,$$

$$a_{ij}^2 = \iint_\Omega r\left(\tilde{\lambda}\left(\frac{\partial\varphi_i}{\partial r}\frac{\partial\varphi_j}{\partial r} + \frac{\partial\varphi_i}{\partial z}\frac{\partial\varphi_j}{\partial z}\right) - c_w v\left(\sum_{i=1}^{N}\alpha_i^1\varphi_i\right)\varphi_i\frac{\partial\varphi_j}{\partial z}\right)drdz +$$

$$+\int_{\Gamma_2} r\alpha\varphi_i(r,z)\ \varphi_j(r,z)d\,\Gamma_2,$$

$$F_i^1(t) = 0, F_i^2(t) = \int_{\Gamma_2} r\alpha T_a\ \varphi_i(r,z)d\,\Gamma_2,$$

$$\bar{H}_{0i}^1 = \iint_\Omega W_0(r,z)\varphi_i(r,z)d\Omega\,, \bar{H}_{0i}^2 = \iint_\Omega T_0(r,z)\varphi_i(r,z)d\Omega\,,\ i,j = \overline{1,N}.$$

To estimate the time-discrete approximate generalized solution, we use the Crank–Nicolson scheme (Deineka et al. 1995).

Let $\tilde{T} = J\tau$ for some integer $J \geq 1$. We are going to find a sequence $\{H^j(r,z)\}_{j=0}^J \subset Z_t^N$, $H^j(r,z) = H(r,z,j\tau)$ such that H^j approximates $h^N \in Z^N$ optimally in $W_2^1(\Omega)$. Let us denote

$$\partial_\tau P^j = \frac{1}{\tau}(P^{j+1} - P^j),\ P^{j+1/2} = \frac{1}{2}(P^{j+1} + P^j), j = \overline{0,J-1}.$$

The Crank–Nicolson scheme for the Cauchy problem [1.25], [1.26] can be written in the form

$$m(\partial_\tau H^j, V) + A\left(H^{j+1/2}; H^{j+1/2}, V\right) = \left(F^{j+1/2}, V\right), j = \overline{0,J-1}, \qquad [1.27]$$

$$(H^0, V) = (h_0, V)\ \forall V \in Z_0^N \qquad [1.28]$$

where $H^{j+1}(r,z) = \sum_{i=1}^{2N}\alpha_i^{j+1}\Phi_i(r,z), j = \overline{0,J-1}$. We can rewrite [1.27], [1.28] in the matrix form as follows:

$$\frac{1}{\tau}\bar{M}(\alpha^{j+1} - \alpha^j) + \frac{1}{2}A\left(\frac{\alpha^{j+1}+\alpha^j}{2}\right)(\alpha^{j+1} + \alpha^j) = \frac{1}{2}(F^{j+1} + F^j),$$

$$\alpha^0 = M^{-1}\bar{H}_0, j = \overline{0,J-1}.$$

Then, the theorem is true.

THEOREM 1.2.– *Let the classical solution of the initial-boundary value problem [1.1]–[1.6] $h(r,z,t)$ satisfies the conditions of Theorem 1.1, and its partial* *derivatives* $\frac{\partial^3 h_i}{\partial t^3}, \frac{\partial^3 h_i}{\partial z \partial t^2}, \frac{\partial^3 h_i}{\partial r \partial t^2}$, $i = 1,2$ *are bounded on* $\bar{\Omega}_{\bar{T}}$. *Then, there exist constants* $\bar{\delta}, \bar{\bar{\delta}}, C > 0$, *such that the following estimate holds*:

$$\left\|h^J - H^J\right\|_{L_2}^2 + \bar{\delta}\tau \sum_{j=0}^{J-1} \left\|h^{j+1/2} - H^{j+1/2}\right\|_{H_0^1}^2 +$$

$$+\bar{\bar{\delta}}\tau \sum_{j=0}^{J-1} \left\|h_2^{j+1/2} - H_2^{j+1/2}\right\|_{L_2(\Gamma_2)}^2 \leq C\left(\bar{h}^{2k} + \tau^4\right). \quad [1.29]$$

PROOF.– Using Taylor's formula, we have

$$\frac{\partial h(r,z,(j+1/2)\tau)}{\partial t} = \partial_\tau h^j + \rho_j,\ h(r,z,(j+1/2)\tau) = h^{j+1/2} + \xi_j,$$

$$F(r,z,(j+1/2)\tau) = F^{j+1/2} + \eta_j,$$

$$\rho_j = O(\tau^2), \xi_j = O(\tau^2), \eta_j = O(\tau^2), \frac{\partial \xi_j}{\partial r} = O(\tau^2), \frac{\partial \xi_j}{\partial z} = O(\tau^2). \quad [1.30]$$

Then, the relation [1.8] for $t = (j+1/2)\tau$ can be presented as

$$m(\partial_\tau h^j + \rho_j, V) + A(h^{j+1/2} + \xi_j; h^{j+1/2} + \xi_j, V) = (F^{j+1/2} + \eta_j, V),$$

$$j = \overline{0, J-1},\ \forall V \in Z_0^N. \quad [1.31]$$

Let $V = \frac{z^{j+1} + z^j}{2} = z^{j+1/2} = (h-H)^{j+1/2} = (h-\tilde{h})^{j+1/2} + (\tilde{h}-H)^{j+1/2}$, where $\tilde{h}(r,z,t)$ is an arbitrary function from Z^N. Then, using the equalities [1.27], [1.31] and taking into account that $(\tilde{h}-H)^{j+1/2} \in Z_0$, we get

$$m(\partial_\tau z^j + \rho_j, (\tilde{h}-H)^{j+1/2}) + A(h^{j+1/2} + \xi_j; h^{j+1/2} + \xi_j, (\tilde{h}-H)^{j+1/2}) -$$

$$-A(H^{j+1/2}; H^{j+1/2}, (\tilde{h}-H)^{j+1/2}) = (\eta_j, (\tilde{h}-H)^{j+1/2}), j = \overline{0, J-1}. \quad [1.32]$$

Let us consider the relation

$$m(\partial_\tau z^j, z^{j+1/2}) + A(H^{j+1/2}; z^{j+1/2}, z^{j+1/2}) = m(\partial_\tau z^j, (h-\tilde{h})^{j+1/2}) +$$

$$+A(H^{j+1/2}; z^{j+1/2}, (h-\tilde{h})^{j+1/2}) + m(\partial_\tau z^j + \rho_j, (\tilde{h}-H)^{j+1/2}) +$$

$$+A(h^{j+1/2} + \xi_j; h^{j+1/2} + \xi_j, (\tilde{h}-H)^{j+1/2}) - A(H^{j+1/2}; H^{j+1/2}, (\tilde{h}-H)^{j+1/2}) -$$

$$-m(\rho_j, (\tilde{h}-H)^{j+1/2}) - A(h^{j+1/2}+\xi_j; h^{j+1/2}+\xi_j, (\tilde{h}-H)^{j+1/2}) +$$

$$+A(H^{j+1/2}; h^{j+1/2}, (\tilde{h}-H)^{j+1/2}), j=\overline{0,J-1}.$$

Taking into account [1.32] and [1.10] from the last equality, we have

$$m(\partial_\tau z^j, z^{j+1/2}) + a(H^{j+1/2}; z^{j+1/2}, z^{j+1/2}) + b_2(z^{j+1/2}, z^{j+1/2}) =$$

$$= m(\partial_\tau z^j, (h-\tilde{h})^{j+1/2}) + b_1(H^{j+1/2}; z^{j+1/2}, z^{j+1/2}) +$$

$$+A(H^{j+1/2}; z^{j+1/2}, (h-\tilde{h})^{j+1/2}) + A(H^{j+1/2}; h^{j+1/2}+\xi_j, (\tilde{h}-H)^{j+1/2}) -$$

$$-A(h^{j+1/2}+\xi_j; h^{j+1/2}+\xi_j, (\tilde{h}-H)^{j+1/2}) - A(H^{j+1/2}; \xi_j, (\tilde{h}-H)^{j+1/2})$$

$$-m(\rho_j, (\tilde{h}-H)^{j+1/2}) + (\eta_j, (\tilde{h}-H)^{j+1/2}), j=\overline{0,J-1}. \qquad [1.33]$$

Let us estimate the terms in [1.33] taking into account the ε-inequality, the triangle inequality and the relations [1.7], [1.10], [1.16]. We get

$$m(\partial_\tau z^j, z^{j+1/2}) \geq \tfrac{c_1}{\tau}\left(\left\|z^{j+1}\right\|_{L_2}^2 - \left\|z^j\right\|_{L_2}^2\right),$$

$$a(H^{j+1/2}; z^{j+1/2}, z^{j+1/2}) \geq c_2\left\|z^{j+1/2}\right\|_{H_0^1}^2, b_2\left(z^{j+1/2}, z^{j+\frac{1}{2}}\right) \geq r_0\alpha\left\|z_2^{j+1/2}\right\|_{L_2(\Gamma_2)}^2,$$

$$\left|b_1(H^{j+1/2}; z^{j+1/2}, z^{j+1/2})\right| \leq c_3\left(\varepsilon_1\left\|z^{j+1/2}\right\|_{L_2}^2 + \tfrac{1}{4\varepsilon_1}\left\|z^{j+1/2}\right\|_{H_0^1}^2\right),$$

$$\left|A(H^{j+1/2}; z^{j+1/2}, (h-\tilde{h})^{j+1/2})\right| \leq c_4\left(\varepsilon_2\left\|z^{j+1/2}\right\|_{H_0^1}^2 + \frac{1}{4\varepsilon_2}\left\|(h-\tilde{h})^{j+1/2}\right\|_{H_0^1}^2\right) +$$

$$+c_3\left(\varepsilon_3\left\|z^{j+1/2}\right\|_{L_2}^2 + \frac{1}{4\varepsilon_3}\left\|\left(h-\tilde{h}\right)^{j+1/2}\right\|_{H_0^1}^2\right) + r_1\alpha\left(\varepsilon_4\left\|z_2^{j+1/2}\right\|_{L_2(\Gamma_2)}^2 +\right.$$

$$\left.+\tfrac{1}{4\varepsilon_4}\left\|\left(h_2-\tilde{h}_2\right)^{j+1/2}\right\|_{L_2(\Gamma_2)}^2\right),$$

$$\left|m(\rho_j, (\tilde{h}-H)^{j+1/2})\right| \leq \left|m(\rho_j, (h-\tilde{h})^{j+1/2})\right| + \left|m(\rho_j, z^{j+1/2})\right| \leq$$

$$\leq (\varepsilon_5+\varepsilon_6)\left\|\rho_j\right\|_{L_2}^2 + \tfrac{1}{4\varepsilon_5}\left\|(h-\tilde{h})^{j+1/2}\right\|_{L_2}^2 + \tfrac{1}{4\varepsilon_6}\left\|z^{j+1/2}\right\|_{L_2}^2,$$

$$\left|(\eta_j, (\tilde{h}-H)^{j+1/2})\right| \leq (\varepsilon_7+\varepsilon_8)\left\|\eta_j\right\|_{L_2}^2 + \tfrac{1}{4\varepsilon_7}\left\|(h-\tilde{h})^{j+1/2}\right\|_{L_2}^2 + \tfrac{1}{4\varepsilon_8}\left\|z^{j+1/2}\right\|_{L_2}^2,$$

$$\big|A\big(H^{j+1/2}; h^{j+1/2}+\xi_j, (\tilde{h}-H)^{j+1/2}\big)$$

$$-A\big(h^{j+1/2}+\xi_j; h^{j+1/2}+\xi_j, (\tilde{h}-H)^{j+1/2}\big)-A\big(H^{j+1/2}; \xi_j, (\tilde{h}-H)^{j+1/2}\big)\big|$$

$$\le c_5\left\|z^{j+1/2}+\xi_j\right\|_{L_2}\ \left\|(\tilde{h}-H)^{j+1/2}\right\|_{H_0^1}+c_6\left\|z^{j+1/2}+\xi_j\right\|_{L_2}\ \left\|(\tilde{h}-H)^{j+1/2}\right\|_{H_0^1}$$

$$+c_4\left\|\xi_j\right\|_{H_0^1}\ \left\|(\tilde{h}-H)^{j+1/2}\right\|_{H_0^1}+c_3\left\|\xi_j\right\|_{L_2}\ \left\|(\tilde{h}-H)^{j+1/2}\right\|_{H_0^1}$$

$$+r_1\alpha\left\|\xi_j\right\|_{L_2(\Gamma_2)}\left\|(\tilde{h}-H)^{j+1/2}\right\|_{L_2(\Gamma_2)}\le\tilde{\tilde{c}}\left(\left\|z^{j+1/2}\right\|_{L_2}^2+\left\|z^{j+1/2}\right\|_{H_0^1}^2\right.$$

$$+\left\|z^{j+1/2}\right\|_{L_2(\Gamma_2)}^2+\left\|(h-\tilde{h})^{j+1/2}\right\|_{H_0^1}^2+\left\|(h-\tilde{h})^{j+1/2}\right\|_{L_2(\Gamma_2)}^2$$

$$\left.+\left\|\xi_j\right\|_{L_2}^2+\left\|\xi_j\right\|_{H_0^1}^2+\left\|\xi_j\right\|_{L_2(\Gamma_2)}^2\right), j=\overline{0,J-1}, \qquad [1.34]$$

$\tilde{\tilde{c}}=\tilde{\tilde{c}}(c_i,\varepsilon_j,r_1,\alpha)$, c_i, $i=\overline{1,6}$ are determined by [1.20], $\varepsilon_j=const>0$, $j=1,2,\ldots$ are determined from ε-inequalities.

Taking into account the estimates [1.34], we obtain the following inequality from the equality [1.33]:

$$\frac{1}{\tau}\left(\left\|z^{j+1}\right\|_{L_2}^2-\left\|z^j\right\|_{L_2}^2\right)+\delta_1\left\|z^{j+1/2}\right\|_{H_0^1}^2+\delta_2\left\|z_2^{j+1/2}\right\|_{L_2(\Gamma_2)}^2$$

$$\le\delta\left(\left\|(h-\tilde{h})^{j+1/2}\right\|_{H_0^1}^2+\left\|(h-\tilde{h})^{j+1/2}\right\|_{L_2}^2+\left\|(h-\tilde{h})^{j+1/2}\right\|_{L_2(\Gamma_2)}^2+\left\|\xi_j\right\|_{L_2}^2\right.$$

$$\left.+\left\|\xi_j\right\|_{H_0^1}^2+\left\|\xi_j\right\|_{L_2(\Gamma_2)}^2+\left\|\rho_j\right\|_{L_2}^2+\left\|\eta_j\right\|_{L_2}^2+\left\|z^{j+1}\right\|_{L_2}^2+\left\|z^j\right\|_{L_2}^2\right)$$

$$+\frac{1}{c_1}m\left(\frac{z^{j+1}-z^j}{\tau},(h-\tilde{h})^{j+1/2}\right), j=\overline{0,J-1}. \qquad [1.35]$$

δ_1, δ_2 are positive due to appropriately chosen constants in ε-inequalities.

Similarly (Rectoris 1977), we multiply the inequality [1.35] for each $j=\overline{0,J-1}$ by the corresponding function $(1+\delta\tau)^{-1}g^j(\tau)$, where $g(\tau)=\frac{1-\delta\tau}{1+\delta\tau}$ is bounded from above and below by some positive numbers less than one. Finding the sums of the left and right parts of the inequalities along j and taking [1.30] into account, we establish the validity of the inequality

$$\|z^J\|_{L_2}^2 + \bar{\delta}\tau \sum_{j=0}^{J-1} \|z^{j+1/2}\|_{H_0^1}^2 + \bar{\bar{\delta}}\tau \sum_{j=0}^{J-1} \|z_2^{j+1/2}\|_{L_2(\Gamma_2)}^2$$

$$\leq C\left(\sum_{j=0}^{J-1} \|(h-\tilde{h})^{j+1/2}\|_{H_0^1}^2 \tau + \sum_{j=0}^{J-1} \left\|\frac{(h-\tilde{h})^{j+1/2} - (h-\tilde{h})^{j-1/2}}{\tau}\right\|_{L_2}^2 \tau\right.$$

$$+\|(h-\tilde{h})^0\|_{L_2}^2 + \|(h-\tilde{h})^{1/2}\|_{L_2}^2 + \|(h-\tilde{h})^{J-1/2}\|_{L_2}^2$$

$$\left.+\tau \textstyle\sum_{j=0}^{J-1} \|(h-\tilde{h})^{j+1/2}\|_{L_2(\Gamma_2)}^2\right) + O(\tau^4). \qquad [1.36]$$

The estimate [1.29] is obtained from [1.36] taking into account the following transformations:

$$\frac{(h-\tilde{h})^{j+1/2} - (h-\tilde{h})^{j-1/2}}{\tau} = \frac{1}{\tau}\left\{(h-\tilde{h})^j + \frac{\tau}{2}\left((h-\tilde{h})^j\right)' + \frac{\tau^2}{8}\left((h-\tilde{h})^j\right)''\right.$$

$$\left.+O(\tau^3) - (h-\tilde{h})^j + \frac{\tau}{2}\left((h-\tilde{h})^j\right)' - \frac{\tau^2}{8}\left((h-\tilde{h})^j\right)'' + O(\tau^3)\right\}$$

$$= \left(\frac{\partial(h-\tilde{h})}{\partial t}\right)^j + O(\tau^2)$$

and estimates of interpolation by FEM polynomials (Deineka et al. 1995). □

Thus, the estimates of the convergence rate for the continuous in time and discrete approximate solutions of the axis-symmetric initial-boundary value problem for the system of filtration and heat transfer equations have been obtained.

1.2. The analysis of numerical modeling of soil mass dynamics in the presence of unsteady pressure filtration

The prediction of the soil hydraulic structure condition largely depends on an integrated, systematic approach that allows taking into account various factors that influence the behavior of these objects in the process of mathematical modeling. Thus, the construction, theoretical justification and numerical implementation of the most adequate model is a complex and topical problem.

In this section, on the example of a profile problem for the filter base of a gravity dam, three initial-boundary value problems are presented in order to determine the dynamic characteristics of soil base under the condition of the upper reservoir water level changes.

For each of these differential models, approximate generalized Galerkin solutions are constructed using the FEM, and a comparative analysis of the calculation results is carried out in Bohaienko et al. (2014).

Mathematical models

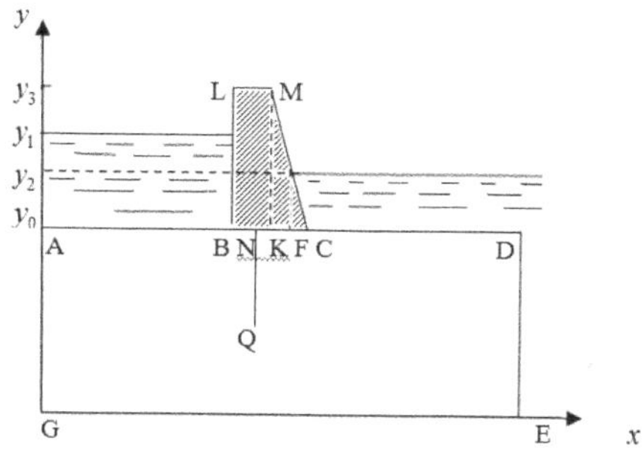

Figure 1.1. *The simulation domain*

All problems presented in this section were solved without taking into account curtain NQ in the domain shown in Figure 1.1. *Model I* is the initial-boundary value problem for the parabolic–hyperbolic system that includes the non-stationary quasi-linear filtration equation and elasticity theory equations describing the stress–strain state of soil massif, described in Marchenko and Samoilenko (2012):

$$\tilde{\mu}\frac{\partial h}{\partial t}-\frac{\partial}{\partial x}\left(K_f(\theta,t)\frac{\partial h}{\partial x}\right)-\frac{\partial}{\partial y}\left(K_f(\theta,t)\frac{\partial h}{\partial y}\right)=0,$$

$$\rho_{ws}\frac{\partial^2 u_{CK}}{\partial t^2}-\mu\Delta u_{sk}-(\lambda+\mu)\frac{\partial}{\partial x}\left(\frac{\partial u_{sk}}{\partial x}+\frac{\partial v_{sk}}{\partial y}\right)-\frac{\partial P}{\partial x}=0,$$

$$\rho_{ws}\frac{\partial^2 v_{CK}}{\partial t^2}-\mu\Delta v_{sk}-(\lambda+\mu)\frac{\partial}{\partial y}\left(\frac{\partial u_{sk}}{\partial x}+\frac{\partial v_{sk}}{\partial y}\right)-\frac{\partial P}{\partial y}=-\rho_{ws}g, \qquad [1.37]$$

where $(x,y,t)\in\Omega_{\tilde{T}}$, $\Omega_{\tilde{T}}=\Omega\times(0,\tilde{T}]$; h is the piezometric head, u_{sk}, v_{sk} are the horizontal and vertical displacement for the soil skeleton, respectively, $P=\rho_w g(h-y)$ is the hydrostatic pressure, $K_f(\theta,t)$ is the filtration coefficient,

$\theta = \frac{\partial u}{\partial x} + \frac{\partial v}{\partial y}$ is the volumetric deformation, λ, μ are the Lame constants, $\tilde{\mu}$ is the water capacity, ρ_{ws} is the density of the water-saturated soil, ρ_w is the density of water and g is the acceleration of gravity.

We set the boundary conditions as follows (see Figure 1.1):

$$h(x,y,t) = y_1 + \tilde{h}(t), (x,y,t) \in AB \times (0,\tilde{T}],$$

$$h(x,y,t) = y_2, (x,y,t) \in CD \times (0,\tilde{T}], \quad [1.38]$$

$$\frac{\partial h}{\partial x} = 0, (x,y,t) \in (GA \cup DE) \times (0,\tilde{T}],$$

$$\frac{\partial h}{\partial y} = 0, (x,y,t) \in (GE \cup BK \cup KF \cup FC) \times (0,\tilde{T}],$$

$$u_{sk} = 0, (x,y,t) \in (GA \cup DE) \times (0,\tilde{T}], \; u_{sk} = v_{sk} = 0,$$

$$(x,y,t) \in GE \times (0,\tilde{T}], \quad [1.39]$$

$$\tau_{xy} = 0, (x,y,t) \in (GA \cup DE) \times (0,\tilde{T}],$$

$$\sigma_n = -\rho_{mat}\, g(y_3 - y_0), \quad (x,y,t) \in BK \times (0,\tilde{T}],$$

$$\sigma_n = -\rho_{mat}\, g(\tilde{y}(x) - y_0), \quad (x,y,t) \in KF \times (0,\tilde{T}],$$

$$\sigma_n = -\rho_{mat}\, g(\tilde{y}(x) - y_0) - \rho_w\, g(y_2 - \tilde{y}(x)), \quad (x,y,t) \in FC \times (0,\tilde{T}],$$

$$\tau_s = 0, (x,y,t) \in (BK \cup KF \cup KC) \times (0,\tilde{T}],$$

$$\sigma_n = 0, \quad \tau_s = 0, \quad (x,y,t) \in (AB \cup CD) \times (0,\tilde{T}]$$

where n is the external normal to the contour, σ_n, τ_s are the normal and tangential components of the stress vector, ρ_{mat} is the density of the material of the dam body, $\tilde{y}(x)$ is the equation of a line that passes through the points M and C.

The physical parameters and the parameters of the domain are as follows: $K_f(x,y,t) = \bar{K}_f e^{\varepsilon(\theta)/(0.18\varepsilon_t - 0.048)}$, $\bar{K}_f = 0.3m/day$, $\varepsilon(\theta) = \varepsilon_0 + \theta(1-\varepsilon_0)$, $\varepsilon_0 = 0.64$, $\varepsilon_t = 1.06$, $\gamma_p = W_t/\gamma_w$, $\gamma_p = 2.54$ is the volumetric weight of soil particles, $W_t = 0.282$ is the total soil moisture content, γ_p is the specific gravity of water, $\rho_{ws} = 1.94\; ton/m^3$, $\rho_{mat} = 2.3\; ton/m^3$, $\tilde{\mu} = 10^{-3}$, $y_0 = 260\; m$, $y_1 = 377\; m$, $y_2 = 266\; m$, $y_3 = 420\; m$, $\tilde{h}(t) = t/86400\; m/s$ (corresponds to the rate of water rising equal to 10 m/day), $E = 5000\ ton/m^3$ is the elastic modulus, $\nu = 0.3$ is Poisson's ratio.

Initial conditions have the form

$$h(x,y,0) = h_0, \quad u_{\text{ск}}(x,y,0) = u_0, \quad v_{sk}(x,y,0) = v_0,$$

$$\frac{\partial u_{sk}}{\partial t}(x,y,0) = \frac{\partial v_{sk}}{\partial t}(x,y,0) = 0, \quad (x,y) \in \bar{\Omega}.$$

Based on the Galerkin method and assuming the existence of an additional condition $\frac{\partial h}{\partial t}(x,y,0) = h_1$, $(x,y) \in \overline{\Omega}$, the following generalized problem is formulated in Marchenko and Samoilenko (2012):

$$m\left(\frac{\partial^2 w}{\partial t^2}, z\right) + \overline{m}\left(\frac{\partial w}{\partial t}, z\right) + a(w,z) = \left(\tilde{F}, z\right),\ w \in Z,\ \forall t \in (0, \tilde{T}], \qquad [1.40]$$

$$(w(\cdot,\cdot,0), z) = (w^0, z), \left(\frac{\partial w}{\partial t}(\,\cdot\,,\,\cdot\,,0),\ z\right) = (w^1,\ z) \quad \forall z \in Z_0 \qquad [1.41]$$

where

$$m\left(\frac{\partial^2 w}{\partial t^2}, z\right) = \rho_{\text{гр}} \iint_\Omega \left(\frac{\partial^2 w_2}{\partial t^2} z_2 + \frac{\partial^2 w_3}{\partial t^2} z_3\right) d\Omega,\ \overline{m}\left(\frac{\partial w}{\partial t}, z\right) = \iint_\Omega \tilde{\mu} \frac{\partial w_1}{\partial t} z_1\, d\Omega,$$

$$a(w,z) = W_1(w,z) + W_2(w,z) + F(w,z),$$

$$W_1(w,z) = \iint_\Omega K_f(\theta,t)\left(\frac{\partial w_1}{\partial x}\frac{\partial z_1}{\partial x} + \frac{\partial w_1}{\partial y}\frac{\partial z_1}{\partial y}\right) d\Omega,$$

$$W_2(w,z) = \iint_\Omega \left((\lambda + 2\mu)\left(\frac{\partial w_2}{\partial x}\frac{\partial z_2}{\partial x} + \frac{\partial w_3}{\partial y}\frac{\partial z_3}{\partial y}\right) + \lambda\left(\frac{\partial w_3}{\partial y}\frac{\partial z_2}{\partial x} + \frac{\partial w_2}{\partial x}\frac{\partial z_3}{\partial y}\right)\right.$$

$$\left. + \mu\left(\frac{\partial w_2}{\partial y} + \frac{\partial w_3}{\partial x}\right)\left(\frac{\partial z_2}{\partial y} + \frac{\partial z_3}{\partial x}\right)\right) d\Omega,$$

$$F(w,z) = \iint_\Omega \rho_w g(w_1 - y)\left(\frac{\partial z_2}{\partial x} + \frac{\partial z_3}{\partial y}\right) d\Omega,$$

$$\left(\tilde{F}, z\right) = -\iint_\Omega \rho_{ws} g z_3\, d\Omega - \int_{BK} \rho_{mat}\ g(y_3 - y_0) z_3\, d\Gamma -$$

$$+\int_{KF} \rho_{mat}\, g(\tilde{y}(x) - y_0) z_3\, d\Gamma - \int_{FC} (\rho_{mat}\, (\tilde{y}(x) - y_0) +$$

$$+\rho_w(y_2 - \tilde{y}(x)))g z_3 d\Gamma + \int_{AB} \rho_w\, g(y_1 - y)) z_3 d\Gamma + \int_{CD} \rho_w\, g(y_2 - y) z_3 d\Gamma,$$

$w^0 = (h_0, u_0, v_0)^T$, $w^1 = (h_1, 0,0)^T$, $(\cdot,\cdot)$ is a scalar product of vectors.

The solution to the problem [1.40], [1.41] is sought on the set Z of vector functions $w(x,y,t) = (h(x,y,t), u_{sk}(x,y,t), v_{sk}(x,y,t))^{\mathrm{T}} = (w_1(x,y,t), w_2(x,y,t), w_3(x,y,t))^{\mathrm{T}}$, the components of which, their time derivatives $\frac{\partial w}{\partial t}(x,y,t)$, $\frac{\partial^2 w}{\partial t^2}(x,y,t)$, $\forall t \in (0,T]$, $\frac{\partial w}{\partial t}(x,y,0)$ and $w(x,y,0)$ belong to the space $W_2^1(\Omega)$. The set Z_0 contains vector functions $z(x,y) = (z_1(x,y), z_2(x,y), z_3(x,y))^T$ with components from the space $W_2^1(\Omega)$ that satisfy homogeneous main boundary conditions [1.38], [1.39].

An approximate generalized solution $w^N(x,y,t) \in Z^N$ of the problem [1.40], [1.41] is sought by the FEM on the subset $Z^N \subset Z$ and satisfies the following integral identities:

$$m\left(\frac{\partial^2 w^N}{\partial t^2}, z^N\right) + \overline{m}\left(\frac{\partial w^N}{\partial t}, z^N\right) + a(w^N, z^N) = \left(\tilde{F}, z^N\right) \forall t \in \left(0, \tilde{T}\right], \qquad [1.42]$$

$$(w^N(\cdot\,,\cdot\,,0), z^N) = (w^0, z^N), \left(\frac{\partial w^N}{\partial t}(\cdot,\cdot\,,0),\ z^N\right) = (w^1,\ z^N)\ \forall z^N \in Z_0^N. \qquad [1.43]$$

To obtain an approximate time-discrete solution $\{W^j(x,y)\}_{j=0}^{J} \subset Z_t^N$ of the problem [1.6], [1.7], the Crank–Nicolson scheme (Marchenko and Samoilenko 2012) is used.

Model II is the model of dynamic consolidation of water-saturated soils (Zaretsky and Lombardo 1983; Skopetsky and Marchenko 2005; Skopetsky et al. 2008), in the construction of which three groups of equations for quasi-two-phase soils were considered:

1) equations describing the law of momentum change in classical mechanics;

2) groups of equations that determine the velocity of fluid in a porous medium based on Navier–Stokes and Darcy–Gersevanov laws;

3) equations that follow from the mass conservation law for soil phases.

The system of the equations is formulated on the basis of the Terzaghi effective stress principle and takes into account elastic deformations only; thus, it has the following form:

$$(1-m)\rho_p \frac{\partial^2 u_{sk}}{\partial t^2} + \frac{\rho_w g m^2}{\bar{K}_f}\left(\frac{\partial u_{sk}}{\partial t} - \frac{\partial u_w}{\partial t}\right) -$$

$$-\left(\mu \Delta u_{sk} + (\lambda+\mu)\frac{\partial}{\partial x}\left(\frac{\partial u_{sk}}{\partial x} + \frac{\partial v_{sk}}{\partial y}\right)\right) - (1-m)\frac{\partial P_w}{\partial x} = 0,$$

$$(1-m)\rho_p \frac{\partial^2 v_{sk}}{\partial t^2} + \frac{\rho_w g m^2}{\bar{K}_f}\left(\frac{\partial v_{sk}}{\partial t} - \frac{\partial v_w}{\partial t}\right) -$$

$$-\left(\mu \Delta v_{sk} + (\lambda+\mu)\frac{\partial}{\partial y}\left(\frac{\partial u_{sk}}{\partial x} + \frac{\partial v_{sk}}{\partial y}\right)\right) - (1-m)\frac{\partial P_w}{\partial y} = -(1-m)\rho_p g,$$

$$m\rho_w \frac{\partial^2 u_w}{\partial t^2} + \frac{\rho_w g m^2}{\bar{K}_f}\left(-\frac{\partial u_{sk}}{\partial t} + \frac{\partial u_w}{\partial t}\right) - m\frac{\partial P_w}{\partial x} = 0,$$

$$m\rho_w \frac{\partial^2 v_w}{\partial t^2} + \frac{\rho_w g m^2}{\bar{K}_f}\left(-\frac{\partial v_{sk}}{\partial t} + \frac{\partial v_w}{\partial t}\right) - m\frac{\partial P_w}{\partial y} = -m\rho_w g,$$

$$\frac{\partial P_w}{\partial t} = \frac{M^w}{m}\left((1-m)\frac{\partial}{\partial t}\left(\frac{\partial u_{sk}}{\partial x} + \frac{\partial v_{sk}}{\partial y}\right) + m\frac{\partial}{\partial t}\left(\frac{\partial u_w}{\partial x} + \frac{\partial v_w}{\partial y}\right)\right) \quad [1.44]$$

where $(x,y,t) \in \Omega_{\tilde{T}}$, $\Omega_{\tilde{T}} = \Omega \times (0,\tilde{T}]$; u_w, v_w are the horizontal and vertical components of the fluid displacement vector, respectively; P_w is the pore pressure; ρ_p, ρ_w are the densities of mineral particles and liquid, respectively ($\rho_{ws} = \rho_p(1-m)$ $+\rho_w m$), $\rho_p = 2.54\ ton/m^3$; $m = 0.39$ is the porosity; $\bar{K}_f$ is the filtration coefficient given in the previous model; $M^w = 203 \cdot 10^3\ tonf/m^3$ is the elastic modulus of the fluid.

The boundary conditions have the form

$$u_{sk} = 0,\ u_w = 0,\ (x,y,t) \in (GA \cup DE) \times (0,\tilde{T}], \quad [1.45]$$

$$\tau_{xy} = 0,\ \frac{\partial P_w}{\partial x} = 0,\ (x,y,t) \in (GA \cup DE) \times (0,\tilde{T}],\ \frac{\partial P_w}{\partial y} = 0,\ (x,y,t) \in GE \times (0,\tilde{T}], \quad [1.46]$$

$$u_{sk} = v_{sk} = 0,\ v_w = 0, w_{w,n} = 0, (x,y,t) \in GE \times (0,\tilde{T}], \quad [1.47]$$

$$\sigma_n + P_w = -P_w^+ = -\rho_w\ g(y_1 + \tilde{h}(t) - y_0),\ \ (x,y,t) \in AB \times (0,\tilde{T}], \quad [1.48]$$

$$\sigma_n + P_w = -P_w^+ = -\rho_w\ g(y_2 - y_0),\ \ (x,y,t) \in CD \times (0,\tilde{T}], \quad [1.49]$$

$$\sigma_n = kmw_{rel,n}, \tau_s = 0{,}0 \le k \le \infty, (x,y,t) \in (AB \cup CD) \times (0,\tilde{T}], \quad [1.50]$$

$$\sigma_n + P_w = -\rho_{mat}\ g(y_3 - y_0), \tau_s = 0, w_{rel,n} = 0,(x,y,t) \in BK \times (0,\tilde{T}], \quad [1.51]$$

$$\sigma_n + P_w = -\rho_{mat}\ g(\tilde{y}(x) - y_0), \tau_s = 0, w_{rel,n} = 0,(x,y,t) \in KF \times (0,\tilde{T}], \quad [1.52]$$

$$\sigma_n + P_w = -\rho_{mat}\ g(\tilde{y}(x) - y_0) - \rho_w\ g\big(y_2 - \tilde{y}(x)\big), \tau_s = 0,$$

$$w_{rel,n} = 0,\ (x,y,t) \in FC \times (0,\tilde{T}] \quad [1.53]$$

where $w_{w,n}$ is the normal component of the fluid displacement vector $w_w = (u_w,\ v_w)^T$, $w_{rel,n}$ is the normal component of the relative water displacement vector $w_{rel} = w_w - w_{sk}$.

Condition [1.50] is based on [1.48], [1.49] and the condition

$$P_{w,n}^+ - P_{w,n} = kmw_{rel,n},\ 0 \le k \le \infty,\ \ (x,y,t) \in (AB \cup CD) \times (0,T]$$

which reflects the proportionality of the difference between the values of water pressure and pore pressure to the level of openness of the boundary at the edge between two media (Seimov et al. 1990). The case when the pores are completely open corresponds to the value of the coefficient $k = 0$. In the absence of connection with the voids of soil medium (e.g. in the presence of an impermeable membrane), conditions [1.48]–[1.50] at $k = +\infty$ correspond to the condition $\sigma_n = -P_w^+$.

Initial conditions are as follows:

$$u_{sk}(x,y,0) = u_w(x,y,0) = u_0, v_{sk}(x,y,0) = v_w(x,y,0) = v_0,$$

$$\frac{\partial u_{sk}}{\partial t}(x,y,0) = \frac{\partial v_{sk}}{\partial t}(x,y,0) = 0, \frac{\partial u_w}{\partial t}(x,y,0) = \frac{\partial v_w}{\partial t}(x,y,0) = 0,$$

$$P_w(x,y,0) = \frac{M^w}{m}\left((1-m)\left(\frac{\partial u_{sk}(x,y,0)}{\partial x} + \frac{\partial v_{sk}(x,y,0)}{\partial y}\right) + \right.$$

$$+m\left(\frac{\partial u_w(x,y,0)}{\partial x}+\frac{\partial v_w(x,y,0)}{\partial y}\right)\Bigg)-$$

$$-\rho_w g(h_0-y)=\frac{M^w}{m}\left(\frac{\partial u_0}{\partial x}+\frac{\partial v_0}{\partial y}\right)-\rho_w g(h_0-y),(x,y)\in\bar{\Omega}. \qquad [1.54]$$

The generalized solution of the initial-boundary value problem [1.44]–[1.54] is the vector function $w(x,y,t)\in Z$, which satisfies the integral relations in the form [1.40], [1.41], where the quadratic forms $m\left(\frac{\partial^2 w}{\partial t^2},z\right)$, $\overline{m}\left(\frac{\partial w}{\partial t},z\right)$, $a(w,z)$ and $(\tilde{F},z)$ are written similarly to Skopetsky et al. (2008) and Skopetsky and Marchenko (2005), Z is the set of vector functions $w(x,y,t)=(u_{sk}(x,y,t),v_{sk}(x,y,t),$ $u_w(x,y,t),v_w(x,y,t))^T=(w_{sk}(x,y,t),w_w(x,y,t))^T=(w_1(x,y,t),\ w_2(x,y,t),$ $w_3(x,y,t),\ w_4(x,y,t))^T$, satisfying the main boundary conditions [1.45], [1.47] and whose components, as well as some of their partial time derivatives, belong to the space $W_2^1(\Omega)$ $\forall t\in(0,T]$.

The set Z_0 contains vector functions $z(x,y)=(z_1(x,y),\ z_2(x,y),\ z_3(x,y),$ $z_4(x,y))^T$, that satisfy the homogeneous main boundary conditions [1.9], [1.11], and their components belong to the space $W_2^1(\Omega)$.

According to *model III*, the stress–strain state of the dam foundation is described by the system of equations of elasticity theory

$$\rho_{ws}\frac{\partial^2 u_{sk}}{\partial t^2}-\mu\Delta u_{sk}-(\lambda+\mu)\frac{\partial}{\partial x}\left(\frac{\partial u_{sk}}{\partial x}+\frac{\partial v_{sk}}{\partial y}\right)=0,$$

$$\rho_{ws}\frac{\partial^2 v_{sk}}{\partial t^2}-\mu\Delta v_{sk}-(\lambda+\mu)\frac{\partial}{\partial y}\left(\frac{\partial u_{sk}}{\partial x}+\frac{\partial v_{sk}}{\partial y}\right)=-\rho_{ws}g,$$

for $(x,y,t)\in\Omega_{\tilde{T}},\ \Omega_{\tilde{T}}=\Omega\times(0,\tilde{T}]$.

The boundary and initial conditions were set as follows:

$$u_{sk}=0,\tau_{xy}=0,(x,y,t)\in(GA\cup DE)\times(0,\tilde{T}],$$

$$u_{sk}=v_{sk}=0,(x,y,t)\in GE\times(0,\tilde{T}],$$

$$\sigma_n=-\rho_w\ g(y_3-y_0),(x,y,t)\in BK\times(0,\tilde{T}],$$

$$\sigma_n=-\rho_{mat}\ g(\tilde{y}(x)-y_0),(x,y,t)\in KF\times(0,\tilde{T}],$$

$$\sigma_n = -\rho_{mat}\ g(\tilde{y}(x) - y_0) - \rho_w\ g(y_2 - \tilde{y}(x)),\ (x, y, t) \in FC \times (0, \tilde{T}],$$

$$\tau_s = 0,\ (x, y, t) \in (BK \cup KF \cup KC) \times (0, \tilde{T}],$$

$$\sigma_n = -\rho_w\ g(y_1 + \tilde{h}(t) - y_0),\ \tau_s = 0,\ (x, y, t) \in AB \times (0, \tilde{T}],$$

$$\sigma_n = -\rho_w\ g(y_2 - y_0),\ \tau_s = 0,\ (x, y, t) \in CD \times (0, \tilde{T}],$$

$$u_{sk}(x, y, 0) = u_0, \qquad v_{sk}(x, y, 0) = v_0, \qquad \frac{\partial u_{sk}}{\partial t}(x, y, 0) = \frac{\partial v_{sk}}{\partial t}(x, y, 0) = 0,$$
$$(x, y) \in \bar{\Omega}.$$

For time-continuous $w^N(x, y, t) = (u_{sk}^N(x, y, t), v_{sk}^N(x, y, t))^T$ and completely discrete $\{W^j(x, y)\}_{j=0}^J$ approximate generalized solutions constructed in a similar way to the previous problems, the following estimates hold (Skopetsky et al. 1995):

$$\|w^N - w\|_{L_\infty(0,T;L_2(\Omega))} \le Cs^k,\ \max_{0\le j\le J}\|W^j - w^j\|_{L_2(\Omega)} \le C(s^k + \tau^2).$$

It is assumed that the functions encountered in the problem's statements have sufficient smoothness.

The initial conditions for solving all three initial-boundary value problems were taken from the solution of stationary problems: to determine the initial stress–strain state, a stationary problem corresponding to *model III* was solved; to determine the initial pressures, the stationary filtration problem corresponding to *model I* was solved.

Figure 1.2(a) shows the deformation of the rectangular grid of the computational domain based on the results of the calculation at the time $t = 100\ hours$ for *model I*; Figure 1.2(b) – for *model II* at $k = 0$ in the boundary condition [1.40], Figure 1.2(c) – for *model III*. The obtained numerical solutions show that *model I* gives the maximum absolute values for the isolines of horizontal displacements among three models and the minimum values for vertical displacements; *model III*: maximum – along the vertical and minimum along the horizontal. *Model II* with the value of $k = 0$ in the condition [1.40] gives the average absolute values for isolines in both horizontal and vertical displacements.

The comparative analysis of the graphs of water pressure isolines for *model I* (see Figure 1.3(a)) and *model II* at $k = 0$ (see Figure 1.3(b)) for the same moment of time $t = 100\ hours$ showed that they practically coincide. A decrease in the permeability of the bottom of the upper reservoir, in our formulation – an increase in the values of the coefficient k in [1.40] (Figure 1.3(b) corresponds to $k = 0$, Figure 1.3(c) – to $k = 100$) – leads to the decrease in pressure values in the section AB of the simulation domain's boundary and in the part of the domain under the upper reservoir, along with the increase in pressure at the boundary CD and in the part of the domain under the lower reservoir.

Analysis of simulation results

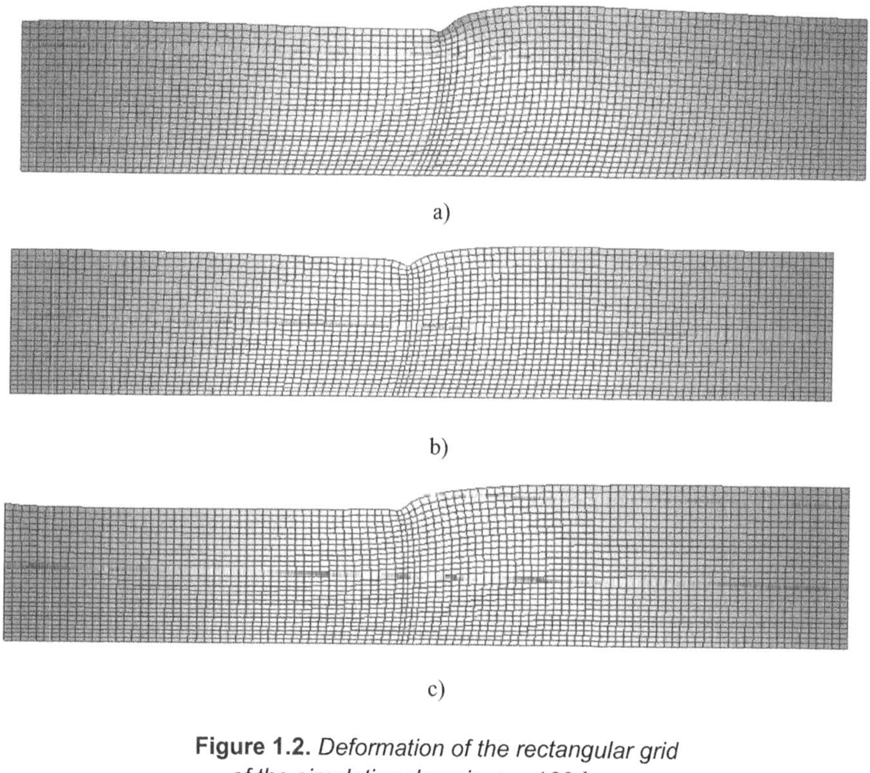

Figure 1.2. *Deformation of the rectangular grid of the simulation domain,* $t = 100\ hours$

Consideration of the permeability of the bottom of reservoirs in the boundary condition [1.40] of *model II* has a significant effect on the displacements of both the skeleton w_{sk} and water w_w in soil. A decrease in permeability leads to a decrease in the absolute values of the components of the vectors w_{sk} and w_w. Figures 1.4 and

1.5 show the isolines of horizontal displacements u_{sk} and u_w, respectively, for the time $t = 100\,hours$. In both figures, option (a) corresponds to $k = 0$, (b) to $k = 100$, (c) to $k = 1000$ as well as the absolute values of the components of the vector w_{rel}.

a)

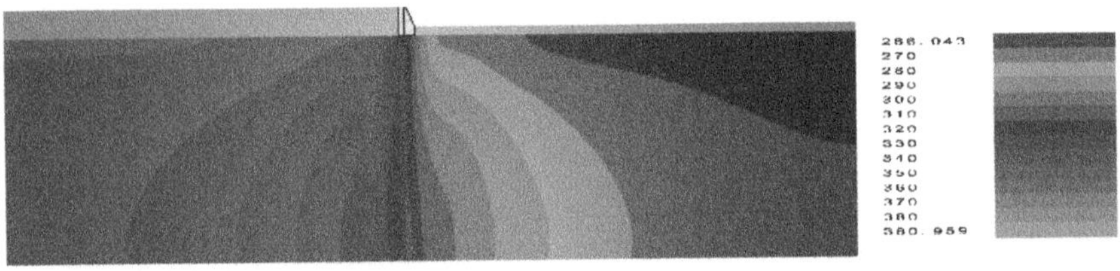

b)

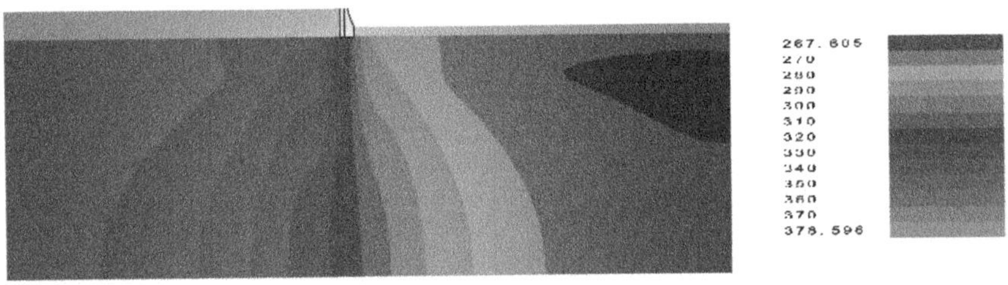

c)

Figure 1.3. *Pressure isolines, $t = 100\,hours$. For a color version of this figure, see www.iste.co.uk/koroliouk/computational2.zip*

The analysis of the results indicates that *model II* of dynamic consolidation of water-saturated soil, in which the Terzaghi effective stress principle was taken into account not only when deriving differential equations, but also in boundary conditions, when the pressure of the water column in the reservoir is redistributed to the stress in the skeleton and pore pressure (which allows the taking into account of the permeability of the bottom of reservoirs), allows getting the most realistic picture of dynamic processes in water-saturated soils. At the same time, the lack of consideration of filtration processes in the study of the stress–strain state (*model III*) of such a medium leads to a significant distortion of the results compared to the first two models, in which these processes are taken into account.

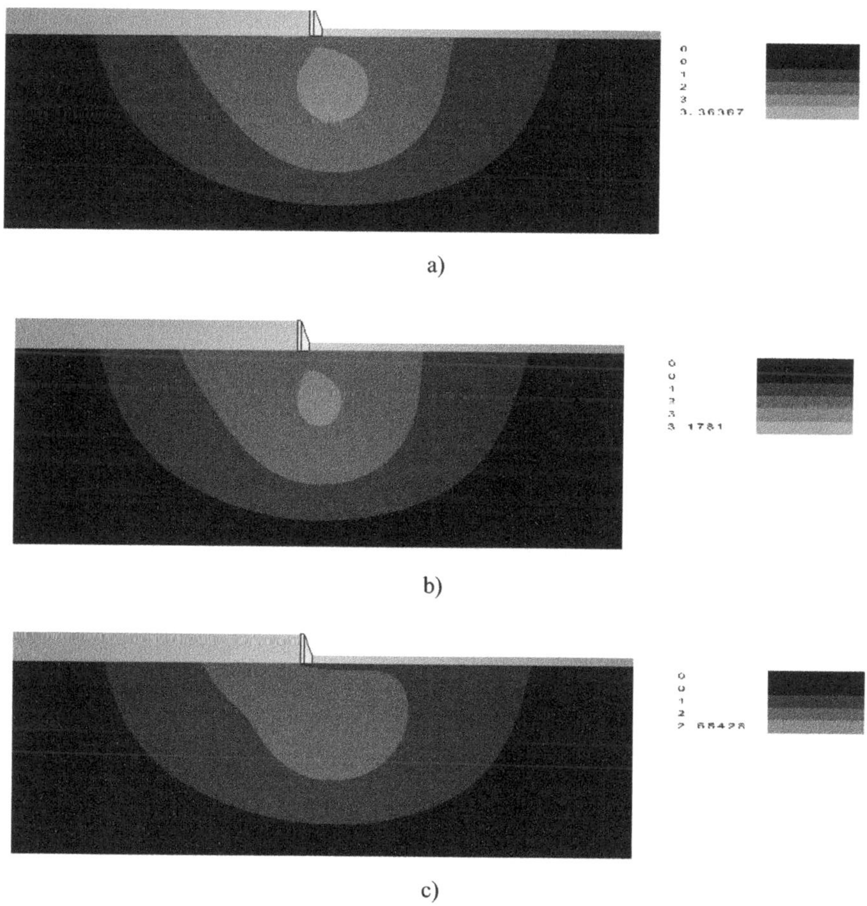

Figure 1.4. *Isolines of horizontal displacements of the skeleton u_{sk}. For a color version of this figure, see www.iste.co.uk/koroliouk/computational2.zip*

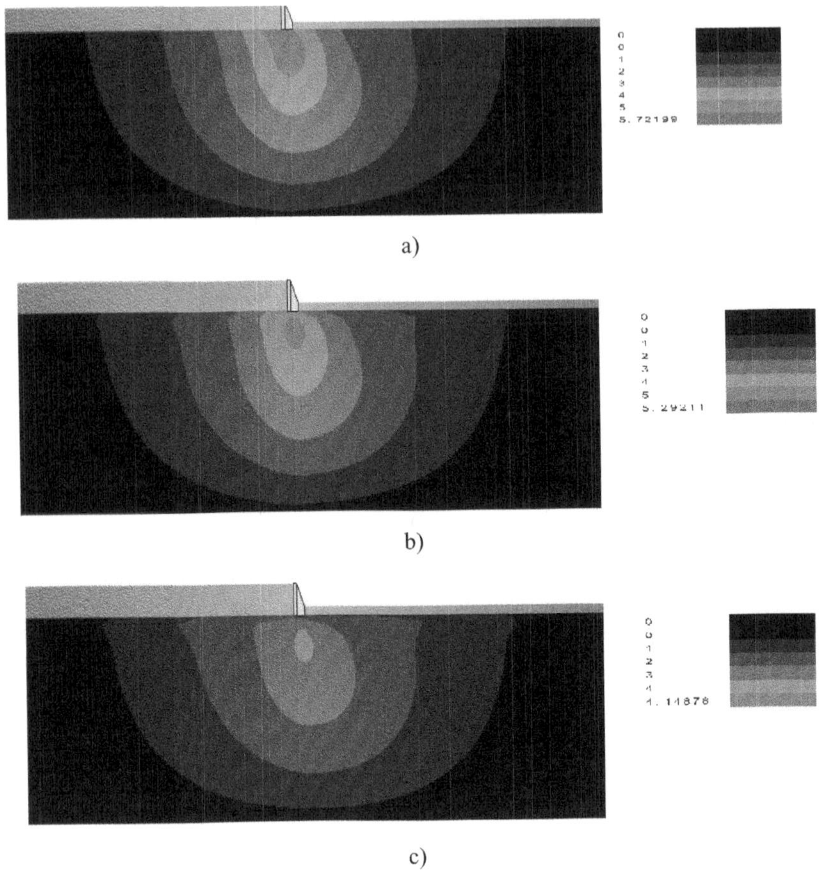

Figure 1.5. *Isolines of horizontal water displacements u_w. For a color version of this figure, see www.iste.co.uk/koroliouk/computational2.zip*

1.3. The analysis of numerical simulation of non-isothermal processes in soil

The problem of water inflow into a pit

In this section, on the example of the problem of water inflow into a pit, the comparative results of modeling non-stationary non-isothermal processes in moisture-saturated soils by means of an approximate solution of initial-boundary value problems for two quasi-linear parabolic–hyperbolic systems are presented. An analysis of calculations was carried out depending on the temperature of the external environment, with a change in the water level in the upper reservoir, and variable depth of a pit (Bohaienko et al. 2016).

The first model is an initial-boundary value problem for a system of differential equations, which includes non-stationary equations of filtration, thermal conductivity and equations describing the stress–strain state of soil. The second is the initial-boundary value problem for the system of equations of dynamic consolidation of moisture-saturated soils and the equation of heat transfer.

Formulation of the problem

By analogy with *model I* from section 1.2, the system of differential equations is written as follows and considered in the domain $\Omega_{\tilde{T}} = \Omega \times (0, \tilde{T}]$ (see Figure 1.6):

$$\tilde{\mu} \frac{\partial h}{\partial t} - \frac{\partial}{\partial x}\left(K_f(T,\theta)\frac{\partial h}{\partial x}\right) - \frac{\partial}{\partial y}\left(K_f(T,\theta)\frac{\partial h}{\partial y}\right) - 0,$$

$$c_T \frac{\partial T}{\partial t} - \frac{\partial}{\partial x}\left(\lambda_T \frac{\partial T}{\partial x} - c_w v_x T\right) - \frac{\partial}{\partial y}\left(\lambda_T \frac{\partial T}{\partial y} - c_w v_y T\right) = 0,$$

$$\rho_{ws}\frac{\partial^2 u_{sk}}{\partial t^2} - \mu \Delta u_{sk} - (\lambda + \mu)\frac{\partial}{\partial x}\left(\frac{\partial u_{sk}}{\partial x} + \frac{\partial v_{sk}}{\partial y}\right) - \frac{\partial P}{\partial x} = 0,$$

$$\rho_{ws}\frac{\partial^2 v_{sk}}{\partial t^2} - \mu \Delta v_{sk} - (\lambda + \mu)\frac{\partial}{\partial y}\left(\frac{\partial u_{sk}}{\partial x} + \frac{\partial v_{sk}}{\partial y}\right) - \frac{\partial P}{\partial y} = -\rho_{ws} g, (x, y, t) \in \Omega_{\tilde{T}} \quad [1.55]$$

where T is the temperature, $K_f(T,\theta)$ is the filtration coefficient, $(v_x, v_y)^T$ is the filtration rate, c_T, c_w are the volumetric heat capacities of the medium and water, respectively, and λ_T is the thermal conductivity coefficient.

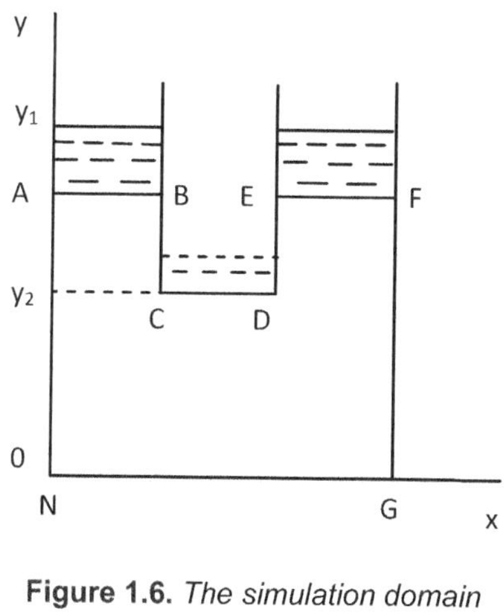

Figure 1.6. *The simulation domain*

Boundary conditions are as follows:

$$h(x,y,t) = y_1, (x,y,t) \in \Gamma^1 \times (0,\tilde{T}],$$

$$h(x,y,t) = y_2 + \tilde{h}(t), (x,y,t) \in D \times (0,\tilde{T}], \quad [1.56]$$

$$\frac{\partial h}{\partial n} = 0, (x,y,t) \in (\Gamma^2 \cup GN) \times (0,\tilde{T}], \ \frac{\partial T}{\partial n} = 0, (x,y,t) \in \Gamma^2 \times (0,\tilde{T}], \quad [1.57]$$

$$T(x,y,t) = T_1, (x,y,t) \in (\Gamma^1 \cup CD) \times (0,\tilde{T}],$$

$$T(x,y,t) = T_2, (x,y,t) \in GN \times (0,\tilde{T}], \quad [1.58]$$

$$u_{sk} = 0, \ (x,y,t) \in \Gamma^2 \times (0,\tilde{T}], \ u_{sk} = v_{sk} = 0, \ (x,y,t) \in GN \times (0,\tilde{T}], \quad [1.59]$$

$$\tau_s = 0, (x,y,t) \in (\partial\Omega \backslash GN) \times (0,\tilde{T}], \sigma_n = 0, (x,y,t) \in (\Gamma^1 \cup CD) \times (0,\tilde{T}], \quad [1.60]$$

$$\Gamma^1 = AB \cup EF, \Gamma^2 = NA \cup BC \cup DE \cup FG.$$

The physical parameters are as follows:

$$K_f(T,\theta) = \bar{K}_f(T)e^{f(\theta)}, \bar{K}_f(T) = \bar{K}_f(10)(0.7 + 0.03T), \bar{K}_f(10) = \begin{cases} \bar{\bar{K}}_f, P \geq 0, \\ \bar{\bar{K}}_f e^{2.48P}, P < 0, \end{cases} \quad \bar{\bar{K}}_f = 0.3\ m/day, \quad \tilde{h}(t) = \frac{1}{|CD|}\int_{CD}\int_0^t v_y(x,y,\tau)d\tau\, d\Gamma,$$

$$v_y(x,y,t) = -K_f(T,\theta)\frac{\partial h}{\partial y},$$

$$\rho_{ws} = 1.94\ ton/m^3, \tilde{\mu} = 10^{-3}, E = 5000\ ton/m^3, \nu = 0.3.$$

Initial conditions have the form

$$h(x,y,0) = H_0(x,y), \quad T(x,y,0) = T_0(x,y), \quad u_{sk}(x,y,0) = U_0(x,y),$$

$$v_{sk}(x,y,0) = V_0(x,y), \quad \frac{\partial u_{sk}}{\partial t}(x,y,0) = \frac{\partial v_{sk}}{\partial t}(x,y,0) = 0, \quad (x,y) \in \bar{\Omega}. \quad [1.61]$$

In addition, we assume that the conditions $\frac{\partial h}{\partial t}(x,y,0) = \tilde{H}_0(x,y)$, $\frac{\partial T}{\partial t}(x,y,0) = \tilde{T}_0(x,y)$, $(x,y) \in \overline{\Omega}$ hold.

Taking this into account, a generalized problem (Marchenko and Samoilenko 2012) is formulated for a set Z of vector functions $w(x,y,t) = (h(x,y,t), T(x,y,t), u_{sk}(x,y,t), v_{sk}(x,y,t))^T = (w_1, w_2, w_3, w_4)^T$ whose components and their time derivatives $\frac{\partial w}{\partial t}(x,y,t)$, $\frac{\partial^2 w}{\partial t^2}(x,y,t)$ $\forall t \in (0,\tilde{T}]$, as well as $\frac{\partial w}{\partial t}(x,y,0)$,

$w(x, y, 0)$, belong to $W_2^1(\Omega)$, satisfy the main boundary conditions [1.56], [1.58], [1.59] and integral identities [1.42], [1.43], where

$$m\left(\frac{\partial^2 w}{\partial t^2}, z\right) = \rho_{ws} \iint_\Omega \left(\frac{\partial^2 w_3}{\partial t^2} z_3 + \frac{\partial^2 w_4}{\partial t^2} z_4\right) d\Omega,$$

$$\overline{m}(\frac{\partial w}{\partial t}, z) = \iint_\Omega \left(\tilde{\mu}\ \frac{\partial w_1}{\partial t} z_1 + c_T\ \frac{\partial w_2}{\partial t} z_2\right) d\Omega,$$

$$a(w, z) = W_1(w, z) + W_2(w, z) + W_3(w, z),$$

$$W_1(w, z) = \iint_\Omega \left(\left(K_f(T, \theta)\left(\frac{\partial w_1}{\partial x}\frac{\partial z_1}{\partial x} + \frac{\partial w_1}{\partial y}\frac{\partial z_1}{\partial y}\right) + \lambda_T \left(\frac{\partial w_2}{\partial x}\frac{\partial z_2}{\partial x} + \frac{\partial w_2}{\partial y}\frac{\partial z_2}{\partial y}\right)\right) d\Omega,$$

$$W_2(w, z) = \iint_\Omega \left((\lambda + 2\mu)\left(\frac{\partial w_3}{\partial x}\frac{\partial z_3}{\partial x} + \frac{\partial w_4}{\partial y}\frac{\partial z_4}{\partial y}\right) + \lambda\left(\frac{\partial w_3}{\partial y}\frac{\partial z_4}{\partial x} + \frac{\partial w_4}{\partial x}\frac{\partial z_3}{\partial y}\right)\right.$$

$$\left.+\mu\left(\frac{\partial w_3}{\partial y} + \frac{\partial w_4}{\partial x}\right)\left(\frac{\partial z_3}{\partial y} + \frac{\partial z_4}{\partial x}\right)\right) d\Omega,$$

$$W_3(w, z) = \iint_\Omega \left(c_w K_f(T, \theta) w_2 \left(\frac{\partial w_1}{\partial x}\frac{\partial z_2}{\partial x} + \frac{\partial w_1}{\partial y}\frac{\partial z_2}{\partial y}\right)\right.$$

$$\left.+\rho_w g(w_1 - y)\left(\frac{\partial z_3}{\partial x} + \frac{\partial z_4}{\partial y}\right)\right)\ d\Omega,$$

$$(F, z) = -\iint_\Omega \rho_{ws} g z_4\, d\Omega,\ w_0 = (H_0, T_0, U_0, V_0)^T,\ \widetilde{w}_0 = (\tilde{H}_0, \tilde{T}_0, 0{,}0)^T.$$

The set Z_0 includes vector functions $z(x, y) = (z_1, z_2, z_3, z_4)^T$ with components from $W_2^1(\Omega)$ that satisfy the zero main boundary conditions. An approximate generalized solution $w^N(x, y, t) \in Z^N$ is sought by the FEM on a subset of $Z^N \subset Z$.

Model II describes the dynamic consolidation of water-saturated soils (section 1.2, Skopetsky et al. 2010), taking into account convective heat transfer

$$(1 - m)\rho_p \frac{\partial^2 u_{sk}}{\partial t^2} + \frac{\rho_w g m^2}{\bar{K}_f(T)}\left(\frac{\partial u_{sk}}{\partial t} - \frac{\partial u_w}{\partial t}\right) -$$

$$-\left(\mu \Delta u_{sk} + (\lambda + \mu)\frac{\partial}{\partial x}\left(\frac{\partial u_{sk}}{\partial x} + \frac{\partial v_{sk}}{\partial y}\right)\right) - (1 - m)\frac{\partial P}{\partial x} = 0,$$

$$(1-m)\rho_p \frac{\partial^2 v_{sk}}{\partial t^2} + \frac{\rho_w g m^2}{\bar{K}_f}\left(\frac{\partial v_{sk}}{\partial t} - \frac{\partial v_w}{\partial t}\right) -$$

$$-\left(\mu \Delta v_{sk} + (\lambda+\mu)\frac{\partial}{\partial y}\left(\frac{\partial u_{sk}}{\partial x} + \frac{\partial v_{sk}}{\partial y}\right)\right) - (1-m)\frac{\partial P}{\partial y} = -(1-m)\rho_p g,$$

$$m\rho_w \frac{\partial^2 u_w}{\partial t^2} + \frac{\rho_w g m^2}{\bar{K}_f}\left(-\frac{\partial u_{sk}}{\partial t} + \frac{\partial u_w}{\partial t}\right) - m\frac{\partial P}{\partial x} = 0,$$

$$m\rho_w \frac{\partial^2 v_w}{\partial t^2} + \frac{\rho_w g m^2}{\bar{K}_f}\left(-\frac{\partial v_{sk}}{\partial t} + \frac{\partial v_w}{\partial t}\right) - m\frac{\partial P}{\partial y} = -m\rho_w g,$$

$$\frac{\partial P}{\partial t} - \frac{M^w}{m}\left((1-m)\frac{\partial}{\partial t}\left(\frac{\partial u_{sk}}{\partial x} + \frac{\partial v_{sk}}{\partial y}\right) + m\frac{\partial}{\partial t}\left(\frac{\partial u_w}{\partial x} + \frac{\partial v_w}{\partial y}\right)\right) = 0,$$

$$c_T \frac{\partial T}{\partial t} - \frac{\partial}{\partial x}\left(\lambda_T \frac{\partial T}{\partial x} - c_w \frac{\partial u_w}{\partial t} T\right) - \frac{\partial}{\partial y}\left(\lambda_T \frac{\partial T}{\partial y} - c_w \frac{\partial v_w}{\partial t} T\right) = 0, (x,y,t) \in \Omega_{\tilde{T}}. \quad [1.62]$$

$\bar{K}_f(T)$ is the filtration coefficient described in the previous model, $\rho_p = 2.54\ ton/m^3$, $M^w = 203 \cdot 10^3\ tonf/m^2$.

Boundary conditions have the form

$$u_{sk} = 0, u_w = 0, \tau_{xy} = 0, \frac{\partial T}{\partial n} = 0, (x,y,t) \in \Gamma^2 \times (0,\tilde{T}],$$

$$u_{sk} = v_{sk} = 0, v_w = 0, T(x,y,t) = T_2, (x,y,t) \in GN \times (0,\tilde{T}],$$

$$\sigma_n = 0, \tau_s = 0, T(x,y,t) = T_1, (x,y,t) \in (\Gamma^1 \cup CD) \times (0,\tilde{T}],$$

$$P = -\rho_w\, g y_1,\ (x,y,t) \in \Gamma^1 \times (0,\tilde{T}],$$

$$P = -\rho_w\, g w_{rel,n},\ (x,y,t) \in CD \times (0,\tilde{T}]. \quad [1.63]$$

Initial conditions are as follows:

$$u_{sk}(x,y,0) = u_w(x,y,0) = U_0(x,y), v_{sk}(x,y,0) = v_w(x,y,0) = V_0(x,y),$$

$$P(x,y,0) = \rho_w g(H_0(x,y) - y), T(x,y,0) = T_0(x,y),$$

$$\frac{\partial u_{sk}}{\partial t}(x,y,0) = \frac{\partial v_{sk}}{\partial t}(x,y,0) = \frac{\partial u_w}{\partial t}(x,y,0) = \frac{\partial v_w}{\partial t}(x,y,0) = 0, (x,y) \in \bar{\Omega}. \quad [1.64]$$

The generalized solution of the initial-boundary value problem [1.62]–[1.64] in the presence of an additional condition $\frac{\partial T}{\partial t}(x,y,0)=\tilde{T}_0(x,y)$, $(x,y)\in\overline{\Omega}$ is a vector function $w(x,y,t)=(u_{sk}(x,y,t),v_{sk}(x,y,t),u_w(x,y,t),v_w(x,y,t),T(x,y,t))^T=$ $=(w_1,\ w_2,\ w_3,\ w_4,w_5)^T\in Z$, satisfying relations of the form [1.42], [1.43], where

$$m\left(\frac{\partial^2 w}{\partial t^2},z\right)=$$

$$=\iint_\Omega\left(\rho_p(1-m)\left(\frac{\partial^2 w_1}{\partial t^2}z_1+\frac{\partial^2 w_2}{\partial t^2}z_2\right)+\rho_w m\left(\frac{\partial^2 w_3}{\partial t^2}z_3+\frac{\partial^2 w_4}{\partial t^2}z_4\right)\right)d\Omega,$$

$$\overline{m}\left(\frac{\partial w}{\partial t},z\right)=$$

$$=\iint_\Omega\left(\frac{\rho_w g m^2}{\bar{K}_f(T)}\left(\frac{\partial(w_1-w_3)}{\partial t}(z_1-z_3)+\frac{\partial(w_2-w_4)}{\partial t}(z_2-z_4)\right)+c_T\frac{\partial w_5}{\partial t}z_5\right)d\Omega,$$

$$a(w,z)=W_1(w,z)+W_2(w,z)+W_3(w,z),$$

$$W_1(w,z)=\iint_\Omega\left((\lambda+2\mu)\left(\frac{\partial w_1}{\partial x}\frac{\partial z_1}{\partial x}+\frac{\partial w_2}{\partial y}\frac{\partial z_2}{\partial y}\right)+\lambda\left(\frac{\partial w_1}{\partial y}\frac{\partial z_2}{\partial x}+\frac{\partial w_2}{\partial x}\frac{\partial z_1}{\partial y}\right)\right.$$

$$\left.+\mu\left(\frac{\partial w_1}{\partial y}+\frac{\partial w_2}{\partial x}\right)\left(\frac{\partial z_1}{\partial y}+\frac{\partial z_2}{\partial x}\right)\right)d\Omega,$$

$$W_2(w,z)-\frac{M^w}{m}\iint_\Omega\left((1-m)\left(\frac{\partial w_1}{\partial x}+\frac{\partial w_2}{\partial y}\right)+\right.$$

$$\left.+m\left(\frac{\partial w_3}{\partial x}+\frac{\partial w_4}{\partial y}\right)\right)\left((1-m)\left(\frac{\partial z_1}{\partial x}+\frac{\partial z_2}{\partial y}\right)+m\left(\frac{\partial z_3}{\partial x}+\frac{\partial z_4}{\partial y}\right)\right)d\Omega,$$

$$W_3(w,z)=\iint_\Omega\left(\lambda_T\left(\frac{\partial w_5}{\partial x}\frac{\partial z_5}{\partial x}+\frac{\partial w_5}{\partial y}\frac{\partial z_5}{\partial y}\right)-c_w\frac{\partial w_3}{\partial t}w_5\frac{\partial z_5}{\partial x}-\right.$$

$$\left.-c_w\frac{\partial w_4}{\partial t}w_5\frac{\partial z_5}{\partial y}\right)d\Omega+\int_{CD}\rho_w g(w_4-w_2)((1-m)z_2+mz_4)d\Gamma,$$

$$(F,z)=\iint_\Omega\left(-(1-m)\rho_p g z_2-m\rho_w g z_4+\left(\frac{M^w}{m}\left(\frac{\partial U_0}{\partial x}+\frac{\partial V_0}{\partial y}\right)-\right.\right.$$

$$-\rho_w g(H_0 - y))\left((1-m)\left(\frac{\partial z_1}{\partial x}+\frac{\partial z_2}{\partial y}\right)+m\left(\frac{\partial z_3}{\partial x}+\frac{\partial z_4}{\partial y}\right)\right)\Bigg)d\Omega -$$

$$-\int_{\Gamma^1} \rho_w g y_1((1-m)z_2 + m z_4)d\Gamma,$$

$$w_0 = (U_0, V_0, U_0, V_0, T_0)^T, \widetilde{w}_0 = (0{,}0{,}0{,}0, \tilde{T}_0)^T.$$

Z is the set of vector functions satisfying the main boundary conditions, the components of which, as well as some of their partial time derivatives, belong to $W_2^1(\Omega)$, $\forall t \in (0,T]$. The vector functions $z(x,y) = (z_1(x,y),\ z_2(x,y),\ z_3(x,y),\ z_4(x,y), z_5(x,y))^{\mathrm{T}}$, with components $z_i \in W_2^1(\Omega)$, $i = \overline{1,5}$ satisfying the homogeneous main boundary conditions and belonging to the set Z_0.

To obtain an approximate time-discrete solution to the problem $\{W^j(x,y)\}_{j=0}^J \subset Z_t^N$ for both models, the Crank–Nicolson scheme (Marchenko and Samoilenko 2012; Skopetsky et al. 2010) is used.

The values of the initial conditions $H_0(x,y)$, $T_0(x,y)$, $U_0(x,y)$, $V_0(x,y)$ at the nodal points are taken from the solution of the stationary problem corresponding to *model I*, $\tilde{H}_0(x,y) \equiv 0$, $\tilde{T}_0(x,y) \equiv 0$.

The analysis of simulation results

The expected result for both models is a faster filling of a pit at a higher temperature of the filtered liquid. The influence of the external temperature T_1 on the dependence of the rate of pit filling on the pressure $y_1 = |NA| + H$ (H is the thickness of the surface water layer) turned out to be insignificant for both models. The temperature influence on the dependence of the pit filling rate on its depth l is significant only in the case of *model I* (see Figure 1.7(a) and (b)). Thus, for the relative difference $d\tilde{h} = (\tilde{h}(25) - \tilde{h}(15))/\tilde{h}(15)$ of water levels in a pit $(\tilde{h}(T_1))(t)$ at $T_1 = 25^oC$ and $T_1 = 15^oC$ ($T_2 = 5^oC$ in all cases), the following dependencies were obtained at the time of $t = 400$ hours: at $l = 2m$, the average $d\tilde{h}$ is equal to 0.07 for different values of H, at $l = 5m$ – to 0.05, at $l = 7m$ – to 0.03, i.e. the dependency of pit filling on the temperature factor decreases with the increase of pit depth. For *model II*, $d\tilde{h}$ is practically independent of H and l (at $t = 400$ hours, its average value is 0.03).

Figures 1.8 and 1.9 show graphs of $\tilde{h}(T_1)$ changes in time for both models: the lines —✱— correspond to *model I* at $H = 0.2m$, —■— at $H = 1.2m$; —✕— correspond to *model II* at $H = 0.2m$, —◆— at $H = 1.2m$.

An analysis of the numerical values of soil displacement vector $(u_{sk}, v_{sk})^T$ indicates that when solving the problem according to *model II* at a higher temperature T_1, both u_{sk} and v_{sk} are less in absolute values than at a lower temperature. So, for u_{sk} obtained at $T_1 = 15^oC$ and $T_1 = 25^oC$, the relative difference $du_{sk} = \frac{u_{sk}(25)-u_{sk}(15)}{u_{sk}(15)}$ is up to (–0.03) for all H, the corresponding $dv_{sk} = \frac{v_{sk}(25)-v_{sk}(15)}{v_{sk}(15)}$ is up to (–0.35) for $H = 0.2m$ and up to (–0.38) for $H = 1.2m$. A decrease in u_{sk} with an increase in T_1 is also characteristic of *model I* (du_{sk} is (–0.05) – (–0.17) for all H, but in this case, the values of dv_{sk} increase from 0.006 for $H = 0.2m$ to 0.025 for $H = 1.2m$).

For all simulation options, the following takes place: the absolute values of both u_{sk} and v_{sk} obtained by *model I* are greater than the corresponding values calculated by *model II* (the relative values reach 60% for u_{sk} and 40% for v_{sk}).

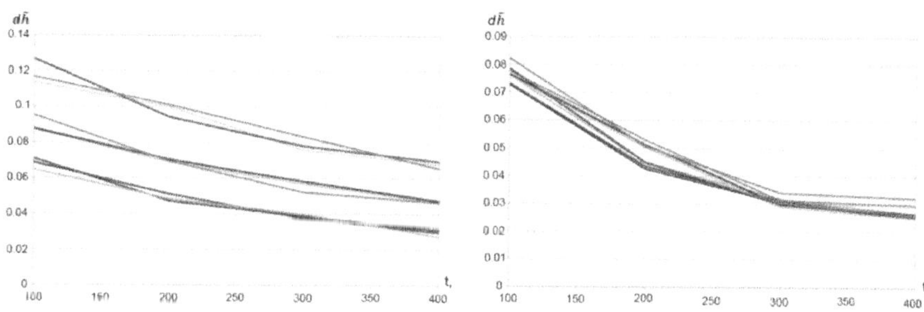

a) b)

Figure 1.7. *The relative difference $d\tilde{h}$: a) for model I, b) for model II (for l =2m — as H =0.2m, — – H =0.6m, — – H =1.2m; for l =5m — – H =0.2m, — – H =0.6m, — – H =1.2m; for l =7m — – H =0.2m, — – H =0.6m, — – H=1.2m). For a color version of this figure, see www.iste.co.uk/koroliouk/computational2.zip*

The results of simulations using *model II* of water displacement vector w_w for a period of 400 hours with an increase of T_1 from 15^oC to 25^oC give the maximum relative increases in the absolute values of the components at the highest calculated water level $H = 1.2m$ and the smallest calculated depth $l = 2m$ (u_w – up to 7%, v_w – up to 4.7%).

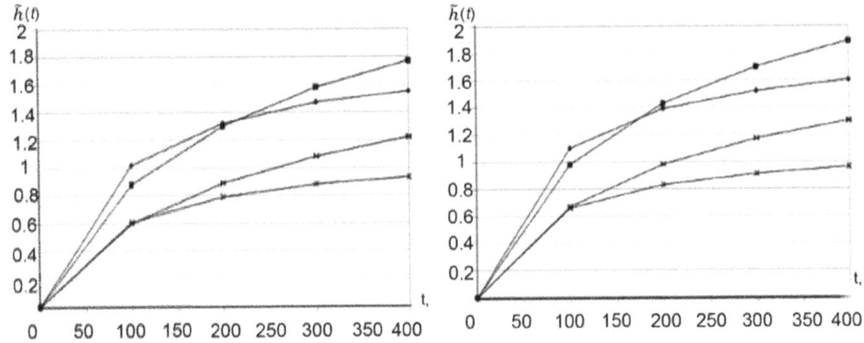

a) $T_1 = 15^{\circ}C$ b) $T_1 = 25^{\circ}C$

Figure 1.8. *Pit depth* $l = 2m$

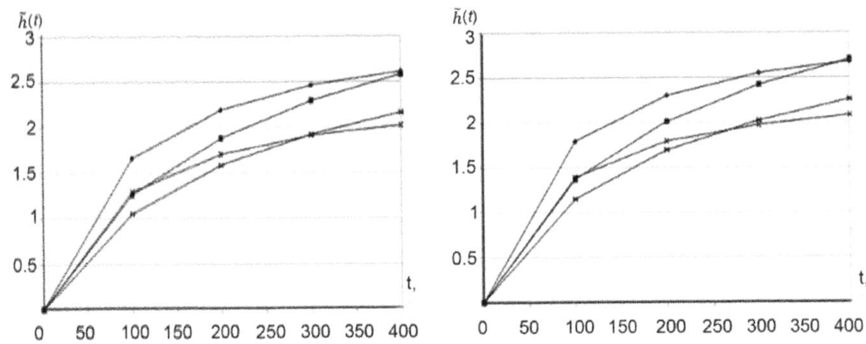

a) $T_1 = 15^{\circ}C$ b) $T_1 = 25^{\circ}C$

Figure 1.9. *Pit depth* $l = 5m$

1.4. The study of soil massif state in the foundations of hydraulic structures

For the system [1.55] in the domain depicted in Figure 1.1, the following initial-boundary value problem was considered. Boundary conditions are as follows:

$$h(x, y, t) = y_1, (x, y, t) \in AB \times (0, \tilde{T}],$$

$$h(x, y, t) = y_2 + \tilde{h}(t), (x, y, t) \in CD \times (0, \tilde{T}],$$

$$\frac{\partial h}{\partial x} = 0, (x,y,t) \in (GA \cup DE) \times (0,\tilde{T}],$$

$$\frac{\partial h}{\partial y} = 0,\ (x,y,t) \in (GE \cup BK \cup KF \cup FC) \times (0,\tilde{T}],$$

$$T(x,y,t) = T_1(t), (x,y,t) \in AB \times (0,\tilde{T}],$$

$$T(x,y,t) = T_2(t), (x,y,t) \in CD \times (0,\tilde{T}],$$

$$T(x,y,t) = T_3(x,t), (x,y,t) \in BC \times (0,\tilde{T}],$$

$$\frac{\partial T}{\partial x} = 0,\ (x,y,t) \in (GA \cup DE) \times (0,\tilde{T}],$$

$$\frac{\partial T}{\partial y} = 0,\ (x,y,t) \in (BK \cup KF \cup FC \cup GE) \times (0,\tilde{T}],$$

$$u_{sk} = 0, (x,y,t) \in (GA \cup DE) \times (0,\tilde{T}], u_{sk} = v_{sk} = 0, (x,y,t) \in GE \times (0,\tilde{T}],$$

$$\tau_{xy} = 0, (x,y,t) \in (GA \cup DE) \times (0,\tilde{T}],$$

$$\sigma_n = -\rho_{mat}\ g(y_3 - y_0), \quad (x,y,t) \in BK \times (0,\tilde{T}],$$

$$\sigma_n = -\rho_{mat}\ g(\tilde{y}(x) - y_0), \quad (x,y,t) \in KF \times (0,\tilde{T}],$$

$$\sigma_n = -\rho_{mat}\ g(\tilde{y}(x) - y_0) - \rho_w\ g(y_2 - \tilde{y}(x)), \quad (x,y,t) \in FC \times (0,\tilde{T}],$$

$$\tau_s = 0, (x,y,t) \in (BK \cup KF \cup KC) \times (0,\tilde{T}],$$

$$\sigma_n = 0, \quad \tau_s = 0, \quad (x,y,t) \in (AB \cup CD) \times (0,\tilde{T}] \qquad [1.65]$$

where $\sigma_n, \tau_s, \rho_{mat}, \tilde{y}(x)$ correspond to the notations given in section 1.2.

Non-homogeneous coupling conditions of non-ideal contact on the thin inclusion $\gamma = NQ$ have the form $((x,y,t) \in \gamma \times (0,\tilde{T}])$:

$$\left[K_f(T,\theta)\left(\frac{\partial h}{\partial x}cos(n,x) + \frac{\partial h}{\partial y}cos(n,y)\right)\right] = 0,$$

$$\left\{K_f(T,\theta)\left(\frac{\partial h}{\partial x}cos(n,x) + \frac{\partial h}{\partial y}cos(n,y)\right)\right\}^{\pm} = r_1(h,T,u_{sk},v_{sk})[h], \qquad [1.66]$$

$$\left[\left(\lambda_T\frac{\partial T}{\partial x} - c_w v_x T\right)cos(n,x) + \left(\lambda_T\frac{\partial T}{\partial y} - c_w v_y T\right)cos(n,y)\right] = 0,$$

$$\left\{\left(\lambda_T \frac{\partial T}{\partial x} - c_w v_x T\right) cos(n,x) + \left(\lambda_T \frac{\partial T}{\partial y} - c_w v_y T\right) cos(n,y)\right\}^{\pm}$$

$$= r_2(h, T, u_{sk}, v_{sk})[T], \quad [1.67]$$

$$[u_n] = 0, \quad [1.68]$$

$$[\sigma_n] = 0,$$

$$[\tau_s] = 0, \quad \{\tau_s\}^{\pm} = R(h, T, u_{sk}, v_{sk})\ [u_s]. \quad [1.69]$$

Here, $r_1(h, T, u_{sk}, v_{sk}), r_2(h, T, u_{sk}, v_{sk}), R(h, T, u_{sk}, v_{sk})$ are the known functions of a thin inclusion γ that characterize its moisture conductivity, thermal conductivity and durability, respectively; n is the normal to γ; u_n, u_s are the normal and tangent components of the displacements, respectively, $[v] = v^+ - v^-$ is the "jump" of function (Deineka et al. 1998).

The physical parameters and parameters of the domain are as follows:

$$\tilde{T} = 180day, K_f(T,\theta) = \bar{K}_f(T)e^{\overline{0.18\varepsilon_t - 0.048}^{\varepsilon(\theta)}}, \varepsilon(\theta) = \varepsilon_0 + \theta(1 - \varepsilon_0),$$

$$\varepsilon_0 = 0.64, \varepsilon_t = 1.06, \bar{K}_f(T) = \bar{K}_f(10)(0.7 + 0.03T), \bar{K}_f(10) = \begin{cases} \bar{\bar{K}}_f, P \geq 0 \\ \bar{\bar{K}}_f e^{2.48P}, P < 0 \end{cases},$$

$$\bar{\bar{K}}_f = 0.3\frac{m}{day}, \rho_{ws} = 1.94\frac{ton}{m^3}, \rho_{mat} = 2.3\frac{ton}{m^3}, \tilde{\mu} = 10^{-3}, y_0 = 260m,$$

$$y_1 = 377\ m, y_2 = 266\ m, y_3 = 420\ m, \tilde{h}(t) = \frac{1}{|CD|}\int_{CD}\int_0^t v_y(x,y,\tau)d\tau\, d\Gamma,$$

$$v_y(x,y,t) = -K_f(T,\theta)\frac{\partial h}{\partial y}, E = 5000\ ton/m^3, \nu = 0.3;$$

$$T_i(t) = \begin{cases} T_i^1, & t \in (0,90], \\ T_i^1 + \frac{T_i^2 - T_i^1}{90 \cdot 86400}t, & t \in (90,180],\ i = 1,2, \end{cases} \quad [1.70]$$

$T_1^1 = 12°$, $T_2^1 = 15°$, $T_1^2 = 1°$, $T_2^2 = 3°$, $T_3(x,t) = T_1(t) + \left(T_2(t) - T_1(t)\right)\frac{x - x_b}{x_c - x_b}, x \in [x_b, x_c]$, $t \in (0,180]$. The last relation reflects the scenario of a gradual decrease in temperature (conditionally summer–autumn period).

Initial conditions have the form

$$h(x,y,0) = H_0(x,y), \quad T(x,y,0) = T_0(x,y), \quad u_{sk}(x,y,0) = U_0(x,y),$$

$$v_{sk}(x,y,0) = V_0(x,y), \quad \frac{\partial u_{sk}}{\partial t}(x,y,0) = \frac{\partial v_{sk}}{\partial t}(x,y,0) = 0, \quad (x,y) \in \bar{\Omega}. \qquad [1.71]$$

On the basis of the Galerkin method and assuming the fulfillment of the additional conditions

$$\frac{\partial h}{\partial t}(x,y,0) = \tilde{H}_0(x,y), \frac{\partial T}{\partial t}(x,y,0) = \tilde{T}_0(x,y), \quad (x,y) \in \overline{\Omega}, \qquad [1.72]$$

the generalized problem is formulated for a set Z of vector functions $w(x,y,t) = (h(x,y,t), T(x,y,t), u_{sk}(x,y,t), v_{sk}(x,y,t))^{\mathrm{T}} = (w_1, w_2, w_3, w_4)^{\mathrm{T}}$ whose components and their time derivatives $\frac{\partial w}{\partial t}(x,y,t)$, $\frac{\partial^2 w}{\partial t^2}(x,y,t)$ $\forall t \in (0,\tilde{T}]$, as well as $\frac{\partial w}{\partial t}(x,y,0)$, $w(x,y,0)$, belong to $W_2^1(\Omega)$, satisfy the main boundary conditions from [1.64], [1.58], [1.59], the main coupling condition [1.68], and the integral identities [1.42], [1.43], where

$$m\left(\frac{\partial^2 w}{\partial t^2}, z\right) = \rho_{ws} \iint_\Omega \left(\frac{\partial^2 w_3}{\partial t^2} z_3 + \frac{\partial^2 w_4}{\partial t^2} z_4\right) d\Omega,$$

$$\overline{m}(\frac{\partial w}{\partial t}, z) = \iint_\Omega \left(\tilde{\mu}\, \frac{\partial w_1}{\partial t} z_1 + c_T\, \frac{\partial w_2}{\partial t} z_2\right) d\Omega\ ,$$

$$a(u; w, z) = W_1(u; w, z) + R_1(u; w\ z) + W_2(u; w, z) +$$

$$+R_2(u; w, z) - W_3(u; w, z) + W_4(w, z),$$

$$W_1(u; w, z) = \iint_\Omega \left(K_f(u_2, \theta^u)\left(\frac{\partial w_1}{\partial x}\frac{\partial z_1}{\partial x} + \frac{\partial w_1}{\partial y}\frac{\partial z_1}{\partial y}\right) +\right.$$

$$\left. + \lambda_T \left(\frac{\partial w_2}{\partial x}\frac{\partial z_2}{\partial x} + \frac{\partial w_2}{\partial y}\frac{\partial z_2}{\partial y}\right)\right) d\Omega,$$

$$R_1(u; w, z) = \int_\gamma r_1(u)[w_1][z_1] d\gamma + \int_\gamma r_2(u)[w_2][z_2] d\gamma,$$

$$W_2(u; w, z) = \iint_\Omega \left((\lambda + 2\mu)\left(\frac{\partial w_3}{\partial x}\frac{\partial z_3}{\partial x} + \frac{\partial w_4}{\partial y}\frac{\partial z_4}{\partial y}\right) + \lambda\left(\frac{\partial w_3}{\partial y}\frac{\partial z_4}{\partial x} + \frac{\partial w_4}{\partial x}\frac{\partial z_3}{\partial y}\right) +\right.$$

$$+\mu\left(\frac{\partial w_3}{\partial y}+\frac{\partial w_4}{\partial x}\right)\left(\frac{\partial z_3}{\partial y}+\frac{\partial z_4}{\partial x}\right)\Bigg)\,d\Omega,$$

$$R_2(u;w,z)=\int_\gamma R(u)[w_s][z_s]d\gamma,$$

$$W_3(u;w,z)=\iint_\Omega\left(b_1(u)w_2\frac{\partial z_2}{\partial x}+b_2(u)w_2\frac{\partial z_2}{\partial y}\right)d\Omega,$$

$$b_1(u)=c_wK_f(u_2,\theta^u)\frac{\partial u_1}{\partial x},\ b_2(u)=c_wK_f(u_2,\theta^u)\frac{\partial u_1}{\partial y},$$

$$W_4(w,z)=\iint_\Omega\rho_w g(w_1-y)\left(\frac{\partial z_3}{\partial x}+\frac{\partial z_4}{\partial y}\right)d\Omega,$$

$$(\tilde{F},z)=-\iint_\Omega\rho_{ws}gz_4\,d\Omega-\int_{BK}\rho_{mat}\ g(y_3-y_0)z_4\,d\Gamma-$$

$$\int_{KF}\rho_{mat}\ g(\tilde{y}(x)-y_0)z_4\,d\Gamma-\int_{FC}\Big(\rho_{mat}\ (\tilde{y}(x)-y_0)+\rho_w\big(y_2-\tilde{y}(x)\big)\Big)gz_4d\Gamma+$$

$$+\int_{AB}\rho_w\ g(y_1-y))z_4d\Gamma+\int_{CD}\rho_w\ g(y_2+\tilde{h}-y)z_4d\Gamma \qquad [1.73]$$

where $u=(u_1,u_2,u_3,u_4)\in Z;\ \ \theta^u=\frac{\partial u_3}{\partial x}+\frac{\partial u_4}{\partial y},\ \ \widetilde{W}^0=(H_0,T_0,U_0,V_0)^T,\ \ \widetilde{W}^1=(\tilde{H}_0,\tilde{T}_0,0,0)^T$, $(\cdot,\cdot)$ is the scalar product of vectors.

To find an approximate generalized solution, we assume that functions given in [1.55], [1.64]–[1.72] are sufficiently smooth on $\Omega_{\tilde{T}}$.

To obtain an approximate time-discrete solution $\{W^j(x,y)\}_{j=0}^J\subset Z_t^N$, the Cauchy problem is written in the form $\forall z^N\in Z_0^N$:

$$\langle W^0,z^N\rangle=\langle\widetilde{W}^0,z^N\rangle,\ \langle\theta^0,z^N\rangle=\langle\widetilde{W}^1,z^N\rangle, \qquad [1.74]$$

$$m(\partial_\tau\theta^j,z^N)+\tilde{m}(\theta^{j+1/2},z^N)+a(W^{j+1/2};W^{j+1/2},z^N)=(\tilde{F}^{j+1/2},z^N), \qquad [1.75]$$

$$W^{j+1}=W^j+\tau\theta^{j+1/2},j=\overline{0,J-1}, \qquad [1.76]$$

$$W^{j+1}(x,y)=\sum_{i=1}^{N'}\alpha_i^{j+1}\,\Phi_i(x,y),\ \ \theta^{j+1}(x,y)=\sum_{i=1}^{N'}d_i{}^{j+1}\Phi_i(x,y)\,,j=\overline{0,J-1},$$

Here, $\{W^j(x,y)\}_{j=0}^{J} \subset Z_t^N$ is the approximate time-discrete solution, $\Phi_i(r,z)$, $i=\overline{1,N}$ are the basic vector functions.

The scheme [1.74]–[1.76] can be rewritten in the form

$$\alpha^0 = \widetilde{M}^{-1}\overline{W}^0, \quad d^0 = \widetilde{M}^{-1}\overline{W}^1,$$

$$\left(M + \frac{\tau}{2}\bar{M} + \frac{\tau^2}{4}L\left(\frac{W^{j+1}+W^j}{2}\right)\right)d^{j+1} = \left(M - \frac{\tau}{2}\bar{M} - \frac{\tau^2}{4}L\left(\frac{W^{j+1}+W^j}{2}\right)\right)d^j -$$

$$-\tau\, L\left(\frac{W^{j+1}+W^j}{2}\right)\alpha^j + \frac{\tau}{2}\left(\tilde{F}^{j+1}+\tilde{F}^j\right),$$

$$\alpha^{j+1} = \alpha^j + \frac{\tau}{2}\left(d^{j+1}+d^j\right), \quad j=\overline{0,J-1}. \qquad [1.77]$$

Matrices $\widetilde{M}$, M, $\bar{M}$ and $L=A+B+R$ have dimension $(N''\times N')$ ($N'=4N$ is the quantity of basis functions corresponding to all N nodal points in the partition of the domain $\overline{\Omega}$, $N''=N'-N^1$, N^1 is the quantity of basis functions corresponding to nodal points on sections with boundary conditions of the first type) and structure

$$\widetilde{M} = \begin{pmatrix} M_1 & 0 & 0 \\ 0 & M_2 & 0 \\ 0 & 0 & M_3 \end{pmatrix}, M = \begin{pmatrix} 0 & 0 & 0 \\ 0 & 0 & 0 \\ 0 & 0 & \rho_{ws}M_3 \end{pmatrix}, \bar{M} = \begin{pmatrix} \tilde{\mu}M_1 & 0 & 0 \\ 0 & c_T M_2 & 0 \\ 0 & 0 & 0 \end{pmatrix},$$

$$A = \begin{pmatrix} A_1 & 0 & 0 \\ 0 & A_2 & 0 \\ 0 & 0 & A_3 \end{pmatrix}, B = \begin{pmatrix} 0 & 0 & 0 \\ B_1 & 0 & 0 \\ 0 & B_2 & 0 \end{pmatrix}, R = \begin{pmatrix} R_1 & 0 & 0 \\ 0 & R_2 & 0 \\ 0 & 0 & R_3 \end{pmatrix},$$

$$M_1 = M_2 = \{m_{si}\}_{s=\overline{1,N''}}^{i=\overline{1,N'}},\; M_3 = \{M_{si}\}_{s=\overline{1,N''}}^{i=\overline{1,N'}},\; M_{si} = \begin{pmatrix} m_{si} & 0 \\ 0 & m_{si} \end{pmatrix},$$

$$m_{si} = \iint_\Omega \varphi_i(x,y)\varphi_s(x,y)d\Omega,\; s=\overline{1,N''}, i=\overline{1,N'};$$

$$A_1 = \{a_1^{si}\}_{s=\overline{1,N''}}^{i=\overline{1,N'}},\; a_1^{si} = \iint_\Omega K_f\left(W^{j+1/2}\right)\left(\frac{\partial\varphi_i}{\partial x}\frac{\partial\varphi_s}{\partial x} + \frac{\partial\varphi_i}{\partial y}\frac{\partial\varphi_s}{\partial y}\right)d\Omega,$$

$$A_2 = \{a_2^{si}\}_{s=\overline{1,N''}}^{i=\overline{1,N'}},\; a_2^{si} = \iint_\Omega \lambda_T\left(\frac{\partial\varphi_i}{\partial x}\frac{\partial\varphi_s}{\partial x} + \frac{\partial\varphi_i}{\partial y}\frac{\partial\varphi_s}{\partial y}\right)d\Omega,$$

$$A_3 = \{A_{si}\}_{s=\overline{1,N''}}^{i=\overline{1,N'}},\; A_{si} = \begin{pmatrix} a_{11}^{si} & a_{12}^{si} \\ a_{21}^{si} & a_{22}^{si} \end{pmatrix},$$

$$a_{11}^{si} = \iint_{\Omega} \left((\lambda + 2\mu) \frac{\partial \varphi_i}{\partial x} \frac{\partial \varphi_s}{\partial x} + \mu \frac{\partial \varphi_i}{\partial y} \frac{\partial \varphi_s}{\partial y} \right) d\Omega,$$

$$a_{12}^{si} = \iint_{\Omega} \left(\lambda \frac{\partial \varphi_i}{\partial y} \frac{\partial \varphi_s}{\partial x} + \mu \frac{\partial \varphi_i}{\partial x} \frac{\partial \varphi_s}{\partial y} \right) d\Omega, \; a_{21}^{si} = \iint_{\Omega} \left(\lambda \frac{\partial \varphi_i}{\partial x} \frac{\partial \varphi_s}{\partial y} + \mu \frac{\partial \varphi_i}{\partial y} \frac{\partial \varphi_s}{\partial x} \right) d\Omega,$$

$$a_{22}^{si} = \iint_{\Omega} \left((\lambda + 2\mu) \frac{\partial \varphi_i}{\partial y} \frac{\partial \varphi_s}{\partial y} + \mu \frac{\partial \varphi_i}{\partial x} \frac{\partial \varphi_s}{\partial x} \right) d\Omega,$$

$$B_1 = \{B_{si}^1\}_{s=\overline{1,N''}}^{i=\overline{1,N'}}, B_2 = \{B_{si}^2\}_{s=\overline{1,N''}}^{i=\overline{1,N'}},$$

$$B_{si}^2 = \begin{pmatrix} \tilde{b}_1^{si} \\ \tilde{b}_2^{si} \end{pmatrix}, \tilde{b}_1^{si} = \iint_{\Omega} \rho_w g \varphi_i \frac{\partial \varphi_s}{\partial x} d\Omega, \tilde{b}_2^{si} = \iint_{\Omega} \rho_w g \varphi_i \frac{\partial \varphi_s}{\partial y} d\Omega,$$

$$R_1 = \{\tilde{r}_1^{si}\}_{s,i=1}^{N}, \tilde{r}_1^{si} = \int_{\gamma} r_1 \left(W^{j+\frac{1}{2}} \right) q_{si} d\gamma, q_{si} = [\varphi_s][\varphi_i],$$

$$R_2 = \{\tilde{r}_2^{si}\}_{s,i=1}^{N}, \tilde{r}_2^{si} = \int_{\gamma} r_2 (W^{j+1/2}) q_{si} d\gamma, \; q_{si} = [\varphi_s][\varphi_i],$$

$$R_3 = \{R_3^{si}\}_{s=\overline{1,N''}}^{i=\overline{1,N'}}, \; R_3^{si} = \begin{pmatrix} \tilde{r}_{11}^{si} & \tilde{r}_{12}^{si} \\ \tilde{r}_{21}^{si} & \tilde{r}_{22}^{si} \end{pmatrix}, \; s = \overline{1, N''}, \; i = \overline{1, N'},$$

$$\tilde{r}_{11}^{si} = \int_{\gamma} R(\widehat{W}^{j+1/2}) m^2 \, q_{si} d\gamma, \tilde{r}_{12}^{si} = \tilde{r}_{21}^{si} = \int_{\gamma} R(\widehat{W}^{j+1/2}) l m q_{si} d\gamma,$$

$$\tilde{r}_{22}^{si} = \int_{\gamma} R(\widehat{W}^{j+1/2}) l^2 \, q_{si} d\gamma, q_{si} = [\varphi_s] \, [\varphi_i].$$

The vectors $\overline{W}^0, \overline{W}^1, \tilde{F}$ have the structure

$$\overline{W}^0 = \begin{pmatrix} \overline{W}_1^0 \\ \overline{W}_2^0 \\ \overline{W}_3^0 \end{pmatrix}, \overline{W}^1 = \begin{pmatrix} \overline{W}_1^1 \\ \overline{W}_2^1 \\ 0 \end{pmatrix}, \tilde{F} = \begin{pmatrix} 0 \\ 0 \\ \tilde{F}^3 \end{pmatrix}$$

where

$$\overline{W}_{1,s}^0 = \iint_{\Omega} H_0(x, y) \varphi_s(x, y) d\Omega, \; \overline{W}_{1,s}^1 = \iint_{\Omega} \tilde{H}_0(x, y) \varphi_s(x, y) d\Omega, s = \overline{1, N},$$

$$\overline{W}_{2,s}^0 = \iint_{\Omega} T_0(x, y) \varphi_s(x, y) d\Omega, \overline{W}_{2,s}^1 = \iint_{\Omega} \tilde{T}_0(x, y) \varphi_s(x, y) d\Omega, s = \overline{1, N''},$$

$$\overline{W}^0_{3,2s-1} = \iint_\Omega U_0(x,y)\varphi_s(x,y)d\Omega\,, \overline{W}^0_{3,2s} = \iint_\Omega V_0(x,y)\varphi_s(x,y)d\Omega, s = \overline{1,N''},$$

$$\tilde{F}^3_{2s-1}(t) = \iint_\Omega \rho_w g \frac{\partial \varphi_s}{\partial x} d\Omega,$$

$$\tilde{F}^3_{2s}(t) = \iint_\Omega g\left(\rho_w \frac{\partial \varphi_s}{\partial y} - \rho_{ws}\varphi_s(x,y)\right) d\Omega - \int_{BK} \rho_{mat}\ g(y_3 - y_0)\varphi_s(x,y)\, d\Gamma -$$

$$-\int_{KF} \rho_{mat} g(\tilde{y}(x) - y_0)\varphi_s(x,y)\, d\Gamma -$$

$$-\int_{FC} (\rho_{mat}(\tilde{y}(x) - y_0) + \rho_w(y_2 - \tilde{y}(x)))g\varphi_s(x,y)d\Gamma +$$

$$+\int_{AB} \rho_w\ g(y_1 - y))\varphi_s(x,y)d\Gamma + \int_{CD} \rho_w\ g\left(y_2 + \tilde{h} - y\right)\varphi_s(x,y)d\Gamma, s = \overline{1,N''},$$

$l = cos(n,x), m = cos(n,y)$, n is the external normal to the contour.

The sought solution vector $\alpha(N'')$ of the problem [1.77] on the j-th time layer, $j = \overline{1,J}$, has the following structure:

$$\alpha(t_j) = (\alpha_1(t_j),\ \alpha_2(t_j),\ \alpha_3(t_j))^T = (\alpha_{1,1}(t_j), \alpha_{1,2}(t_j), \ldots, \alpha_{1,N}(t_j),$$

$$\alpha_{2,1}(t_j), \alpha_{2,2}(t_j), \ldots \alpha_{2,N}(t_j), \alpha_{3,1}(t_j), \ldots, \alpha_{3,2N}(t_j))^T.$$

We can supplement the system with rectangular matrices to the system with square matrices of dimension $(N' \times N')$ according to the standard procedure in accordance with the global numbering of nodes in the partitioned area, taking into account the main boundary conditions for N^1 basis functions corresponding to the nodal points on the sections with boundary conditions of the first kind.

Calculation of parameter-criteria for the durability of soil structures

The local durability of the soil is determined by some combination of acting stresses, at which the soil passes into the limit state. The limit state is characterized by the fact that soil deformations are stabilized. The limit state is accompanied by a progressive process of accumulation of plastic deformations.

It is believed that in the case of a flat problem, the destruction occurs due to the displacement of one part of the rock over another, and the limit state at the displacement sites is characterized by Coulomb's law (Zaretsky and Lombardo 1983)

$$\tau_\alpha^{lim} = \sigma_\alpha tg\varphi + c$$

where α shows the direction of landslide sites (the most dangerous sites) to the main sites; τ_α^{lim} is the tangential stress; σ_α is the normal stress, φ is the friction angle and c is the shear resistance characteristic (coefficient of soil adhesion).

The durability of the rock is determined by the size

$$\eta = \frac{\tau_\alpha^{lim}}{\tau_\alpha} \quad [1.78]$$

(the margin of safety factor). At $\eta > 1$, the rock has a margin of strength; at $\eta = 1$, it is in a state of limit equilibrium; at $\eta < 1$, the strength of the rock is broken.

Areas inclined to the main ones at an angle $\pi/4 \pm \varphi/2$ are considered, according to the Coulomb–Mohr criterion, to be the most dangerous directions, in which, at the limit ratio between the main stresses, non-damping slips will occur. Then, the Coulomb equation corresponds to the following limit equilibrium condition:

$$\frac{\sigma_1 - \sigma_2}{\sigma_1 + \sigma_2 + 2c \cdot ctg\varphi} = sin\,\varphi,$$

where $\sigma_1,\ \sigma_2$ are the principal stresses,

$$\tau_\alpha = 1/2(\sigma_1 - \sigma_2)\,cos\,\varphi,$$

$$\sigma_\alpha = 1/2(\sigma_1 + \sigma_2) - 1/2(\sigma_1 - \sigma_2)\,sin\,\varphi,$$

$$\tau_\alpha^{lim} = 1/2(\sigma_1 + \sigma_2)tg\,\varphi - 1/2(\sigma_1 - \sigma_2)\,sin\,\varphi\ \ tg\,\varphi + c,$$

$$\eta = \frac{\tau_\alpha^{lim}}{\tau_\alpha} = \frac{(\sigma_1 + \sigma_2 + 2c \cdot ctg\varphi)}{(\sigma_1 - \sigma_2)\,cos^2\,\varphi} - \frac{sin^2\,\varphi}{cos^2\,\varphi}.$$

The values of principal stresses and principal elastic deformations are determined by the formulas

$$\sigma_{1,2} = \frac{c}{2} \pm \frac{1}{2}\sqrt{(\sigma_x - \sigma_y)^2 + 4\tau_{xy}^2},$$

$$\varepsilon_{1,2} = \frac{\varepsilon_x + \varepsilon_y}{2} \pm \frac{1}{2}\sqrt{(\varepsilon_x - \varepsilon_y)^2 + \varepsilon_{xy}^2}$$

where the components of deformations and stresses are calculated based on Hooke's law through the components of displacements of the soil skeleton as follows:

$$\varepsilon_x = \frac{\partial u_{sk}}{\partial x}, \varepsilon_y = \frac{\partial v_{sk}}{\partial y}, \gamma_{xy} = \frac{\partial u_{sk}}{\partial y} + \frac{\partial v_{sk}}{\partial x}, \sigma_x = \lambda\left(\frac{\partial u_{sk}}{\partial x} + \frac{\partial v_{sk}}{\partial y}\right) + 2\mu\frac{\partial u_{sk}}{\partial x},$$

$$\sigma_y = \lambda\left(\frac{\partial u_{sk}}{\partial x} + \frac{\partial v_{sk}}{\partial y}\right) + 2\mu\frac{\partial v_{sk}}{\partial y}, \tau_{xy} = \mu\left(\frac{\partial u_{sk}}{\partial y} + \frac{\partial v_{sk}}{\partial x}\right).$$

The analysis of simulation results includes the problem [1.55], [1.64]–[1.72], which was solved both without and with the consideration of the thin inclusion (curtain) NQ in the domain shown in Figure 1.1.

a) Absence of the anti-filtration curtain

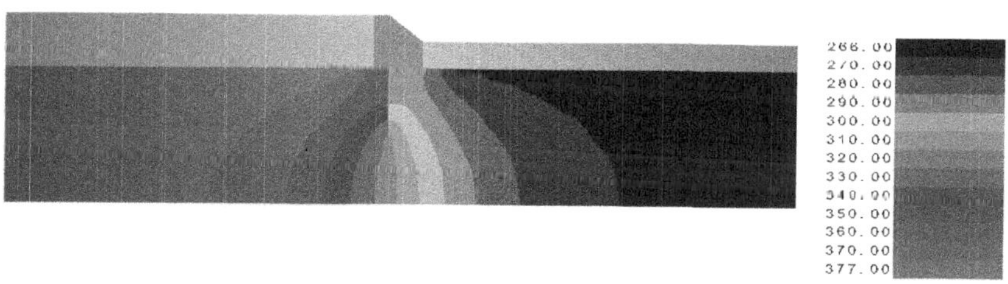

b) Presence of the anti-filtration curtain NQ, $r_1 = 0$:

Figure 1.10. *Pressure isolines, t* = 90 *days. For a color version of this figure, see www.iste.co.uk/koroliouk/computational2.zip*

The initial conditions are calculated with a filtration coefficient that corresponds to a temperature of 13.5°C.

Test simulations were carried out for the case of the presence of a weakly durable ($R = 10$) anti-filtration curtain NQ ($r_1 = 0$), with high heat-insulating properties ($r_2 = 0$) for a period of 135 days.

Figures 1.10 and 1.11 show the effect of a waterproof curtain on the distribution of pressure and water level in the lower bay.

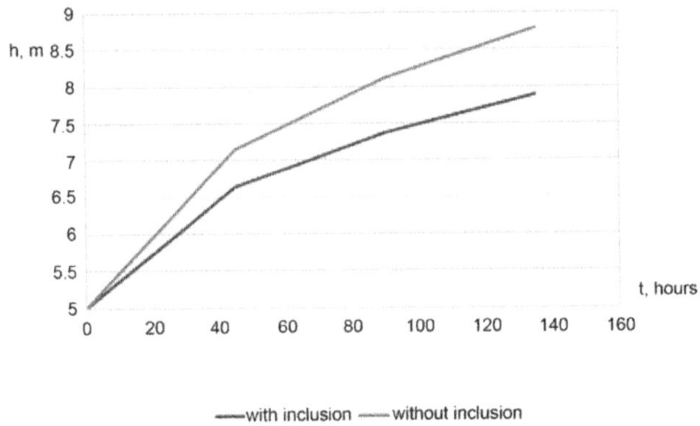

Figure 1.11. *Change in water level in the lower bay. For a color version of this figure, see www.iste.co.uk/korolioúk/computational2.zip*

a) Absence of the curtain, $t = 0$ days

b) Absence of the curtain, $t = 90$ days

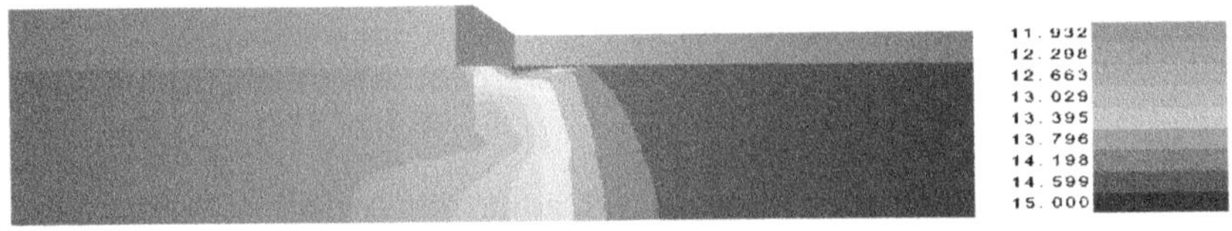

c) Presence of the curtain NQ, $r_2 = 0$, $t = 90$ days

Figure 1.12. *Temperature isolines. For a color version of this figure, see www.iste.co.uk/korolioúk/computational2.zip*

Figure 1.12 shows temperature isolines in the absence and the presence of the curtain NQ with the parameter $r_2 = 0$.

a) Absence of the curtain

b) Presence of the curtain NQ, R = 10

Figure 1.13. *Isolines of displacements along y, $t = 90$ days. For a color version of this figure, see www.iste.co.uk/koroliouk/computational2.zip*

a) Absence of the curtain

b) Presence of the curtain NQ, R = 10

Figure 1.14. *Isolines of η, $t = 90$ days. For a color version of this figure, see www.iste.co.uk/koroliouk/computational2.zip*

Figure 1.13 shows the isolines of vertical displacements of the soil skeleton at the time of 90 days in cases of the absence and the presence of a thin weakly persistent inclusion NQ ($R = 10$), at which a gap in tangential (in our case – vertical) displacements is allowed (condition [1.23] of the problem statement).

Figure 1.14 shows the isolines of the safety factor η [1.78] under the same conditions. It is clearly visible that the most stable zone under the dam body has much smaller dimensions in the presence of weakly persistent inclusion compared to its absence.

1.5. Parallel algorithm for AMLI preconditioner and its application to model soil massif state

The problem of solving systems of linear algebraic equations [1.77] at every step is fundamental in the application of the considered models of the stress–strain state of soil massifs. In many cases, such linear systems have ill-conditioned sparse matrices of large size. In such situations, iterative methods, particularly conjugate gradient method (CG) or stabilized biconjugate gradient method (BiCGstab) in Saad (2003), are mostly used for their solution.

The main technique used to improve convergence and accuracy while solving ill-conditioned linear systems is the use of preconditioners (Saad 2003; Chen 2005) – the matrices, whose multiplication on linear system matrix leads to the decrease of condition number. Since the ideal preconditioner is the inverse matrix, many of them are built as its approximation. The most common methods for constructing preconditioners are incomplete matrix decomposition (e.g. ILU decomposition) and incomplete matrix inversion (e.g. polynomial preconditioners). Among others, it is worth mentioning multigrid, such as algebraic multigrid (AMG) in Shapira (2003) and algebraic multilevel iteration (AMLI) in Chen (2005), and wavelet preconditioners.

Basis matrix method (Kudin et al. 2007) is one of the methods developed to analyze ill-conditioned and rectangular linear systems. This method iteratively constructs an inverse matrix according to orthogonalization procedure. On its base, incomplete inversion algorithm that can be used as a preconditioner was created. Further, it was combined (Bohaienko 2014) with the AMLI preconditioner, whose application for solving elasticity problems was studied in Margenov (1998).

Since the efficiency of preconditioners in most cases cannot be proved theoretically, experimental studies (Bohaienko 2014) have been carried out for AMLI combined with incomplete inversion by basis matrix method. They showed its efficiency for accelerating convergence rate of iterative algorithms while modeling stress–strain state by FEM. Among the known preconditioners, similar results could be achieved only with multigrid ones.

The speed of the AMLI preconditioner, similarly to multigrid preconditioners, in particular AMG, is low and its application takes most of the time spent on linear system solving. This makes the development of parallel schemes crucial for these algorithms. Parallel modifications of AMLI are poorly studied with existing works (Axelsson and Neythceva 1994; Neytcheva 1995) mainly focusing on the effect of restricting its V-cycle.

While forming linear systems in the process of solving the problem [1.55], [1.64]–[1.72], rows were grouped by the unknown functions, thus forming matrices with the following properties:

– block structure;

– high fill-in factor;

– high condition number due to different speeds of simulated processes.

The use of iterative methods for solving linear systems without preconditioners for the considered problems did not allow obtaining solutions of acceptable accuracy. Taking into account the features of linear system matrices, we have chosen the AMLI preconditioner with incomplete inversion by the basis matrix method to improve iterative algorithm convergence rate and performance.

Preconditioner on the base of AMLI and basis matrix method

AMLI preconditioner (Axelsson and Neythceva 1994) is based on the representation of input matrix as having 2x2 blocks with its subsequent LU factorization:

$$A = \begin{bmatrix} A_{11} & A_{12} \\ A_{21} & A_{22} \end{bmatrix} = \begin{bmatrix} I & \\ A_{21}A_{11}^{-1} & I \end{bmatrix} \begin{bmatrix} A_{11} & A_{12} \\ & S \end{bmatrix}, \qquad [1.79]$$

where $S = A_{22} - A_{21}A_{11}^{-1}A_{12}$ is the Schur complement of A.

After approximating the Schur complement with polynomial and recursive application of the expansion [1.79] to the matrix A_{22}, the following algorithm for constructing a preconditioner that approximates A^{-1} matrix is obtained:

$$\begin{cases} S^{(k-1)} = A^{(k-1)}\left[I - P_v\left({M^{(k-1)}}^{-1} A^{(k-1)}\right)\right]^{-1} \\ M^{(k)} = \begin{bmatrix} I & \\ A_{21}^{(k)} {A_{11}^{(k)}}^{-1} & I \end{bmatrix} \begin{bmatrix} A_{11}^{(k)} & A_{12}^{(k)} \\ & S^{(k-1)} \end{bmatrix} \end{cases}. \qquad [1.80]$$

In the simplest case, when $P_v(t) = P_1(t) = 1 - t$, recursive scheme [1.80] is simplified to

$$M^{(k)} = \begin{bmatrix} I & \\ A_{21}^{(k)} A_{11}^{(k)^{-1}} & I \end{bmatrix} \begin{bmatrix} A_{11}^{(k)} & A_{12}^{(k)} \\ & M^{(k-1)} \end{bmatrix}. \qquad [1.81]$$

It is worth noting that $A_{11}^{(k)^{-1}}$ matrices included in [1.81] may be approximated by incomplete decomposition methods such as ILU factorization (Li et al. 2003; Saad 2007).

We propose to approximate these matrices by incomplete decomposition using the basis matrix method (Bohaienko and Kudin 2014). Since the accuracy of the approximation may vary on the stages of recursive procedure, we limit it from below. When the approximation $\tilde{A}$ of inverse matrix A^{-1} is not precise enough, complete inversion ($\tilde{A} = A^{-1}$) is performed. Since the estimation of approximation accuracy as $\varepsilon_1 = \|A\tilde{A}\|$ is computationally complex, we will approximate it as $\varepsilon_2 = max_i(A\tilde{A})_{ii}$.

The computational scheme for the preconditioning procedure [1.81] is as follows (Bohaienko 2020).

The matrix A is divided into $(N + 1) \times (N + 1)$ blocks in the following way:

$$A = \begin{bmatrix} A_{11}^{(0)} & A_{121}^{(0)} & \cdots & A_{12N}^{(0)} \\ A_{211}^{(0)} & \cdots & \cdots & \cdots \\ \cdots & \cdots & A_{11}^{(N-1)} & A_{12N}^{(N-1)} \\ A_{21N}^{(0)} & \cdots & A_{21N}^{(P-1)} & A_{11}^{(N)} \end{bmatrix},$$

$$A_{12}^{(i)} = \left(A_{12,i+1}^{(i)}, \dots, A_{12N}^{(i)}\right),\ A_{21}^{(i)} = \left(A_{21,i+1}^{(i)}, \dots, A_{21N}^{(i)}\right).$$

Considering an application of preconditioner [1.80] as a solution of linear system $M^{(N)}x = b,\ \ x = (x_0, \dots, x_N),\ \ b = (b_0, \dots, b_N)$, it can be performed by a two-stage procedure, V-cycle, consisting of "direct" and "reverse" parts.

On the "direct" part of the procedure, values of the following vectors are sequentially computed:

$$w_0 = \tilde{A}_{11}^{(0)} b_0, w_i = b_i - A_{21}^{(i)} \tilde{A}_{11}^{(i)} w_{i-1}, i = 1, \dots, N. \qquad [1.82]$$

The "reverse" part consists of the sequential computation of the resulting vector:

$$x_N = \tilde{A}_{11}^{(N)} w_N,\ x_i = w_i - \tilde{A}_{11}^{(i)} A_{12}^{(i)} (x_{i+1},\ldots,x_N), i = 0,\ldots,N-1. \qquad [1.83]$$

Algorithm of incomplete inversion by basis matrix method

Incomplete transformations restrict the set of changeable matrix elements in the process of its decomposition, for example, by LU (Bohaienko 2020), or inversion, for example, by the Gramm–Schmidt method (Liu and Zhang 2013). The matrix that approximates the inverse one is thus obtained and can be used to calculate $\tilde{A}_{11}^{(i)}$ matrices in the above-described modification of AMLI algorithm.

The algorithm of incomplete inversion of matrix A, $dim\, A = [n,n]$ by basis matrix method (Bohaienko and Kudin 2014), taking into account its similarity to the Gramm–Schmidt method, can be stated as follows:

Let $A_b^{(i)}$, $dim\, A_b^{(i)} = [n,n]$ be a basis matrix on the i-th iteration of the algorithm, $A_b^{(i)}[j]$ is a column of matrix $A_b^{(i)}$, a_i is a row of matrix A, $A_b^{(0)} = A_b^0$, and R is a set of nonzero element indices for matrix A.

Then, the i-th iteration of the algorithm is as follows:

1) $v = diag\left(AA_b^{(i-1)}\right)$.

2) Set k to be the index of the element of vector v with maximal absolute value.

3) $\alpha = a_k A_b^{(i-1)}$.

4) $A_b^{(i)}[k] = A_b^{(i-1)}[k]/\alpha_k$.

5) $A_b^{(i)}[l]_j = A_b^{(i-1)}[l]_j - \alpha_l A_b^{(i)}[k]_j, l \neq k, (l,j) \in R$.

The algorithm yields $A_b^{(n)} = \tilde{A} \approx A^{-1}$ after completing n iterations.

The following condition must be fulfilled on iterations of the basis matrix method:

$$a_k A_b^{(i)}[l] = \begin{cases} 1, & l = k \\ 0, & l \neq k \end{cases}.$$

Fulfillment of this condition is violated when doing incomplete transformations. The algorithm of incomplete decomposition can be modified using the correction procedure which ensures its fulfillment.

To do this, the following step is added to the algorithm:

6) If $(l,j) \notin R$, $a_{kj} \neq 0$, and $\exists m = \max_{j}|a_{kj}|, A_b^{(i-1)}[l]_j \neq 0$, do the following transformation: $A_b^{(i)}[l]_m = A_b^{(i)}[l]_m - \alpha_l A_b^{(i)}[k]_j a_{kj}/a_{km}$.

Parallel implementation of the preconditioner

To make AMLI algorithm with incomplete inversion by basis matrix method run in parallel on distributed memory systems, the following data partitioning scheme can be applied (Bohaienko 2020). Each of i row blocks $(A_{21i}^{(0)}, \ldots, A_{21i}^{(i-1)}, A_{11}^{(i)}, A_{12,i+1}^{(i)}, \ldots, A_{12P}^{(i)})$ is distributed on the system of P processes by blocks of rows. Thus, each process stores blocks of rows of $A_{11}^{(i)}$, $A_{12}^{(i)}$, $A_{21j}^{(i)}$ submatrices and can perform multiplications that arise while applying the preconditioner [1.80] in parallel. Computational scheme used for multiplying sparse matrix A of $M \times M$ size on vector v has the following form:

1) Process p stores block $B_p = [r_{p0}, r_{p1}]$, $\cup_{i=0}^{P} B_i = [0, \ldots, M]$, $B_i \cap B_j = \emptyset \forall i,j$ of matrix A rows and corresponding elements of vector v.

2) Denote a set of matrix A columns in which nonzero elements are present in row block B_p as C_p.

3) Process p sends values of vector v elements from an index set $C_i \cap B_p$ to processes i, $C_i \cap B_p \neq \emptyset$ and receives values of elements from an index set $C_p \cap B_j$ from processes j, $C_p \cap B_j \neq \emptyset$.

4) After synchronization is performed on step 3, each process independently multiplies a block of matrix A rows on a vector v. The result of multiplication has the same partitioning as vector v has.

Performing "direct" part [1.82] of the AMLI algorithm, when vector multiplications on $A_{21}^{(i)}$ matrix are organized as successive multiplications on $A_{21j}^{(i)}$ submatrices, synchronizations are done once according to the fill-in pattern of a complete $A_{21}^{(i)}$ matrix.

Similarly, when non-changeable component x_i of the result vector x is multiplied on $A_{12i}^{(j)}$, $j < i$ submatrices during "reverse" part [1.83], synchronization can be performed only once according to the combined fill-in pattern of $A_{12i}^{(j)}$ submatrices.

Matrices $\tilde{A}_{11}^{(i)}$ can be obtained in parallel the following way (Bohaienko 2020):

1) Process p handles the sequence $k_{p2} > k > k_{p1}$ of $A_{11}^{(k)}$ matrices. The sequences are constructed to make processes be evenly loaded as far as it is possible. This is ensured by approximately the same number of non-zero elements in $A_{11}^{(k)}$ matrices that are processed by each of the processes.

2) Processes exchange the values of $A_{11}^{(i)}$ matrices' elements to make $A_{11}^{(k)}$, $k_{p2} > k \geq k_{p1}$ matrices completely stored in process p memory. Process p sends row blocks B_p of $A_{11}^{(k)}$, $k_{i2} > k > k_{i1}$ matrices to processes $i, j \neq p$ and receives row blocks B_i of $A_{11}^{(k)}$, $k_{p2} > k \geq k_{p1}$ matrices from them.

3) Each process independently performs complete or incomplete inversion of the matrices.

4) Processes exchange the values of $\tilde{A}_{11}^{(i)}$ matrices' elements to restore the initial data partitioning. Process p sends row blocks B_i of $\tilde{A}_{11}^{(k)}$, $k_{p2} > k \geq k_{p1}$ matrices to processes $i, j \neq p$ and receives row blocks B_p of $\tilde{A}_{11}^{(k)}$, $k_{i2} > k \geq k_{i1}$ matrices from them.

Since data partitioning of blocks of rows within $A_{12}^{(i)}$, $A_{21i}^{(j)}$ matrices may differ for different values of i, additional synchronization should be performed when vectors are multiplied on these matrices. Denote a block of rows from $A_{12}^{(i)}$, $A_{21i}^{(j)}$, submatrices that is handled by process p as B_{pi}.

Then, before multiplying a vector v with a partitioning $B_i = (B_{0i}, \ldots, B_{Pi})$ on a matrix $A_{21}^{(i)}$, we need to synchronize its elements to make each process store the block $B_{pi} \cup \ldots \cup B_{pN}$. To do this, process p sends block $B_{pi} \cap B_{rj}$ of vector elements to process r if $B_{pi} \cap B_{rj} \neq \emptyset, j \neq i$. Then, process p receives a corresponding block from a process r if $B_{pj} \cap B_{ri} \neq \emptyset, j \neq i$.

Similar procedures should be performed when vector $v = (v_{i+1}, \ldots, v_N)$ is multiplied on a matrix $A_{12}^{(i)}$ for each of its components. When non-changeable component x_j is multiplied on $A_{12}^{(i)}$ matrices during the "reverse" part [1.83] of the AMLI algorithm, the synchronization can be performed once before calculating a result of the first multiplication operation.

When the preconditioner is used with the iterative method for solving the linear algebraic system, calculation of matrices $\tilde{A}_{11}$ and sets of elements that need synchronization, is performed once on the first iteration. The execution of these operations will be further denoted as "initialization procedure".

Estimation of algorithm performance includes estimating algorithm performance on the basis of the following assumptions:

– $A_{11}^{(i)}$ matrices have the same fill-in factor k_{11}, and $A_{12j}^{(i)}$, $A_{21j}^{(i)}$ matrices – the same fill-in factor k_2 which do not depend on block size.

– The percentage k_f of $A_{11}^{(i)}$ matrices for which full inverse procedure is applied does not depend on block size.

– Inverse matrices of $A_{11}^{(i)}$ are completely filled.

– The number of elements that each process must synchronize before performing a parallel multiplication is considered proportional (with the coefficient of proportionality equal to k_s) to the number of non-zero elements in the block of matrix rows. We assume that each process performs only one pair of data exchange operations doing the synchronization.

– Submatrices are considered evenly distributed across the system.

Performing calculations on [1.82], [1.83], it is necessary to do $2N$ matrix–vector multiplications for matrix $\tilde{A}_{11}^{(i)}$, $N(N-1)$ multiplications for $A_{12j}^{(i)}$, $A_{21j}^{(i)}$ submatrices, and $2N$ per-component operations with vectors. All of these operations are carried out with vectors of size $\frac{n}{N}$. Computation time of the sequential algorithm spent to perform calculations on [1.82], [1.83] can thus be estimated as in Bohaienko (2020):

$$T_1(n,N) = k_p\left[\frac{n^2}{N}\left(2\left(k_f + (1-k_f)k_{11}\right) + (N-1)k_2\right) + 4n + n(N-1)\right] \quad [1.84]$$

where k_p is the performance coefficient.

Minimal time, in case the condition $2\left(k_f + (1-k_f)k_{11}\right) - k_2 \geq 0$ is fulfilled, is achieved here for

$$N_m = \sqrt{\left(2\left(k_f + (1-k_f)k_{11}\right) - k_2\right)n}. \quad [1.85]$$

Time spent on $\tilde{A}_{11}^{(k)}$ matrices' computation with the assumption that computational complexity of one step of incomplete basis matrix method is proportional to the maximal number of non-zero elements in rows of $A_{11}^{(k)}$ matrices (Bohaienko 2020) can be estimated as follows:

$$T_{i1}(n,N) = Nk_{p2}\left(k_f\frac{n^3}{N^3} + k_{11}\frac{n^2}{N^2}\right) \quad [1.86]$$

where k_{p2} is a performance coefficient.

We evaluate the computation time of parallel algorithm using the linear model of pairwise data exchange operation:

$$T_s(n) = k_e n + k_a \quad [1.87]$$

where k_e is the bandwidth, n is the data size and k_a is the latency.

Estimating parallel computations' performance, we assume that the execution of arithmetic operations is parallelized linearly. Each operation of matrix–vector multiplication in this case needs data synchronization. With the above-mentioned assumptions, the following data synchronizations should be performed:

– $2k_f N$ synchronizations when vectors are multiplied on inverted $A_{11}^{(i)}$ matrices. Considering such matrices are completely filled, such operations require each process to perform $P-1$ exchanges with data block size equal to $\frac{n}{NP}$ elements.

– $2(1-k_f)N$ synchronizations when vectors are multiplied on approximately inverted $A_{11}^{(i)}$ matrices, and $2(N-1)$ synchronizations when vectors are multiplied on $A_{12}^{(i)}$, $A_{21}^{(i)}$ matrices. Taking into account the above assumptions, such operations require each process to perform one exchange with data block size equal to $k_s k_{11}\frac{n}{NP}$ or $k_s k_2\frac{n}{NP}$ correspondingly.

Taking into account [1.87], the total time spent by the parallel algorithm on performing calculations on [1.82], [1.83] can be estimated as in Bohaienko (2020):

$$T_P(n,N) = \frac{1}{P}T_1(n,N) + 2nk_e\left[\frac{k_s}{P}\left((1-k_f)k_{11} + \frac{N-1}{N}k_2\right) + \frac{P-1}{P}k_f\right]$$

$$+(2N(2 + k_f P - 2k_f) - 2)k_a \quad [1.88]$$

where P is the number of processes.

Minimal time spent on data exchanges for a fixed N is achieved for the number of processes is equal to

$$P_m(N) = \sqrt{\frac{2nk_e k_s((1-k_f)k_{11} + \frac{N-1}{N}k_2)}{2k_a k_f N}}. \quad [1.89]$$

When doing parallel computations, the time spent to calculate $\tilde{A}_{11}^{(k)}$ matrices can be estimated with the assumption that each process performs incomplete or complete inversion of $\frac{N}{P}$ submatrices (we consider this number to be integer). Before processing each submatrix, each process executes $P-1$ exchange operations with data block size equal to $k_{11}\frac{n^2}{N^2P}$. After processing, $P-1$ exchange operations are done with data block size equal to $\frac{n^2}{N^2P}$ in the case of complete inversion and $k_{11}\frac{n^2}{N^2P}$ in the case of incomplete inversion.

Thus, the time spent on computing $\tilde{A}_{11}^{(k)}$ matrices in parallel can be estimated as in Bohaienko (2020):

$$T_{iP}(n,N) = \frac{1}{P}T_{i1}(n,N) + \frac{N}{P}(P-1)\left(\left(k_{11}\frac{n^2}{N^2P}k_e + k_a\right)\right.$$

$$\left. + \left(((1-k_f)k_{11} + k_f)\frac{n^2}{N^2P}k_e + k_a\right)\right). \qquad [1.90]$$

Algorithm testing includes a series of computational experiments being performed on the SCIT-3 cluster of the Institute of Cybernetics of National Academy of Sciences of Ukraine to study performance and convergence rate of the considered algorithms (Bohaienko 2020). Eight-core nodes based on Intel Xeon 5345 processors with 2GB RAM per core were used.

In these tests, we solved the linear systems that arise while modeling the stress–strain state of soil by model [1.55], [1.65]–[1.72], taking filtration into account in the domain depicted in Figure 1.15 using FEM. The matrix of corresponding linear system, further denoted as SD, has the size equal to 107856×107856 with an average number of non-zero elements in a row equal to 23.42. Its fill-in pattern is shown in Figure 1.15.

In all tests, linear systems were solved using the BiCGStab algorithm with the proposed modification of AMLI preconditioner. The submatrices $A_{11}^{(k)}$ were completely inverted when $\varepsilon_2 < 0.05$. This resulted in a complete inversion of ~50% of the submatrices.

As an acceptable accuracy of calculations, the 10^{-9} order was taken. The use of the BiCGStab algorithm itself, or with the procedure of matrix row normalization, did not allow the obtaining of solutions of the required accuracy. Computation time without the preconditioner usage was equal to 140 ms per iteration.

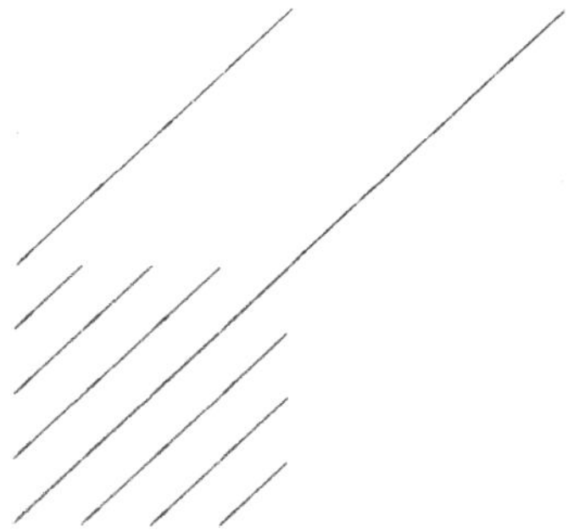

Figure 1.15. *Fill-in pattern of the matrix SD*

Experimental testing of the estimates of computation performance

In the first series of tests, linear system SD was solved using a single processor core with a variable size of blocks on which the matrix is split. Time spent on the preconditioner application was estimated by [1.84]. The values of the estimate's coefficients were, here and further, found using the least-squares method on the base of the measured time with block size ranging from 25 to 4,000. Some of the obtained data are given in Table 1.1 and Figure 1.16.

Block size which minimizes computation time was between 125 and 500, which corresponds to the calculated value of theoretical estimate [1.87] equal to 179.

Obtained experimental data confirm the properties of the algorithm reflected in the estimate [1.84]. The accuracy of the estimate [1.84] was not worse than 25% in the conducted experiments.

Block size	Time spent on calculation upon [1.82], [1.83], ms	Estimate of time spent on calculation upon [1.82], [1.83], ms	Estimation error
50	220	252	-14.63%
75	160	189	-18.32%
100	130	162	-24.63%
125	145	148	-2.73%
500	225	209	7.05%
1,000	380	364	4.19%

Table 1.1. *Characteristics of the sequential algorithm*

In the second series of tests, performance estimate [1.88] for the parallel algorithm was tested measuring the time spent on data exchanges between processes. Values of the estimate's [1.88] coefficients were found on the base of the measured time with block size ranging from 125 to 4,000 and the number of processes ranging from 4 to 40.

Time spent by the parallel algorithm on exchanges while doing calculations upon [1.82], [1.83] and its estimates with block size equal to 500 and 1,000 are shown in Figure 1.18. Estimation error here remains lower than 30% for block size ranging from 250 to 3,000 and the number of processes ranging from 12 to 24. In other cases, the error increased.

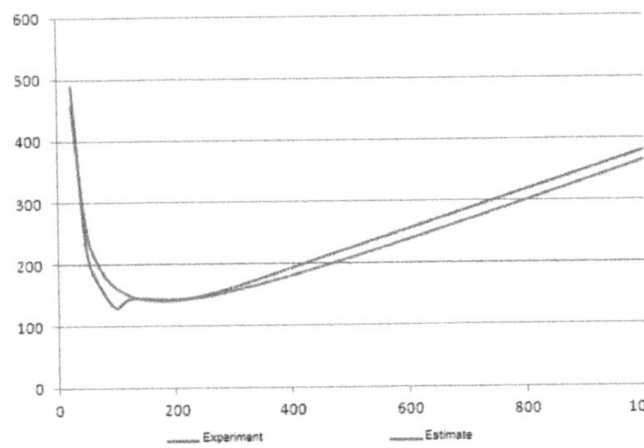

Figure 1.16. *Time spent on computations upon [1.82], [1.83]. For a color version of this figure, see www.iste.co.uk/koroliouk/computational2.zip*

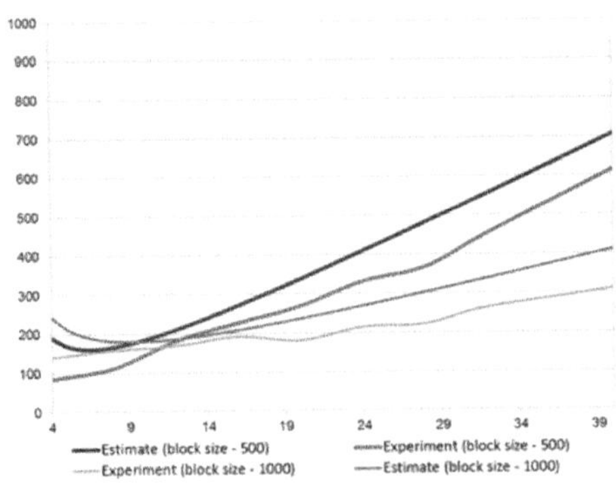

Figure 1.17. *Time spent on exchanges during calculations upon [1.82], [1.83]. For a color version of this figure, see www.iste.co.uk/koroliouk/computational2.zip*

The optimal number of processes measured during computational experiments and the number estimated according to [1.89], under which, with a fixed block size, we achieve the smallest amount of time spent on exchanges when performing calculations, is given in Table 1.2. Here, in most cases, the experimentally obtained values of the optimal number of processes were smaller than those calculated by the estimate [1.89].

Block size	Estimate of optimal number of processes according to [1.89]	Range of process number under which minimal execution time was achieved
250	4	2–4
500	6	2–4
1,000	9	4–8
2,000	12	8–12
4,000	18	16–20

Table 1.2. *Optimal number of processes*

Experimental testing of execution time of preconditioner's initialization

The time spent on computing matrices $\tilde{A}_{11}^{(k)}$ on one core and its estimate according to [1.86] depending on the block size are shown in Figure 1.18. Coefficients' values for the estimate [1.86] were obtained using the measured execution time with block size varying from 125 to 4,000.

The accuracy of estimation is lower than 12% for block size less than 3000, while block size increase leads to the increase of the error. These inaccuracies here may be caused by a change in the percentage of filly inverted submatrices.

In the case of the parallel algorithm, the corresponding time was measured and estimated according to [1.90] for block size ranging from 125 to 4,000 and the number of processes that ranges from 4 to 40.

The general behavior of the estimate [1.90] coincided with the measured time but estimation errors were significant. Time spent on calculation of $\tilde{A}_{11}^{(k)}$ matrices and its estimate depending on the number of processes for block size equal to 1,000 is shown in Figure 1.19.

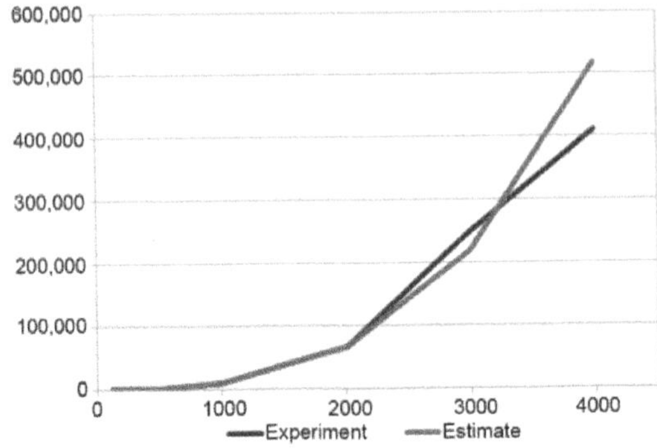

Figure 1.18. *Time, ms, spent on computing the values of* $\tilde{A}_{11}^{(k)}$ *and its estimate in the case of the sequential algorithm*

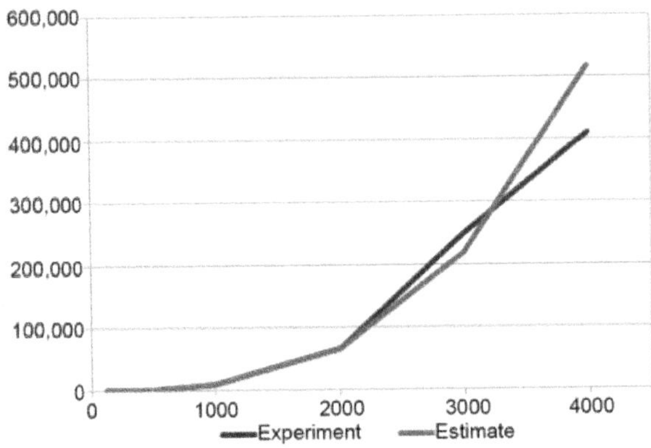

Figure 1.19. *Time, ms, spent on calculating* $\tilde{A}_{11}^{(k)}$ *matrices and its estimate in the case of the parallel algorithm*

Influence of preconditioner's parameters of the convergence of BiCGstab algorithm

The number of iterations required to achieve 10^{-9} order of accuracy was measured along with the overall time spent by BiCGstab algorithm and the proposed preconditioner running on a single processor core. Obtained data are presented in Table 1.3. Here, an increase of block size accelerates the convergence of the iterative algorithm along with the significant decrease of its performance. In this case, time spent on incomplete inversion of submatrices dominates in the total time.

Block size for which the required accuracy is reached in the minimal time was in the range from 75 to 125 elements.

A series of computational experiments were conducted to evaluate the impact of fully inverted submatrices $A_{11}^{(k)}$ percentage k_f on the algorithm's convergence rate and performance. The logarithm of residual at iterations of BiCGstab algorithm with the use of the considered preconditioner for block size equal to 1,000 with different number of completely inverted submatrices is shown in Figure 1.20. The convergence rate of the algorithm here decreases when the number of completely inverted submatrices is reducing and, starting from a particular value, it becomes impossible to obtain solutions of the required accuracy.

Block size	Number of iterations needed to acquire 10^{-9} order of accuracy	Time spent on incomplete $A_{11}^{(k)}$ matrices inversion, ms	Total time spent for acquiring 10^{-9} order of accuracy, ms
50	401	4,050	200,540
75	325	5,910	126,160
100	309	8,460	104,250
125	284	14,480	111,040
250	243	52,110	137,160
500	169	195,980	280,480
1,000	160	875,600	1,005,200
2,000	115	3,492,550	3,666,200

Table 1.3. *Total linear system solving time for the sequential algorithm*

The decrease in the number of fully inverted $A_{11}^{(k)}$ submatrices reduces the time spent on preconditioner's initialization and thus significantly increases the algorithm's performance in the case when a small number of computational resources is used (see Figure 1.22). On the other hand, initialization operations' speed-up is higher for higher values of k_f, so the time spent on initialization becomes close to equal for different values of k_f when the number of processes increase.

Time spent on parallel calculations upon [1.82], [1.83] behaves likewise (see Figure 1.22). The smaller number of completely inverted submatrices, in general, accelerates this operation, but speed-up decreases with the increase in the number of involved computational resources.

All this, given that convergence is faster for a larger number of completely inverted submatrices, leads to the situation when k_f reduction is effective only when a small number of processes are involved (see Figure 1.21) and in the ranges of algorithm's convergence (see Figure 1.20).

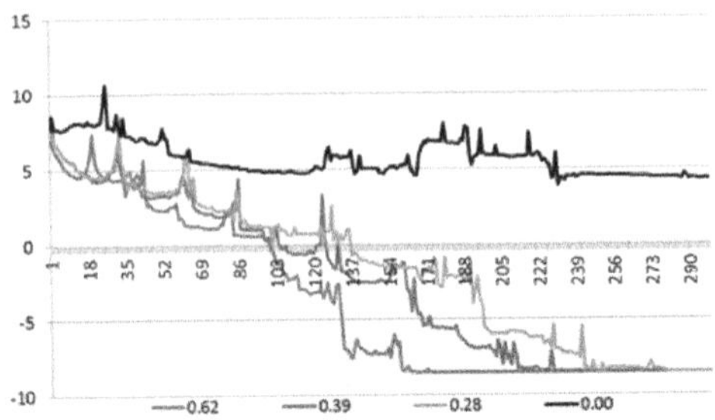

Figure 1.20. *The logarithm of residual at the iterations of the algorithm for different per cents of completely inverted submatrices. For a color version of this figure, see www.iste.co.uk/koroliouk/computational2.zip*

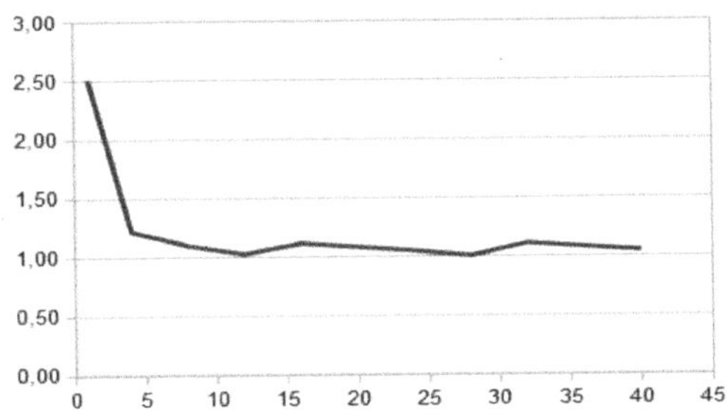

Figure 1.21. *Acceleration due to the decrease of the number of fully inverted matrices ($k_f = 0.39$ compared to $k_f = 0.62$ with respect to the number of processes)*

Acceleration from parallel algorithm's usage includes speed-up of calculations upon [1.82], [1.83] for different block sizes depending on the number of involved processes, which is given in Figure 1.23. Total speed-up while obtaining the linear system's solution with a 10^{-9} order of accuracy in comparison with execution on one processor core with the same block size is given in Figures 1.24 and 1.25, which show the total speed-up comparing parallel algorithm computation time with the best time obtained by the sequential algorithm.

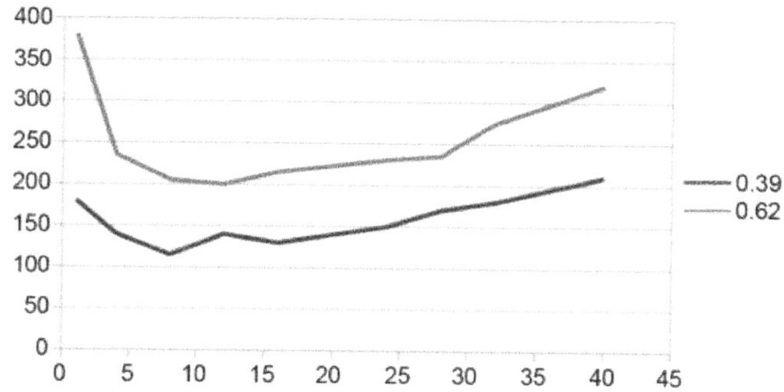

Figure 1.22. *Time, ms, spent on computing according to [1.82], [1.83] with respect to the number of processes for different per cent of fully inverted matrices. For a color version of this figure, see www.iste.co.uk/koroliouk/computational2.zip*

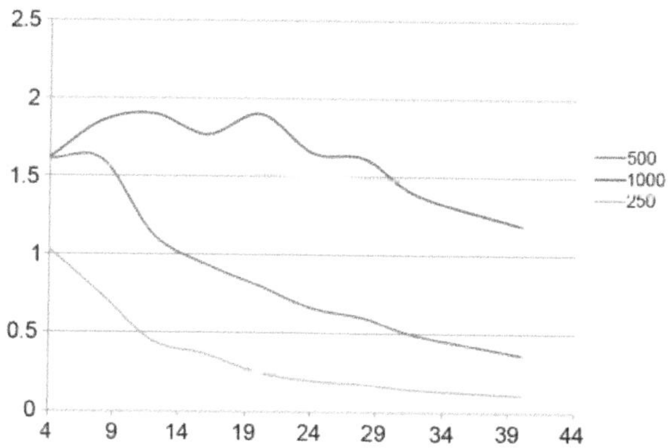

Figure 1.23. *Speed-up of the parallel algorithm performing computation upon [1.82], [1.83]. For a color version of this figure, see www.iste.co.uk/koroliouk/computational2.zip*

Comparing with the best time for which the solution of the required accuracy is obtained on one processor core, the speed-up was achieved only for block size equal to 250 and 500 elements. In this case, the maximal speed-up was equal to 24% when eight processes were used and block size was equal to 500 elements.

Maximal speed-up of calculations upon [1.82], [1.83] increases with block size increase, but the scalability of this operation is low due to a large number of exchanges that must be performed before each matrix–vector multiplication. Block size increase reduces the number of exchanges and only slightly changes their total volume that, as can be seen from Figure 1.23, leads to an increase in the efficiency of parallel computations upon [1.82], [1.83].

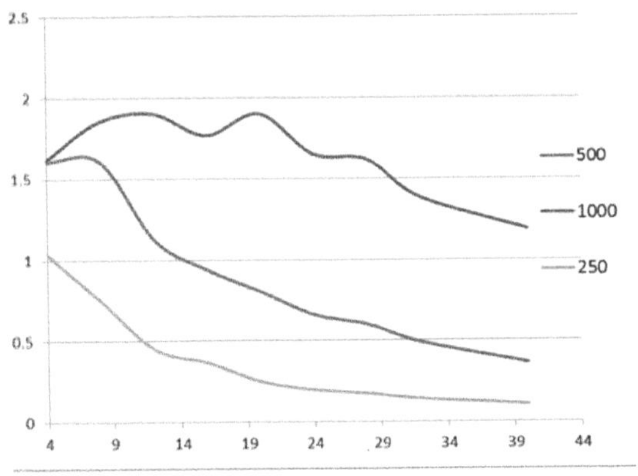

Figure 1.24. *Total speed-up of linear system's solutions. For a color version of this figure, see www.iste.co.uk/koroliouk/computational2.zip*

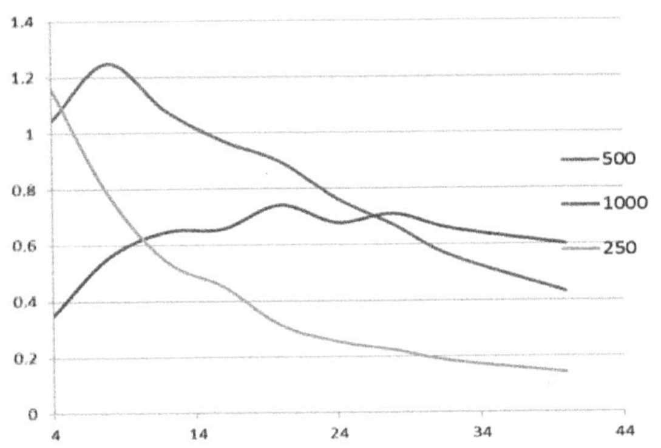

Figure 1.25. *Total speed-up comparing the parallel algorithm computation time with the best time obtained by the sequential algorithm. For a color version of this figure, see www.iste.co.uk/koroliouk/computational2.zip*

Total speed-up of the linear system's solution, as can be seen from Figure 1.24, is significantly greater than the acceleration of the calculations on [1.82], [1.83]. Its main part forms the speed-up of initialization procedures, first of all – the computation of $\tilde{A}_{11}^{(i)}$ matrices.

On the other hand, the time spent on these operations increases with the increase of block size. In the performed tests with block size of more than 1,000 elements, time spent on $\tilde{A}_{11}^{(i)}$ calculations was the dominant factor, which made the speed-up absent in comparison with the best time when performing calculations on one core (see Figure 1.25). By reducing block size, this time becomes smaller, but at the same time, the need for a larger number of synchronizations when doing matrix–vector multiplications reduces algorithm performance and scalability.

Comparison with known algorithms

Efficiency of the considered parallel preconditioner was compared with the efficiency of known algorithms implemented in hypre software (Shapira 2003). We solved linear systems that arise in the problem for the model [1.55], [1.65]–[1.74] (linear system SD) and its non-isothermal counterpart (this linear system will be further denoted as SDT) along with the linear systems *2cubes_sphere* and *offshore* that arise during electromagnetic processes simulation and are present in Tim Davis Matrix Collection (Davis 1997).

The SDT matrix obtained after discretizing the problem of non-isothermal stress–strain state modeling taking into account filtration processes has the size equal to 665434 × 665434 and the fill-in factor equal to 23.52. The matrix *2cubes_sphere* has the size equal to 101492 × 101492 and the fill-in factor equal to 8.62, the matrix *offshore* – the size equal to 259789 × 259789 and the fill-in factor equal to 8.67.

Condition numbers of the matrices were estimated by the LSMR algorithm (Fong and Saunders 2011). Their estimates have the following order: 10^{12} for SD and SDT matrices, 10^{8} for the matrix *2cubes_sphere* and 10^{9} for the matrix *offshore*.

In the case of the first two linear systems, the only combination of iterative method and preconditioner implemented in hypre, which allowed obtaining solution of 10^{-9} accuracy order, was the combination of Gmres (Saad 2003) and AMG preconditioner (Chen 2005) (we will further denote this pair as Gmres+AMG). The *2cubes_sphere* and *offshore* matrices were also solved by other algorithms; however, in order to achieve the accuracy of the order 10^{-9}, the use of preconditioners was necessary in any case. In the performed computational experiments, these matrices were used to compare the efficiency of the considered preconditioner and the AMG preconditioner as those that are effectively solved by it from the point of convergence (other preconditioners, such as Parasails (Chow 2001) and Ilut (Saad 2003), allow us to obtain the necessary solutions faster) and have comparable characteristics with SD and SDT matrices.

The total time needed to obtain SD linear system's solution with 10^{-9} order of accuracy is given in Figure 1.26. Here, the considered parallel preconditioner was used with a block size equal to 250 and 500. Previous testing showed that these sizes are optimal for this linear system. The considered preconditioner was for SD matrix faster than Gmres+AMG when the number of processes was less than 20 and block size equal to 500. When executed on one and four processor cores, the fastest solution was obtained with the block size equal to 250. The scalability of the proposed parallel algorithm here and in the other experiments was significantly lower compared to Gmres+AMG.

In the case of calculating the non-isothermal stress–strain state, taking into account filtration processes (matrix SDT that is significantly larger compared to the matrix SD and with similar fill-in factor) (see Figure 1.27), the algorithm's performance and scalability was worse than in the previous case when compared to Gmres+AMG. Linear system SDT was solved faster by the considered algorithm only in the case of block size equal to 5,000 when all $A_{11}^{(k)}$ submatrices were incompletely inverted and one or four processor cores was used.

The deterioration of the characteristics of the AMLI algorithm with incomplete inversion by the basis matrix method can be explained by the increase of matrix size compared with the previous experiment. Since complete matrix inversion has a third-order complexity, the block size increase significantly increases the time spent on preconditioner's initialization. On the other hand, with the increase of the number of blocks and, consequently, the reduction of their size, the number of data exchanges increases, thus reducing the scalability of the algorithm.

In the case of SDT matrix, optimal block size when performing complete inversions was equal to 1,000 elements, but the algorithm's performance was lower than of Gmres+AMG regardless of the number of involved computational resources. The use of only incomplete inversions here reduces the initialization time of the preconditioner and the time spent on calculations upon [1.82], [1.83]. On the other hand, this significantly decreases convergence rate. For block size equal to 1,000 elements, 10^{-9} accuracy order was achieved after 185 iterations using complete inversions and after 1,454 iterations when only incomplete inversions were performed. At the same time, when block size increases, the convergence worsens and the time of initialization and execution of calculations according to [1.82], [1.83] increases. The optimal block size for the matrix SDT in this case was equal to 5,000 elements.

In the next series of computational experiments, linear systems *2cubes_sphere* and *offshore*, which arise in the simulation of electromagnetic processes, were solved. These matrices have sizes comparable with the previous matrices and three times lower fill-in factor. Algorithms' execution time subject to the number of involved processes for the matrix *offshore* is shown in Figure 1.28, and for the matrix *2cubes_sphere* – in Figure 1.29.

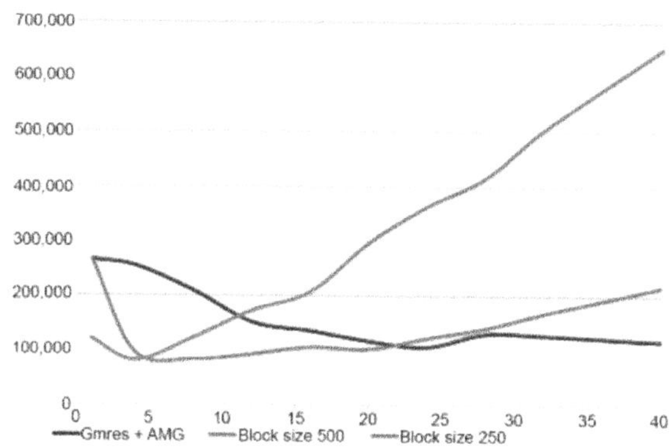

Figure 1.26. *Computation time depending on the number of processes for the dynamic problem of modeling stress–strain state taking filtration into account. For a color version of this figure, see www.iste.co.uk/koroliouk/computational2.zip*

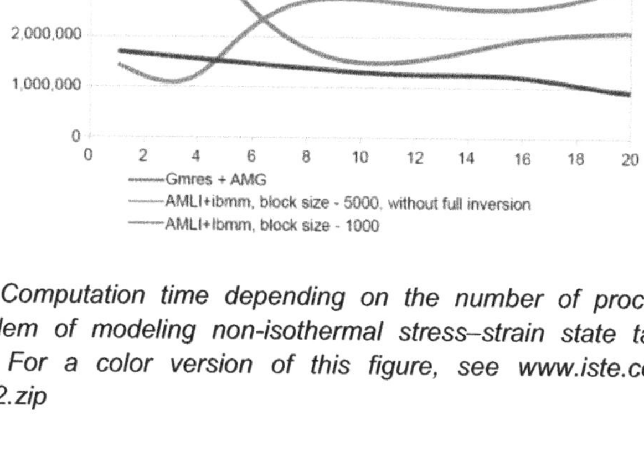

Figure 1.27. *Computation time depending on the number of processes for the dynamic problem of modeling non-isothermal stress–strain state taking filtration into account. For a color version of this figure, see www.iste.co.uk/koroliouk/computational2.zip*

Reduction of fill-in factor and, correspondingly, of the ratio of arithmetic operation number to data exchange volumes leads in the case of these matrices to ineffectiveness of the proposed parallel algorithm – its execution time increases with the increase of the number of involved computational resources. Compared to the previous computational experiments, the scalability of Gmres+AMG algorithm also worsened here.

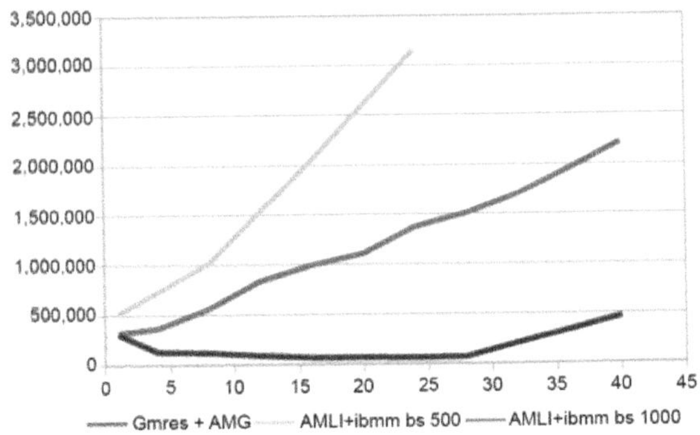

Figure 1.28. *Execution time subject to the number of processes for the matrix offshore. For a color version of this figure, see www.iste.co.uk/koroliouk/computational2.zip*

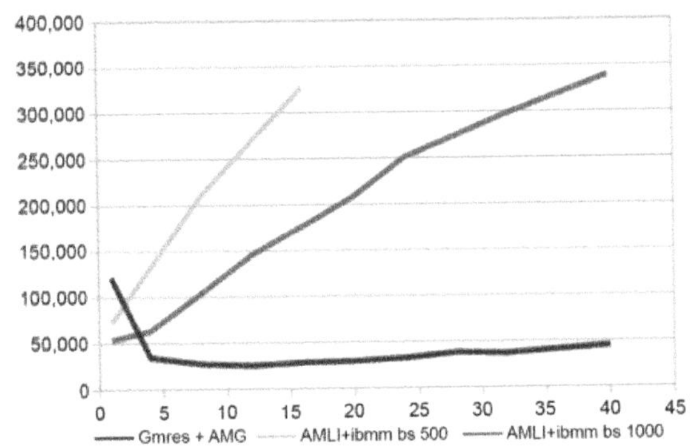

Figure 1.29. *Execution time subject to the number of processes for the matrix 2cubes_sphere. For a color version of this figure, see www.iste.co.uk/koroliouk/computational2.zip*

The performance of AMLI algorithm with incomplete inversion by the basis matrix method when doing sequential computations was greater compared with

Gmres+AMG only for the *2cubes_sphere* matrix, which has a smaller size than the matrix *offshore*.

Summarizing, we can note that the constructed parallel computation scheme that uses the AMLI preconditioner with incomplete inversion by the basis matrix method is effective in modeling strain–stress state of soils, taking into account filtration processes including the case of non-isothermal conditions, which, however, is accompanied by its low speed and scalability. As in the case of multigrid preconditioners, most time when solving linear systems is spent on its application.

Theoretical estimates of the proposed algorithm's performance were obtained and experimentally verified separately for initialization procedures and the application of the preconditioner. In the case of executing the preconditioner on a single processor core, the accuracy of its performance estimate was lower than 25% in the conducted tests. For the case of the parallel algorithm, the error was lower than 30% for block size ranging from 250 to 3,000 elements and the number of processes ranging from 12 to 24. For the other values of parameters, the error was greater. Errors of initialization procedure time estimates were significant. However, within the general behavior tendencies, they coincided with the trends in the measured computation time.

The analysis of the influence of the preconditioner parameters on the convergence of the iterative algorithm showed an acceleration of the convergence with an increase in block size with a significant increase in total execution time. When the number of fully inverted matrices decreases, the convergence of the algorithm decreases too. On the other hand, this decrease significantly reduces the time spent on the preconditioner's initialization operations, which leads to faster algorithm operation when a small amount of computational resources are involved. When they increase, this positive effect is leveled off.

For the linear system SD that has the size equal to 107856×107856 and a condition number of 10^{12} order, the maximal achieved speed-up was 24% when scaling the algorithm on eight processes for block size equal to 500 elements. Low scalability of the algorithm here can be explained by the large number of data exchange operations that must be performed before each matrix–vector multiplication. Increasing the block size reduces it and slightly changes the total volume of exchanges. This leads to an increase in the effectiveness of the preconditioner application, as well as to an increase in the time spent on its initialization, which, as a result, affects the scalability of the algorithm.

The characteristics of the proposed preconditioner were compared with the Gmres algorithm and the AMG preconditioner implemented in the hypre software package. The scalability of the proposed algorithm was generally worse; however, in the case of the matrix SD, the solution with the accuracy order of 10^{-9} was obtained faster with its use when the number of involved processes was less than 20 and block size was equal to 500. In the case of the matrix SDT, which occurs when simulating the effect of temperature changes on strain–stress state, the proposed algorithm was faster than those implemented in the hypre package only when no more than four processor cores were used.

1.6. References

Axelsson, O. and Neythceva, M. (1994). Parallel implementations of the algebraic multilevel iteration method. In *Proceedings of the Fifth SIAM Conference on Applied Linear Algebra*. SIAM, Philadelphia, 372–376.

Bohaienko, V.O. (2014). Using basis matrix method with AMLI preconditioner for solving elasticity problems. Theoretical and applied aspects of cybernetics. In *Proceedings of the 4th International Scientific Conference of Students and Young Scientists*, Kyiv, 55–62.

Bohaienko, V.O. (2020). On parallel algebraic multilevel preconditioner with approximate inversion by basis matrix method. In *Proceedings of The Third International Workshop on Computer Modeling and Intelligent Systems (CMIS-2020)*, Zaporizhzhia, 621–632.

Bohaienko, V.O. and Kudin, V.I. (2014). Building preconditioners using basis matrix method. *International Journal "Information Content and Processing"*, 1(2), 182–187.

Bohaienko, V.A., Marchenko, O.A., Samoilenko, T.A. (2014). Analysis of numerical simulation of soil massif dynamics in the presence of unsteady pressure filtration. *Control Systems and Computers*, 4, 34–41 (in Russian).

Bohaienko, V.A., Marchenko, O.A., Samoilenko, T.A. (2016). Analysis of numerical modeling of non-isothermal processes in the soil massif. *Computer Mathematics*, 2, 3–11 (in Russian).

Chen, K. (2005). *Matrix Preconditioning Techniques and Applications*. Cambridge University Press, Cambridge.

Chow, E. (2001). Parallel implementation and performance characteristics of sparse approximate inverse preconditioners with a priori sparsity patterns. *International Journal of High Performance Computing Applications*, 15, 56.

Davis, T. (1997). University of Florida sparse matrix collection. NA Digest [Online]. Available at: http://www.cise.ufl.edu/research/sparse/matrices.

Deineka, V.S., Sergienko, I.V., Skopetsky, V.V. (1995). *Mathematical Models and Methods for Calculating Problems with Discontinuous Solutions*. Naukova Dumka, Kyiv (in Russian).

Deineka, V.S., Sergienko, I.V., Skopetsky, V.V. (1998). *Models and Methods for Solving Problems with Conjugation Conditions*. Naukova Dumka, Kyiv (in Russian).

Falgout, R.D. and Yang, U.M. (2002). A library of high performance preconditioners. In *Computational Science — ICCS 2002*. Springer, Berlin, Heidelberg.

Fong, D. and Saunders, M. (2011). LSMR: An iterative algorithm for sparse least-squares problems. *SIAM Journal on Scientific Computing*, 33(5), 2950–2971.

Kislitsyn, A.A., Shastunova, U.Yu. Yanbikova, Yu.F. (2018). Experimental study and a mathematical model of the processes in frozen soil under a reservoir with a hot heat-transfer agent. *Journal of Engineering Physics and Thermophysics*, 91(2), 507–514.

Kudin, V.I., Lyashko, S.I., Yatsenko, Yu.P., Khritonenko, N.V. (2007). Method of artificial basis matrices. *Dopovidi NAS of Ukraine*, 9, 30–34 (in Ukrainian).

Li, Z., Saad, Y., Sosonkina, M. (2003). pARMS: A parallel version of the algebraic recursive multilevel solver. *Numerical Linear Algebra with Applications*, 10, 485–509.

Liu, Q.H. and Zhang, F.D. (2013). Incomplete hyperbolic Gram–Schmidt-based preconditioners for the solution of large indefinite least squares problems. *Journal of Computational and Applied Mathematics*, 250, 210–216.

Marchenko, O.A. and Lezhnina, N.A. (2002). Approximate solution of the problem of moisture and heat transfer in freezing soils by the finite element method. *Computer Mathematics*, 1, 24–33 (in Russian).

Marchenko, O.A. and Samoilenko, T.A. (2012). Analyzing an approximate solution to a quasilinear parabolic-hyperbolic problem. *Cybernetics and System Analysis*, 48, 762–773.

Marchenko, O.O. and Samoilenko, T.A. (2020). Construction of the approximate solution of the axisymmetric problem of the dynamics of non-isothermal moisture transfer. *Cybernetics and Computer Technologies*, 1, 41–52 (in Ukrainian).

Margenov, S. (1998). Semicoarsening AMLI algorithms for elasticity problems. *Numerical Linear Algebra with Applications*, 5, 347–362.

Neytcheva, M. (1995). Experience in implementing the algebraic multilevel iteration method on a SIMD-type computer. *Applied Numerical Mathematics*, 19, 71–90.

Pavlov, A.R. and Permyakov, P.P. (1983). Mathematical model and algorithms for an electronic computer analysis of the heat and mass transfer in freezing the soil. *Journal of Engineering Physics*, 44, 220–225.

Rectoris, K. (1977). *Variational Methods in Mathematics, Science and Engineering*. Springer, Berlin, Heidelberg.

Saad, Y. (2003). *Iterative Methods for Sparse Linear Systems*. Society for Industrial and Applied Mathematics, Philadelphia, PA.

Saad, Y. (2007). Schur complement preconditioners for distributed general sparse linear systems. In *Domain Decomposition Methods in Science and Engineering XVI*, 127–138. Springer, Berlin, Heidelberg.

Seimov, V.M., Trofimchuk, A.N., Savitsky, O.A. (1990). *Oscillations and Waves in Layered Media.* Naykova Dumka, Kyiv (in Russian).

Shapira, Y. (2003). *Matrix-Based Multigrid: Theory and Applications.* Springer, Berlin, Heidelberg.

Skopetsky, V.V. and Marchenko, O.A. (2005). Formulation and analysis of problems for dynamic systems of inhomogeneous two-phase media. *Cybernetics and System Analysis*, 41, 865–878.

Skopetsky, V.V., Deineka, V.S., Marchenko, O.A. (1995). Modeling the dynamic equilibrium of objects weakened by thin low-strength inclusions. *Cybernetics and System Analysis*, 31, 261–267.

Skopetsky, V.V., Marchenko, O.A., Samoilenko, T.A. (2008). Approximate solution of a nonlinear system of equations for two-phase media. *Cybernetics and System Analysis*, 44, 524–536.

Skopetsky, V.V., Marchenko, O.A., Samoilenko, T.A. (2010). Construction and analysis of a nonlinear differential model for two-phase media. *Cybernetics and System Analysis*, 46, 252–263.

Zaretsky, Yu.K. and Lombardo, V.N. (1983). *Statics and Dynamics of Soil Dams.* Energoatomizdat, Moscow (in Russian).

2

Computer Simulation of Transdermal Drug Delivery Using Soluble Microneedles

D.A. KLYUSHIN, S. LYASHKO, V.V. ONOTSKYI and O.S. BONDAR
Taras Shevchenko National University of Kyiv, Ukraine

2.1. Introduction

Currently, microneedle systems are increasingly used to penetrate the epidermis and provide controlled transdermal drug delivery (Bhatnagar et al. 2017; Mitra et al. 2017; Phatale et al. 2022; Wu et al. 2023). Microneedles are divided into solid, hollow and soluble, and can also have a complex shape: pyramidal, conical, cylindrical and others. Solid microneedles create a microchannel and are coated with a drug on the outside, hollow microneedles contain active substances and are injected under the skin, and soluble microneedles are a polymeric soluble shell containing the drug (Jin et al. 2018; Parhi and Supriya 2019; Plamadeala et al. 2019; Benbrook and Zhan 2025). Soluble microneedles represent a highly promising strategy for transdermal drug delivery (Pires et al. 2020; Jung and Jin 2021; Kim et al. 2025).

The process of transdermal drug transfer from soluble microneedles is described by the initial boundary value problem of convective diffusion. We will consider one-, two- and three-dimensional cases, and interpret the microneedles as point sources whose intensity depends on the solution of an additional Cauchy problem. Now, related problems are studied intensively. Yadav et al. (2020) published an extensive review of works devoted to mathematical modeling and optimization of

Computational Methods and Mathematical Modeling in Cyberphysics and Engineering Applications 2,
coordinated by Dmitri KOROLIOUK, Sergiy LYASHKO and Nikolaos LIMNIOS.

drug transfer processes from microneedle systems. Recently, a skin transport model developed by Calcutt and Anissimov (2021) and further expanded by Calcutt et al. (2022) was used to study the dependence of dermal drug concentrations on physiological parameters. Newell and Zhan (2023, 2024) applied mathematical modeling to investigate the process of transdermal delivery under various skin models.

Separately, we would like to note the paper (Ronander et al. 2018) where the authors studied the dynamics of changes in the intensity of drug injection with a pyramidal microneedle with a base in the form of a parallelepiped that dissolves together with the drug. However, in this work, the convection and diffusion of drug distribution is largely ignored. In our paper, numerical modeling of the distribution of drugs into the skin, introduced into the skin using an array of pyramidal-, cylindrical-, bar- and cone-shaped microneedles, which self-dissolve in the skin with consideration of convection and diffusion for one-, two-, and three-dimensional cases, was carried out. Thus, we propose to integrate the model of Ronander et al. (2018) with our approach (Klyushin and Onotskyi 2016; Lyashko et al. 2018, 2024; Sandrakov et al. 2019, 2020) to the mathematical simulation of transdermal delivery considering microneedles as point sources, with intensity depending on the solution of a Cauchy problem describing the influence of microneedle parameters (form, size, etc.).

2.2. Mathematical model

Let $\overline{\Omega}=\{0\le x\le a,\ 0\le y\le a, 0\le z\le d_s\}$ be a parallelepiped of skin, into which an array of microneedles located at the nodes of a square grid of a circular shape of radius R (there are M microneedles in total) is inserted to a depth of h_c. The axis Oz is directed downwards. The shape and thickness of the needle are neglected. Thus, we formulate a problem with concentrated point sources. In the domain $\overline{Q}=\overline{\Omega}\times[0,T]$, we consider the initial boundary value problem

$$Lu=\frac{\partial u}{\partial t}-\sum_{i,j=1}^{N}\frac{\partial}{\partial x_i}\left(k_D\frac{\partial u}{\partial x_j}\right)+\sum_{i=1}^{N}k_L\frac{\partial u}{\partial x_i}=f(x,t), \qquad [2.1]$$

$(x=(x_1,x_2,x_3),t)\in\Omega\times(0,T)=\text{int}(\overline{\Omega})\times(0,T)$, $k_D>0,\ k_L>0$ are constant diffusion and convection coefficients, respectively, N is the space dimension (1D, 2D, or 3D), $f(x,t)=\sum_{m=1}^{M}Q_m F(x-x_m, y-y_m)$ is the intensity of point sources, and $F(x-x_m, y-y_m) \in L_2\left((0,T)\times\Omega\right)$ is a function that specifies the effect on the system of a source located at a point (x_m, y_m).

The initial, boundary and periodical conditions are

$$u(x,0) = g(x), \quad x \in \Omega; \tag{2.2}$$

$$u\big|_{x_3=0} = u\big|_{x_3=d_s} = 0, \; u\big|_{x_n=0} = u\big|_{x_n=a}, \left.\frac{\partial u}{\partial x_n}\right|_{x_1=0} = \left.\frac{\partial u}{\partial x_n}\right|_{x_1=a}, \; n=1,2, \text{ or } 3. \tag{2.3}$$

We consider the microneedles and the amount of drugs in them to be the same: $q^m(t) = q(t), m = 1..600$, the function $q(t)$ is defined by the formula (Ronander et al. 2018)

$$q(t) = \beta\rho v_{c,0} - \left(\beta\rho v_c(t) + c(t)\left(v_0 + v_{c,0} - v_c(t)\right)\right) \tag{2.4}$$

We consider the cases of microneedle form: pyramidal, cone, cylinder and timber.

The function v_c for pyramidal microneedles is defined by the formulas (Ronander et al. 2018):

$$v_c(t) = \frac{4}{3}h^3(t)\tan^2\theta,$$

The functions $c(t)$ and $h(t)$ are defined as solutions of the Cauchy problem (Ronander et al. 2018):

$$\frac{dc(t)}{dt} = -(k_L c) + 4\left[\frac{k_D \tan\theta}{\rho\cos\theta}\right]h^2\left[\frac{\beta\rho - c}{v_0 + v_{c,0} - v_c}\right]\left[c_s - \left(\frac{1-\beta}{\beta}\right)c(t)\right],$$

$$c\big|_{t=0} = 0,$$

$$\frac{dh(t)}{dt} = -\frac{k_D}{\rho\sin\theta}\left[c_s - \left[\frac{1-\beta}{\beta}\right]c(t)\right],$$

$$h\big|_{t=0} = h_0.$$

The mass of one cone microneedle is determined by the formula (Ronander et al. 2018):

$$\rho v_c(t) = \rho\left(\frac{\pi}{3}h^3 \tan^2\theta\right),$$

Divide by ρ and differentiate by t:

$$\frac{dv_c}{dt} = \pi h^2(t)\tan^2\theta\frac{dh}{dt}.$$

The change in volume over time Δt is

$$\Delta v_c = -\frac{2\pi R}{\cos\theta}\int_0^h \frac{(h-z)}{h}\Delta n dz = -\frac{\pi R}{\cos\theta}h\Delta n,$$

$$\frac{dv_c}{dt} = -\frac{\pi hR}{\cos\theta}\frac{dn}{dt},$$

$$\frac{dn}{dt} = \frac{k_D}{\rho}\left[c_s - \left[\frac{1-\beta}{\beta}\right]c\right],$$

$$\frac{dv_c}{dt} = -\frac{\pi hR}{\cos\theta}\frac{k_D}{\rho}\left[c_s - \left[\frac{1-\beta}{\beta}\right]c\right],$$

$$\frac{dh(t)}{dt} = -\frac{k_D}{\rho\sin\theta}\left[c_s - \left[\frac{1-\beta}{\beta}\right]c(t)\right],$$

$$h\big|_{t=0} = h_0.$$

The volume of the skin layer is determined from the formula

$$v(t) = v_0 + (v_{c,0} - v_c(t)).$$

Formulate the law of conservation of mass for drugs:

$$\frac{d(vc)}{dt} = -k_L vc + \beta\rho\left(-\frac{dv_c}{dt}\right).$$

This yields the Cauchy problem for c:

$$\frac{dc}{dt} = -(k_L c) + \frac{\pi k_D \tan\theta}{\rho\cos\theta} h^2 \left(\frac{\beta\rho - c}{v_0 + v_{c,0} - v_c(t)} \right) \left(c_s - \frac{1-\beta}{\beta} c \right),$$

$$c\big|_{t=0} = 0.$$

Thus, the functions $c(t)$ and $h(t)$ are defined as solutions of the following Cauchy problem:

$$\frac{dc(t)}{dt} = -(k_L c) + \pi \left[\frac{k_D \tan\theta}{\rho\cos\theta} \right] h^2 \left[\frac{\beta\rho - c}{v_0 + v_{c,0} - v_c} \right] \left[c_s - \left(\frac{1-\beta}{\beta} \right) c(t) \right],$$

$$c\big|_{t=0} = 0,$$

$$\frac{dh(t)}{dt} = -\frac{k_D}{\rho\sin\theta} \left[c_s - \left[\frac{1-\beta}{\beta} \right] c(t) \right],,$$

$$h\big|_{t=0} = h_0.$$

The mass of one cylindrical microneedle is determined by the formula (Ronander et al. 2018):

$$\rho v_c(t) = \rho\left(\pi r^2(t) h(t) \right)$$

Divide by ρ and differentiate with respect to t:

$$\frac{dv_c}{dt} = 2\pi r(t) h(t) \frac{dr}{dt} + \pi r^2(t) \frac{dh}{dt}$$

We assume that the needle dissolves through the side surface and the bottom cover; then, the change in volume over time Δt is

$$\Delta v_c = -2\pi r(t) h(t) \Delta n - \pi r^2(t) \Delta n,$$

$$\frac{dv_c}{dt} = -\pi\left(2r(t)h(t) + r^2(t) \right) \frac{dn}{dt}.$$

Using this fact and the relation $r(t) = h_0 \tan\theta_0 - (h_0 - h(t))$, we obtain

$$\frac{dv_c}{dt} = -\pi\left(2r(t)h(t) + r^2(t)\right)\frac{k_D}{\rho}\left[c_s - \left[\frac{1-\beta}{\beta}\right]c\right].$$

Therefore, we have to solve the following equation with respect to $h(t)$:

$$\frac{dh(t)}{dt} = -\frac{k_D}{\rho}\left[c_s - \left[\frac{1-\beta}{\beta}\right]c\right],$$

$$h\big|_{t=0} = h_0.$$

Using the formula for the skin layer and the law of conservation of mass for drugs, we obtain the following Cauchy problem for c:

$$\frac{dc}{dt} = -(k_L c) + \frac{\pi k_D}{\rho}h^2\left(\frac{\beta\rho - c}{v_0 + v_{c,0} - v_c(t)}\right)\left(2r(t)h(t) + r^2(t)\right)\left(c_s - \frac{1-\beta}{\beta}c\right),$$

$$c\big|_{t=0} = 0.$$

Thus, the functions $c(t)$ and $h(t)$ are defined as solutions of the following Cauchy problem:

$$\frac{dc(t)}{dt} = -(k_L c) + \frac{\pi k_D}{\rho}\left(\frac{\beta\rho - c}{v_0 + v_{c,0} - v_c(t)}\right)\left(2r(t)h(t) + r^2(t)\right)\left(c_s - \frac{1-\beta}{\beta}c\right),$$

$$c\big|_{t=0} = 0,$$

$$\frac{dh(t)}{dt} = -\frac{k_D}{\rho}\left[c_s - \left[\frac{1-\beta}{\beta}\right]c\right],$$

$$h\big|_{t=0} = h_0.$$

The mass of one timber microneedle is determined by the formula (Ronander et al. 2018):

$$\rho v_c(t) = \rho\left(L^2(t)h(t)\right).$$

Divide by ρ and differentiate with respect to t:

$$\frac{dv_c}{dt} = 2L(t)h(t)\frac{dL}{dt} + L^2(t)\frac{dh}{dt}.$$

We assume that the needle dissolves through the side surface and the bottom cover. Therefore, the change in volume over time Δt is

$$\Delta v_c = -4L(t)h(t)\Delta n - L^2(t)\Delta n,$$

$$\frac{dv_c}{dt} = -\left(4L(t)h(t) + L^2(t)\right)\frac{dn}{dt}.$$

Using this fact and the relation $L(t) = l_0 - 2(h_0 - h(t))$, we have

$$\frac{dv_c}{dt} = -\left(4L(t)h(t) + L^2(t)\right)\frac{k_D}{\rho}\left[c_s - \left[\frac{1-\beta}{\beta}\right]c\right].$$

Therefore, the equation with respect to $h(t)$ has the following form:

$$\frac{dh(t)}{dt} = -\frac{k_D}{\rho}\left[c_s - \left[\frac{1-\beta}{\beta}\right]c\right],$$

$$h\big|_{t=0} = h_0.$$

Using the formula for the skin layer of the law of conservation of mass for drugs, we obtain the following Cauchy problem for c:

$$\frac{dc}{dt} = -\left(k_L c\right) + \frac{k_D}{\rho}\left(\frac{\beta\rho - c}{v_0 + v_{c,0} - v_c(t)}\right)\left(4L(t)h(t) + L^2(t)\right)\left(c_s - \frac{1-\beta}{\beta}c\right),$$

$$c\big|_{t=0} = 0.$$

Thus, the functions $c(t)$ and $h(t)$ are defined as solutions of the following Cauchy problem:

$$\frac{dc(t)}{dt} = -(k_L c) + \frac{k_D}{\rho}\left(\frac{\beta\rho - c}{v_0 + v_{c,0} - v_c(t)}\right)\left(4L(t)h(t) + L^2(t)\right)\left(c_s - \frac{1-\beta}{\beta}c\right),$$

$$c\big|_{t=0} = 0.$$

$$\frac{dh(t)}{dt} = -\frac{k_D}{\rho}\left[c_s - \left[\frac{1-\beta}{\beta}\right]c\right],$$

$$h\big|_{t=0} = h_0.$$

2.3. Mathematical methods

Consider the discrete grid on $\overline{Q}$:

$$\overline{Q}_{h\tau} = \left\{\left(x_{1,i} = ih_1, x_{2,j} = jh_2, x_{3,k} = kh_3\right),\ i = \overline{0, N_x},\ j = \overline{0, N_y},\ k = \overline{0, N_z}\right\} \times \\ \times \left\{t_n = \tau n,\ n = \overline{0, N_t}\right\}.$$

Find the discrete solution $u^n_{i,j,k}$ on the grid $\overline{Q}_{h\tau}$. To solve the Cauchy problems for functions $c(t)$ and $h(t)$, we use the Runge–Kutta method. And to solve the initial-boundary value problem [2.1]–[2.3], we use an absolutely stable two-step symmetrized algorithm (DS-algorithm) with the order of approximation $O(\tau^2 + h_1^2 + h_2^2 + h_3^2)$ (Klyushin and Onotskyi 2016).

Decompose the grid $\overline{Q}_{h\tau}$ into two subsets

$$\Omega^{(1)}_{h\tau} = \left\{\left(x_{1,i}, x_{2,j}, x_{3,k}; t_n\right) \mid (i + j + k + n) \text{ is an odd number}\right\},$$

$$\Omega^{(1)}_{h\tau} = \left\{\left(x_{1,i}, x_{2,j}, x_{3,k}; t_n\right) \mid (i + j + k + n) \text{ is an even number}\right\},$$

and take into account the initial conditions:

$$u_{i,j,k}^{0}=0,\ i=\overline{0,N_x},\ j=\overline{0,N_y},\ k=\overline{0,N_z}$$

First, for every $n=\overline{0,N_t-1}$, we use the following explicit finite-difference scheme for $(x_{1,i},x_{2,j},x_{3,k};t_n)\in\Omega_{h\tau}^{(1)}$:

$$u_{i,j,k}^{n+1}=u_{i,j,k}^{n}-\tau k_L\left(\frac{u_{i+1,j,k}^{n}-u_{i-1,j,k}^{n}}{2h_1}\right)+\tau k_D\left(\frac{u_{i+1,j,k}^{n}-2u_{i,j,k}^{n}+u_{i-1,j,k}^{n}}{h_1^2}\right)-$$

$$-\tau k_L\left(\frac{u_{i,j+1,k}^{n}-u_{i,j-1,k}^{n}}{2h_2}\right)+\tau k_D\left(\frac{u_{i,j+1,k}^{n}-2u_{i,j,k}^{n}+u_{i,j-1,k}^{n}}{h_2^2}\right)$$

$$-\tau k_L\left(\frac{u_{i,j,k+1}^{n}-u_{i,j,k-1}^{n}}{2h_3}\right)+\tau k_D\left(\frac{u_{i,j,k+1}^{n}-2u_{i,j,k}^{n}+u_{i,j,k-1}^{n}}{h_3^2}\right)+\tau f_{i,j,k}^{n}.$$

Then, for $(x_{1,i},x_{2,j},x_{3,k};t_n)\in\Omega_{h\tau}^{(2)}$ using founded values, we apply the following implicit finite-difference scheme:

$$u_{i,j,k}^{n+1}=\frac{A}{B},$$

where:

$$A=u_{i,j,k}^{n}-\tau k_L\left(\frac{u_{i+1,j,k}^{n+1}-u_{i-1,j,k}^{n+1}}{2h_1}\right)+\tau k_D\left(\frac{u_{i+1,j,k}^{n+1}+u_{i-1,j,k}^{n+1}}{h_1^2}\right)-$$

$$-\tau k_L\left(\frac{u_{i,j+1,k}^{n+1}-u_{i,j-1,k}^{n+1}}{2h_2}\right)+\tau k_D\left(\frac{u_{i,j+1,k}^{n+1}+u_{i,j-1,k}^{n+1}}{h_2^2}\right)$$

$$-\tau k_L\left(\frac{u_{i,j,k+1}^{n+1}-u_{i,j,k-1}^{n+1}}{2h_3}\right)+\tau k_D\left(\frac{u_{i,j,k+1}^{n+1}+u_{i,j,k-1}^{n+1}}{h_3^2}\right)+\tau f_{i,j,k}^{n+1},$$

$$B = 1 + 2\tau k_D \left(\frac{1}{h_1^2} + \frac{1}{h_2^2} + \frac{1}{h_3^2} \right).$$

$$f_{i,j,k}^n = \begin{cases} Q(\tau n)/(h_1 h_2 h_3), \exists m_0 : (ih_1, jh_2, kh_3) = r_{m_0}, \\ 0, \ \forall m \ (ih_1, jh_2, kh_3) \neq r_m. \end{cases}$$

Also, we take into account the boundary conditions [2.3]. The solution is the set of values $u_{i,j,k}^{n+1}$ at the even time steps. The advantage of this difference scheme is that there is no need to solve a system of linear algebraic equations, since the formulas are explicit.

2.4. Numerical experiments

We conducted three computational experiments at different spatial dimensions: 1D, 2D and 3D, using the parameters from (Ronander et al. 2018):

- $M = 600$;
- $a = 2.1$ is the length and width of the skin area, cm;
- $d_s = 0.08$ is the depth of the skin area, cm;
- $T = 1.5$ is the duration of simulation, hr;
- $h_0 = 0.05$ is the initial height of a microneedle;
- $l_0 = 0.03$ is the length of the base of a pyramidal microneedle;
- $p_w = 3.4 \times 10^{-2}$, cm;
- $c_s = 1.0$ is the coefficient of solvability of matrix polymer in solvent, g/cm^3;
- $v_0 = p_w^2 d_s$ is the initial volume of the drug in a microneedle;
- $\theta = \text{atan}\left(\frac{l_0}{2h_0} \right)$ is a half-angle at the apex of the needle (degree);
- $k_L = 5.16 \times 10^{-2}$ is the elimination rate constant, hr-1;
- $k_D = 1.43 \times 10^{-2}$ is the dissolution rate constant, cm/hr^{-1};

– $\rho = 1.15$ is the density of the drug in a microneedle, g/cm^3;

– $\beta = 0.2142$ is the mass fraction of an active pharmaceutical ingredient of the drug in a microneedle.

In the experiments, we consider every microneedle as point sources buried at the depth $h_c = h_0/4$, i.e. z-coordinate of the mass center of the pyramidal microneedle. The form of a microneedle is taken into account via the functions $v(t)$ and $h(t)$.

The parameters of the Runge–Kutta method and the DS-algorithms are the following: $N_t = 1000, N_x = N_y = 60, N_z = 10$. The parameters of the grid region are chosen so that the positions of the microneedles are located at the grid nodes.

2.5. Results and discussion

Figures 2.1–2.8 present the distributions of drug concentrations and computation errors at the moment $t_0 = 845\tau = 1.2675$ hr when dissolution of the microneedle with the drug at depth h_c was completed.

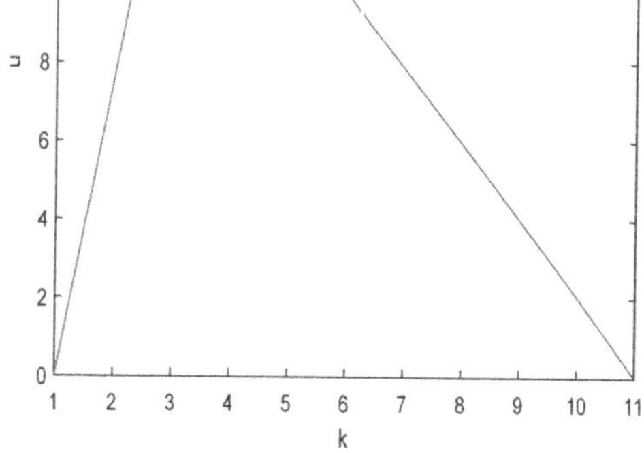

Figure 2.1. *Concentration distribution in depth (1D case) at time $t_0 = 845\tau = 1.2675$ hr; k is the z-coordinate index*

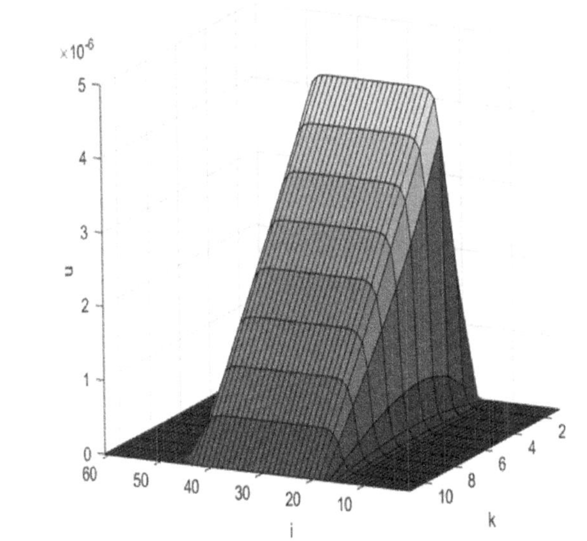

Figure 2.2. *Concentration distribution along the z-axis on the Oxz plane (2D case) at time* $t_0 = 845\tau = 1.2675$ *hr. For a color version of this figure, see www.iste.co.uk/koroliouk/computational2.zip*

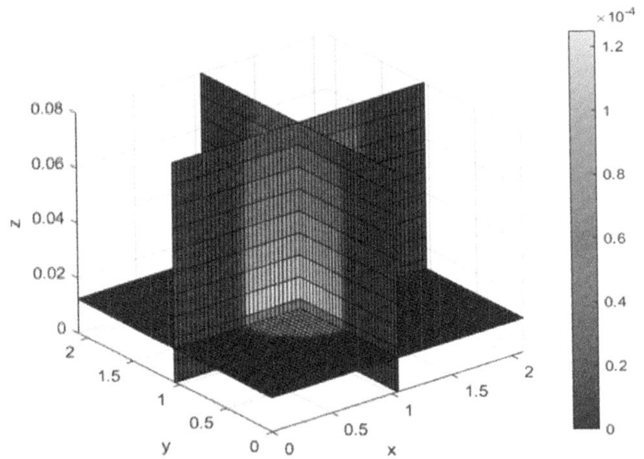

Figure 2.3. *Concentration distribution on the Oxz plane (3D case) at time* $t_0 = 845\tau = 1.2675$ *hr. For a color version of this figure, see www.iste.co.uk/koroliouk/computational2.zip*

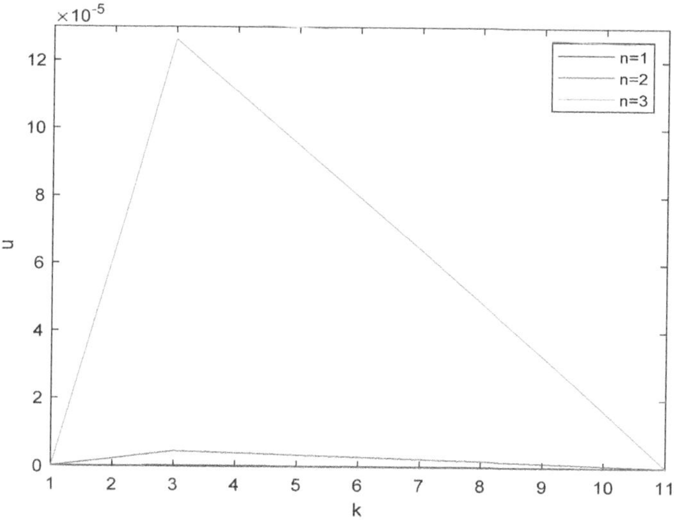

Figure 2.4. *Comparison of the concentration distribution in depth in 1D, 2D and 3D cases at time* $t_0 = 845\tau = 1.2675$ *hr. For a color version of this figure, see www.iste.co.uk/koroliouk/computational2.zip*

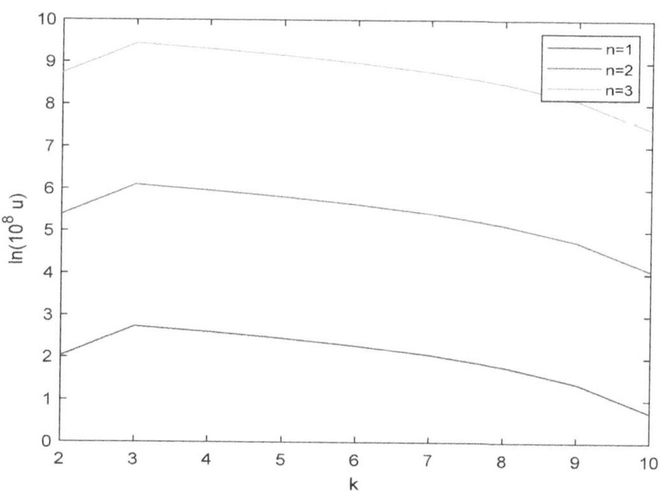

Figure 2.5. *Comparison of logarithms of the concentration distribution in depth in 1D, 2D and 3D cases at time* $t_0 = 845\tau = 1.2675$ *hr. For a color version of this figure, see www.iste.co.uk/koroliouk/computational2.zip*

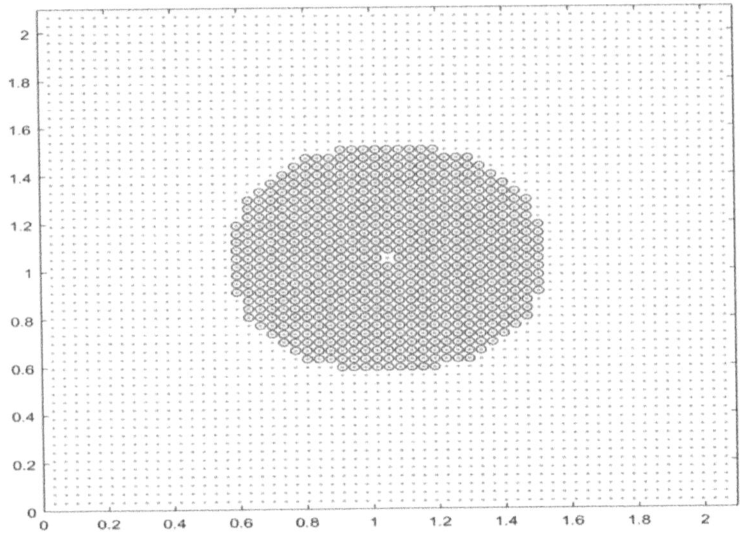

Figure 2.6. *Placement of microneedles (marked in blue) on the Oxy plane. For a color version of this figure, see www.iste.co.uk/koroliouk/computational2.zip*

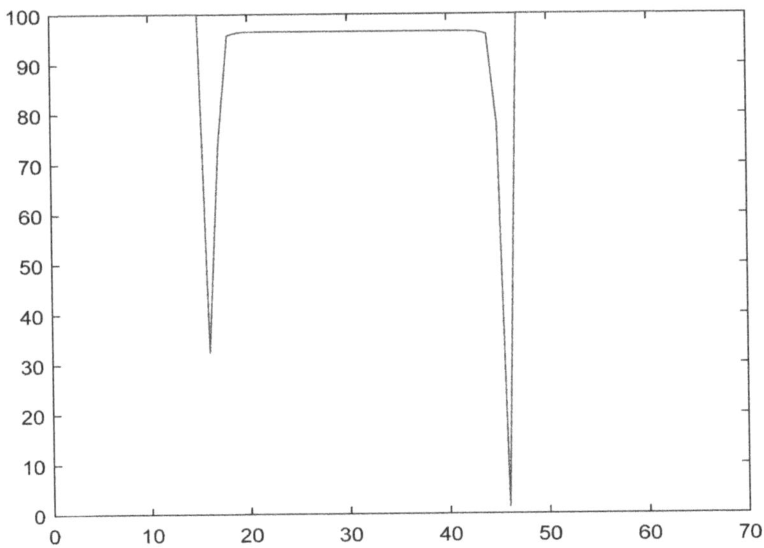

Figure 2.7. *Relative error between the interpolated 1D- and 2D-solutions at time* $t_0 = 845\tau = 1.2675$ *hr*

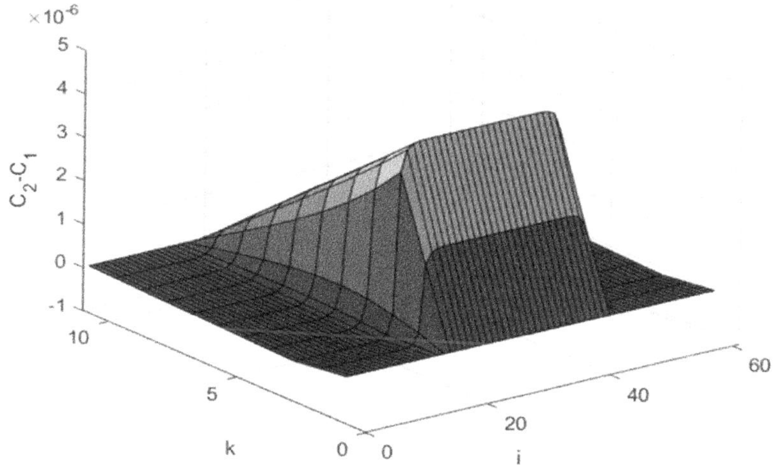

Figure 2.8. *Difference between the interpolated 1D- and 2D-solutions at time* $t_0 = 845\tau = 1.2675$ *hr. For a color version of this figure, see www.iste.co.uk/korolioukcomputational2.zip*

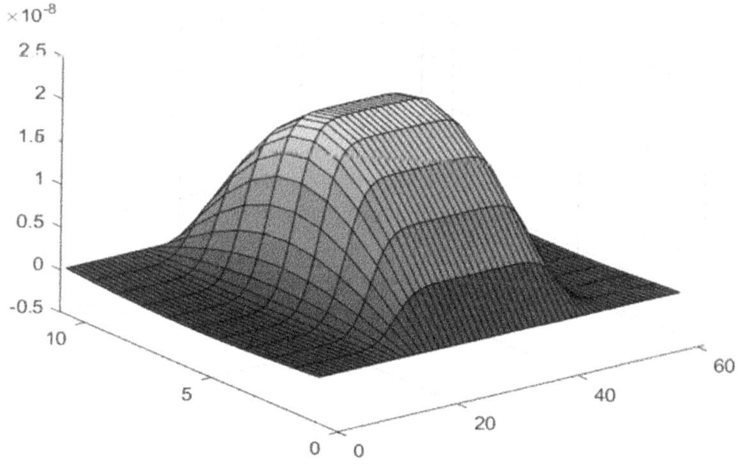

Figure 2.9. *Difference between the interpolated 1D- and 2D-solutions at time* $t_0 = 1.5$ *hr. For a color version of this figure, see www.iste.co.uk/korolioukcomputational2.zip*

The following conclusions can be drawn from the experimental results:

1) Taking into account convection and diffusion using numerical modeling allows us to determine the solution concentration distribution at each point in the spatial region, at each moment in time, with sufficient accuracy and to track the dynamics of this distribution over time (see Figures 2.1–2.3).

2) Experiments with lower dimensionality (1D and 2D) provide significantly worse accuracy compared to the full-size experiment (see Figures 2.4 and 2.5).

3) When comparing a one-dimensional solution interpolated onto a 2D space with a solution to a two-dimensional problem (see Figures 2.6–2.9), it is clear that the error is significant.

The difference in results for different dimensions is due to the fact that the sources influence each other, and this is taken into account to a fuller extent only for the 3D case.

Generalizing experiments were conducted with microneedles in the forms of 1) a pyramid with a square base; 2) a cone; 3) a cylinder; and 4) a bar. Experiments were also conducted with the same number of cobs and the same number of microneedles. The results of the experiments are shown in Figures 2.10–2.14.

The results of the calculations, namely the approximation of the drug concentrations in the microneedle (see Figure 2.10), the heights of the microneedles (see Figure 2.11) and solutions obtained by solving the Cauchy problems (see Figures 2.12 and 2.13) allow us to conclude that drugs are distributed more quickly and evenly in the case of self-dissolving microneedles in the form of bars and cylinders.

This is explained by the fact that the surface area of the microneedles through which dissolution occurs is larger for these shapes. The cone-shaped microneedles took the longest to dissolve.

For the future, we can consider variations of these shapes, namely: pyramids with bases in the form of a regular triangle and polygon, as well as a cylinder and a bar with an oblique cut at a certain angle, or with two symmetrical cuts.

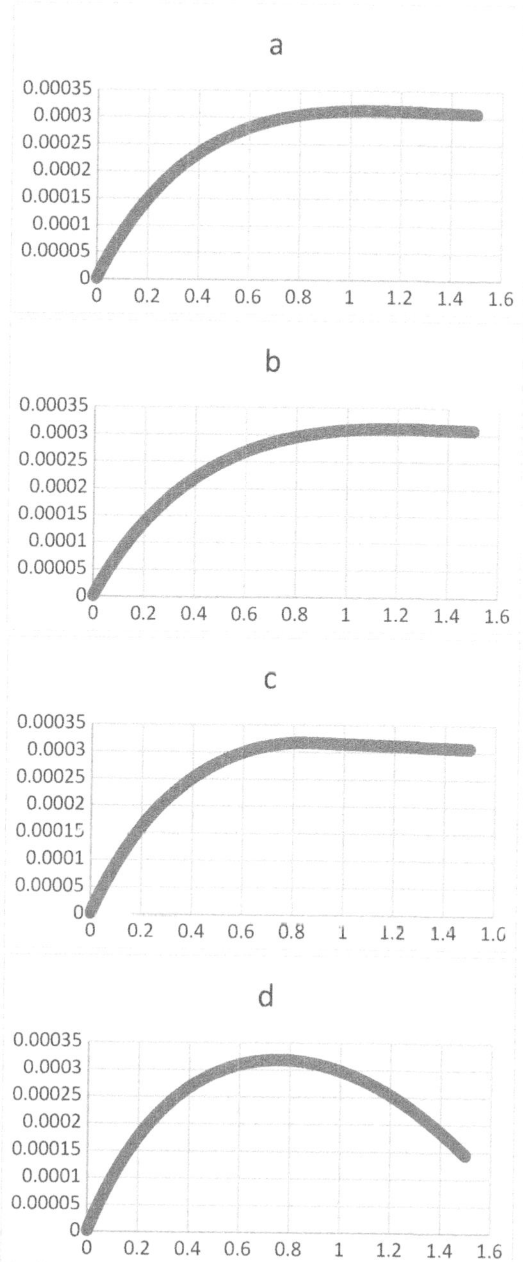

Figure 2.10. *Dynamics of drug concentration: a) pyramid; b) cone; c) cylinder; d) bar*

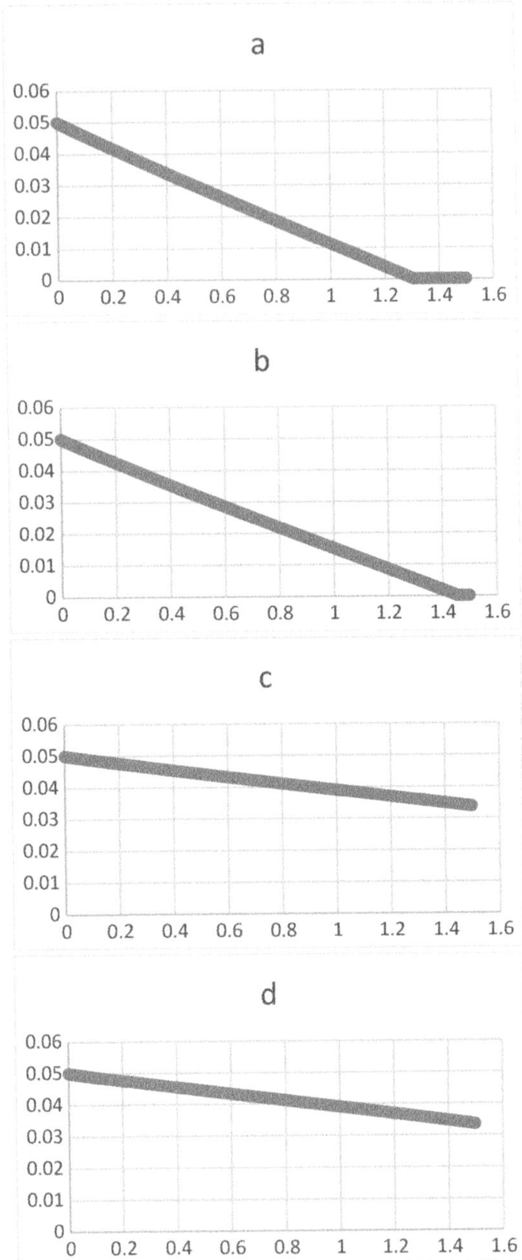

Figure 2.11. *Dynamics of a height microneedle: a) pyramid; b) cone; c) cylinder; d) bar*

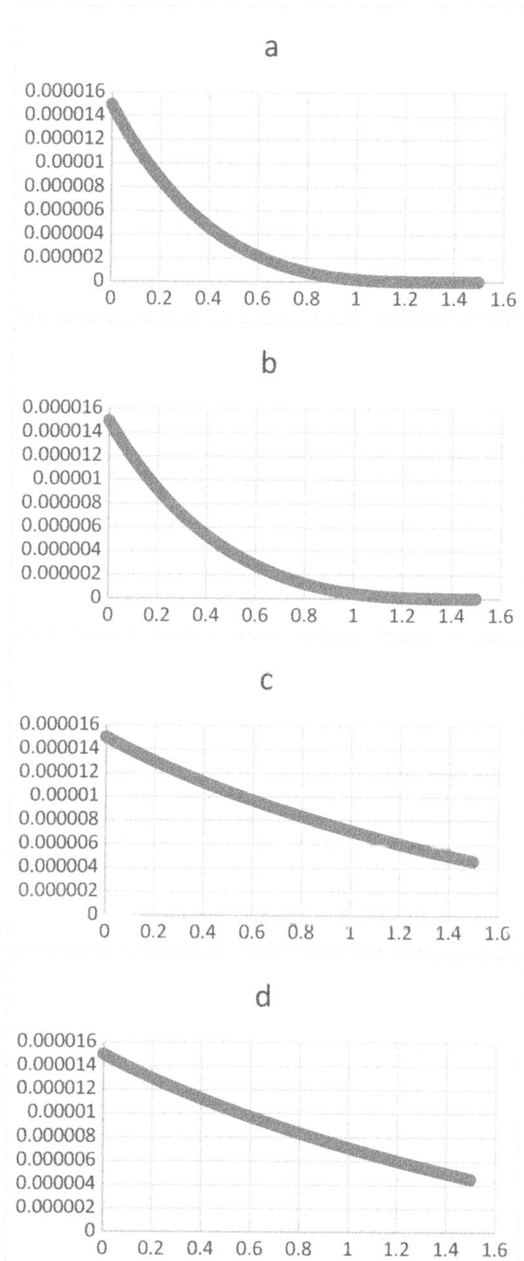

Figure 2.12. *Dynamics of volume: a) pyramid; b) cone; c) cylinder; d) bar*

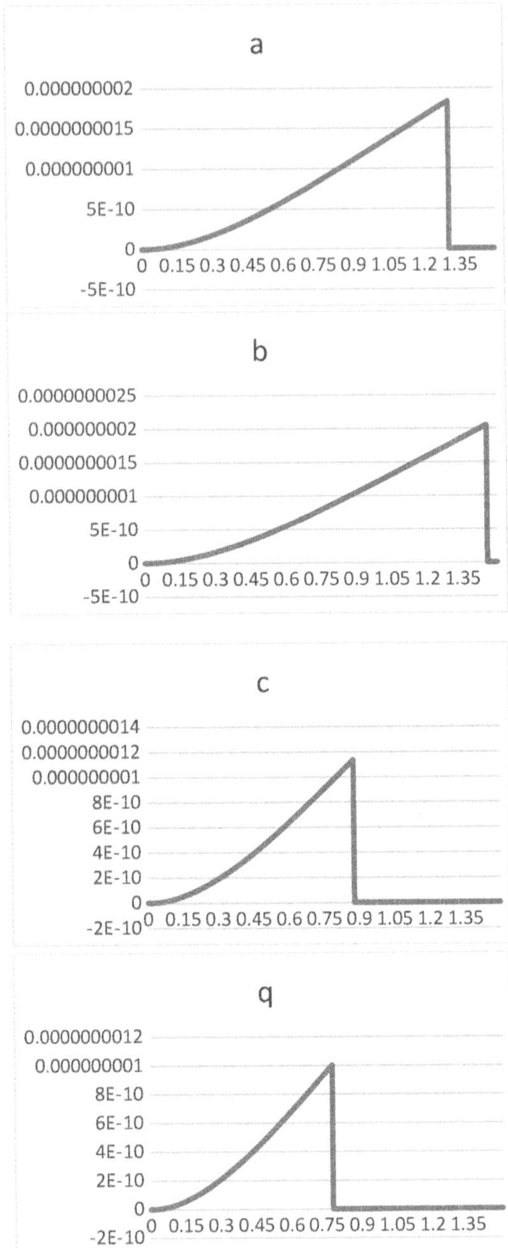

Figure 2.13. *Dynamics of intensity: a) pyramid; b) cone; c) cylinder; d) bar*

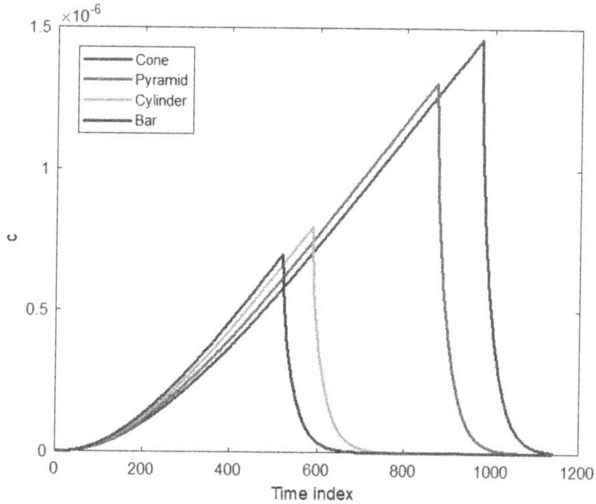

Figure 2.14. *Equalization of the maximum concentration value in the skin for different forms of microneedles (pyramid, cone, cylinder, beam). For a color version of this figure, see www.iste.co.uk/koroliouk/computational2.zip*

2.6. Conclusion

The paper proposes and experimentally substantiates the use of a two-step symmetrized difference algorithm for computer modeling of convection and diffusion in the dermis using microneedles with known coordinates. The influence of the shape and size of the microneedles, as well as physiological parameters on the distribution of drug concentration, is considered. It is shown that drugs are distributed more quickly and evenly in the case of self-dissolving microneedles in the form of bars and cylinders. The proposed method allows us to not only predict the distribution of the drug in the dermis, but also to ensure the target distribution of drugs at a given time, by selecting the parameters of the microneedles.

2.7. References

Benbrook, N. and Zhan, W. (2025). Mathematical modelling of hollow microneedle-mediated transdermal drug delivery. *Drug Deliv. Transl. Res.*, 15, 3226–3251.

Bhatnagar, S., Dave, K., Venuganti, V.V.K. (2017). Microneedles in the clinic. *J. Control. Release*, 260, 164–182.

Calcutt, J.J. and Anissimov, Y.G. (2021). Predicting viable skin concentration: Diffusional and convective drug transport. *J. Pharm. Sci.*, 110, 2823–2832.

Calcutt, J.J., Roberts, M.S., Anissimov, Y.G. (2022). Predicting viable skin concentration: Modelling the subpapillary plexus. *Pharm. Res.*, 39, 783–93.

Jin, X. Zhu, D.D., Chen, B.Z., Ashfaq, M., Guo, X.D. (2018). Insulin delivery systems combined with microneedle technology. *Adv. Drug Deliv. Rev.*, 127, 119–137.

Jung, J.H. and Jin, S.G. (2021). Microneedle for transdermal drug delivery: Current trends and fabrication. *J. Pharm. Investig.*, 51, 503–517.

Kim, J., Jeong, J., Jo, J.K., So, H. (2025). Hollow microneedles as a flexible dosing control solution for transdermal drug delivery. *Mater. Today Bio*, 32, 101754.

Klyushin, D. and Onotskyi, V. (2016). Numerical simulation of 3D unsaturated infiltration from point sources in porous media. *J. Coupled Syst. Multiscale Dyn.*, 4(3), 187–193.

Lyashko, S.I., Klyushin, D.A., Onotskyi, V.V., Lyashko, N.I. (2018). Optimal control of drug delivery from microneedle systems. *Cybern. Syst. Anal.*, 54, 357–365.

Lyashko, S.I., Klyushin, D.A., Onotskyi, V.V., Bondar, O.S. (2024). Optimization of targeted drug delivery from systems of microneedles. *Rep. Nat. Acad. Sci. Ukraine*, 11, 16–23.

Mitra, A.K. Cholkar, K., Mandal, A. (eds) (2017). *Emerging Nanotechnologies for Diagnostics, Drug Delivery, and Medical Devises.* Elsevier, Amsterdam, Oxford, Cambridge.

Newell, B. and Zhan, W. (2023). Mathematical modelling of microneedle-mediated transdermal delivery of drug nanocarriers into skin tissue and circulatory system. *J. Control. Release*, 360, 447–467.

Newell, B.B. and Zhan, W. (2024). Numerical simulation of transdermal delivery of drug nanocarriers using solid microneedles and medicated adhesive patch. *Int. J. Heat Mass Transf.*, 223, 125291.

Parhi, R. and Supriya, N.D. (2019). Review of microneedle based transdermal drug delivery systems. *Int. J. Pharm. Sci. Nanotechnol.*. 12(3), 4511–4523.

Phatale, V., Vaiphei, K.K., Jha, S., Patil, D., Agrawal, M., Alexander, A. (2022). Overcoming skin barriers through advanced transdermal drug delivery approaches. *J. Control. Release*, 351, 361–380.

Pires, L.R., Amado, I.R., Gaspar, J. (2020). Dissolving microneedles for the delivery of peptides – Towards tolerance-inducing vaccines. *Int. J. Pharm.*, 586, 119590.

Plamadeala, C., Gosain, S., Hischen, F., Buchroithner, B., Puthukodan, S., Jacak, J., Bocchino, A., Whelan, D., O'Mahony C., Baumgartner, W., Heitz, J. (2019). Bio-inspired microneedle design for efficient drug/vaccine coating. *Biomed. Microdevices*, 22, 8–19.

Ronnander, P., Simon, L., Spilgies, H., Koch, A. (2018). Modelling the in-vitro dissolution and release of sumatriptan succinate from polyvinylpyrrolidone-based microneedles. *Eur. J. Pharm. Sci.*, 125(1), 54–63.

Sandrakov, G.V., Lyashko S.I., Bondar E.S., Lyashko, N.I. (2019). Modeling and optimization of microneedle systems. *J. Autom. Inf. Sci.*, 51(6), 1–11,

Sandrakov, G.V., Lyashko, S.I., Bondar, E.S., Lyashko, N.I., Semenov V.V. (2020). Modeling of configurations formed under using microneedle systems. *PC&I*, 65(6), 38–47.

Wu, Y., Hutton, A.R.J., Pandya, A., Patravale, V.B., Donnelly, R.F. (2023). Microneedle and polymeric films: Delivery of proteins, peptides and nucleic acids. In *Drug Delivery and Targeting*, Schäfer-Korting, M. and Schubert, U.S. (eds). Springer, Cham, 93–111.

Yadav, P.R., Han, T., Olatunji, O., Pattanayek, S.K., Das, D.B. (2020). Mathematical modelling, simulation and optimisation of microneedles for transdermal drug delivery: Trends and progress. *Pharmaceutics*, 12, 693.

3

Homogenization and Modeling of Processes in Composites Similar to Photonic Crystals

G. V. SANDRAKOV
Taras Shevchenko National University of Kyiv, Ukraine

3.1. Introduction

Understanding the processes in crystals is useful, for example, for designing the processing core of computers. Such processes are modeled based on the well-known Schrödinger equation. As a rule, ideal crystals have a periodic structure, which is characterized by the periodicity of the Schrodinger equation potential. By analogy with such crystals, physicists are developing periodic photonic crystals, in which light waves propagate at the highest speed. It is hoped that mathematical and numerical modeling will lead to an understanding and description of the processes occurring in these photonic crystals. Probably, this approach is the only one possible, since it is quite difficult to conduct experiments in such complex devices. Thus, modeling can help to interpret and even possibly predict new phenomena. However, a relevant mathematical model must be chosen. There are books which propose modeling processes in photonic crystals based on the Schrodinger equation. This approach is not entirely suitable, since light wave processes are modeled by Maxwell's equations, which form a system and can sometimes be reduced to a wave equation with damping, which is significantly different from the Schrödinger equation and has significantly different solution properties.

Computational Methods and Mathematical Modeling in Cyberphysics and Engineering Applications 2,
coordinated by Dmitri KOROLIOUK, Sergiy LYASHKO and Nikolaos LIMNIOS.

The speed of the waves in photonic crystals is much higher than the speed of waves in ordinary crystals, and this can be useful for some applications and uses. The periodicity of photonic crystals is also natural technologically for production. Indeed, having chosen a cell of periodicity, we can continue the cell in a periodic way to obtain a photonic crystal, similar to how walls and squares are covered with tiles. Usually, there are a lot of such cells and the crystal has two scales, which are characterized by its size and the number of cells or equivalently the cell size. Thus, composite two-scale continuous media arise with a periodic structure, which are similar to composite materials with microscale. And mathematical and numerical modeling, for example, electromagnetic and wave processes in such multiscale continuous media becomes relevant and natural. In fact, useful tools for this modeling have been developed as homogenization theory.

Homogenization methods have proven to be a convenient approach for studying problems for multiscale periodic continuous media at a very small microscale. Initially, about 50 years ago, the homogenization methods were developed by N. S. Bakhvalov, I. Babushka, A. Bensoussan, J. L. Lions and G. Papanicolaou to simplify computer simulation problems for composite two-scale media with a periodic structure; details can be found, for example, in Bahvalov (1974, 1975) and Bensoussan et al. (1978). According to the theory, for example, modeling the heat processes in the composite media can be replaced with guaranteed accuracy by modeling the heat process in homogeneous (homogenized) media. In a mathematical sense, this means that the solution of the heat conduction problem for the equation with periodic rapidly oscillating coefficients is close to the solution of the problem for the equation with constant coefficients. The corresponding theorems and accuracy estimates were proved, for example, in Bensoussan et al. (1978) and Bakhvalov and Panasenko (1989) for sufficiently regular data.

Artificial composite materials have shown their usefulness and application in many fields of science and technology, and perhaps the creation of artificial photonic crystals will also be useful for applied purposes. However, it is unnatural to create artificial composites from materials with similar properties. But, if the properties of the composite components differ greatly, then this difference should be compared with the microscale parameter, since the models can be completely different, which will be clearly explained further. For that reason, when multiscale periodic media are composed of materials with very different properties, additional small and large parameters may appear in models. Thus, there is a need to homogenize problems with several parameters that more accurately model the considered composite materials and take into account which can lead to new homogenized models.

Homogenization methods for some initial-boundary value problems with several parameters were developed in Sandrakov (1996, 1997a, 1997b, 1998a, 1998b, 2000), where homogenized problems with constant coefficients and time convolutions are obtained and approximation theorems were proved. The results were obtained using asymptotic and energy methods. The presence of time convolutions in models is called the *memory effect* (Duvaut and Lions 1972). The effect is far from trivial, since the operator spectrum has gaps like those of semiconductors in the detailed sense explained in Sandrakov (2006b, 2007). In the absence of convolutions, the gaps do not arise as noted in the works. Note that such homogenized time convolution problems were previously obtained in Bensoussan et al. (1978) and Sanchez-Palencia (1980) using asymptotic methods. These methods have proven to be effective and allow for an asymptotically controlled reduction of the problem residual on a small microscale, which leads to accurate estimates for the obtained asymptotic approximations through the use of energy methods.

Over time, it turned out to be much easier to prove convergence or multiscale convergence of solutions to solutions of certain homogenized multiscale problems, starting essentially with Allaire (1992). This approach, usually, does not require additional data regularity which is essential for nonlinear problems as, for example, in Piat and Sandrakov (2004) and Sandrakov (2004d, 2005, 2006a) for variational inequalities whose solutions are not regular as a rule. Nevertheless, when studying such variational inequalities in these works, an asymptotic expansion method for linearized problems was implicitly used; this subsequently made it possible to apply convergence methods. Multiscale convergence theorems, which are entirely based on a priori estimates of solutions, can lead to incorrect homogenized multiscale problems, since for problems with additional parameters, such estimates may not be exact. So, inferring the problems from estimates requires high mathematical skills.

In contrast to the method of asymptotic expansions, since the construction of the asymptotics leads to a small residual and therefore to accuracy estimates for the asymptotic approximation, which implies convergence of solutions, at least for sufficiently regular solutions with relevant energy estimates. This approach also allows us to separate fast and slow variables; similar methods are used, for example, in Diaz et al. (2019a, 2019b, 2022a, 2022b). But, for homogenized multiscale problems, such variables are not separated and the type of the corresponding problems is not obvious. In addition, the accuracy of the approximations was not clarified and proved in the multiscale case; further details and relevant references can be found, for example, in Jager and Woukeng (2021), Garttnerat et al. (2022), Ray and Schulz (2023), Garttnerat et al. (2024) and Mascheroni et al. (2025). An approach also leading to homogenized multiscale problems based on the periodic unfolding method is presented, for example, in Cioranescu et al. (2018, 2023).

Here, we will use a method of asymptotic solution expansions for elliptic equations of arbitrary order, the application of which does not require advanced skills as detailed in Sandrakov (1991). This approach is applied to the initial-boundary value problem for the wave equation with damping, which is translated into an elliptic equation using the Laplace transform. The translation is an isomorphism according to Agranovich and Vishik (1964), which allows us to apply the inverse Laplace transform to the resulting homogenized problems. It is this approach that leads to homogenized problems with time convolution, which can form systems of equations for multiphase homogenized models (Sandrakov 2004a, 2004b, 2004c, 2007). The exact formulation of the relevant problem for the wave equation with parameters is given in the next section. Then, asymptotic expansions will be constructed for the regular case of non-degenerate coefficients. After that, the singular case is considered in the next section, which leads to homogenized problems for equations with convolutions and surprisingly homogenized initial conditions.

3.2. Composite media with periodic structures and contrast properties

The purpose here is to investigate models of wave processes in a composite media with a periodic structure, which occupies a bounded domain $\Omega \subset \mathbb{R}^n$ and is composed of two materials with very different properties. For an integer $n \geq 2$, periodic media with the small microscale ε are defined as follows. Let the set F_1 be an open connected 1-periodic (periodic with period 1 in each of the variables $x_1, \ldots, x_n$) subset of $\mathbb{R}^n$ with the locally Lipschitz boundary and $F_0 = \mathbb{R}^n \setminus F_1$ is a set with the locally Lipschitz boundary. For a positive ε, we denote:

$$F_1^\varepsilon = \varepsilon F_1 = \{\varepsilon x : x \in F_1\}, \qquad F_0^\varepsilon = \varepsilon F_0 = \{\varepsilon x : x \in F_0\}.$$

Therefore, the sets $F_1 = F_1^1$ and $F_0 = F_0^1$ with a common boundary ∂F_1 are defined by the sets $Y_1 = F_1 \cap Y$ and $Y_0 = F_0 \cap Y$ with the boundary $\Gamma = \partial F_1$, where $Y = (0,1)^n$ denotes the *periodicity cell*. Thus, the given Y_1 and Y_0 partition the periodicity cell Y into two sets, which correspond to two materials with very different properties separated by a common boundary Γ. It is assumed that Y_1 and Y_0 are the sets of the positive Lebesgue measure $|Y_1|$ and $|Y_0|$ in the space $\mathbb{R}^n$.

Let also the bounded domain Ω be Lipschitz. The sets F_1^ε and F_0^ε for a small enough fixed ε naturally define models for composite media with ε-periodic structure $\Omega_1^\varepsilon = \Omega \cap F_1^\varepsilon$ and $\Omega_0^\varepsilon = \Omega \cap F_0^\varepsilon$, which correspond to different materials in the domain and are bounded by the boundary $\partial\Omega$ of this domain. Examples of models of such periodic media are shown in Figure 3.1.

Figure 3.1. *Models for composite periodic media*

For such composite media, we define the real coefficients of density, damping and conductivity matrices of materials in the domain Ω using the following equalities:

$$m_\mu^\varepsilon = \mu\, m_0(x/\varepsilon),\;\; s_\mu^\varepsilon = \mu\, s_0(x/\varepsilon),\;\; r_\mu^\varepsilon = \mu\, r_0(x/\varepsilon) \quad in \quad \Omega_0^\varepsilon,$$

$$A_\sigma^\varepsilon = \sigma A_0(x/\varepsilon),\;\; b^\varepsilon = b_0(x/\varepsilon) \quad in \quad \Omega_0^\varepsilon, \qquad [3.1]$$

$$m_\mu^\varepsilon = m_1(x/\varepsilon),\;\; s_\mu^\varepsilon = s_1(x/\varepsilon),\;\; r_\mu^\varepsilon = r_1(x/\varepsilon) \quad in \quad \Omega_1^\varepsilon,$$

$$A_\sigma^\varepsilon = A_1(x/\varepsilon),\;\; b^\varepsilon = b_1(x/\varepsilon) \quad in \quad \Omega_1^\varepsilon,$$

where μ and σ are the positive parameters, and ε is a small parameter. For definiteness, let us assume $0<\mu\le 1$, $0<\sigma\le 1$ and $0<\varepsilon\le\varepsilon_0$ for a positive ε_0.

It is supposed that all functions $m_0(y)$, $s_0(y)$, $r_0(y)$, $m_1(y)$, $s_1(y)$, .. and the elements of the matrices $A_0(y)$, $A_1(y)$, $b_0(y)$, $b_1(y)$ from these equalities are defined and bounded for $y \in Y$. In addition, the matrices $A_0(y)$ and $A_1(y)$ are symmetric and elliptic in [3.1]. Here, ellipticity means that there are positive constants α and β such that in the matrix sense the inequalities:

$$0 < \alpha E \le A_0(y) \le \beta E, \quad 0 < \alpha E \le A_1(y) \le \beta E$$

are satisfied, where $y \in Y$ and E are the identity matrix. Hence, the elements of the matrix A_σ^ε may be discontinuous at the boundary $\partial\Omega_1^\varepsilon \setminus \partial\Omega$ separating Ω_0^ε and Ω_1^ε, since the restrictions of the matrix elements on Ω_0^ε and Ω_1^ε may not coincide. Suppose also that density and damping are bounded away from zero and:

$$\alpha \le m_0(y) \le \beta, \quad \alpha \le m_1(y) \le \beta, \quad \alpha \le s_0(y) \le \beta, \quad \alpha \le s_1(y) \le \beta.$$

Let the functions $f \in C_0^\infty([0,T]\times\Omega)$, $u_0 \in C_0^\infty(\Omega)$, $u_1 \in C_0^\infty(\Omega)$ and the vector function $g \in C_0^\infty([0,T]\times\Omega)^n$ be given, where the positive T is fixed. Here and below, the spaces of real functions will be used, the definition of which is given, for example, in Duvaut and Lions (1972) and Sanchez-Palencia (1980).

Define $u = u(t,x)$ as a solution in the sense of distributions of the initial-boundary value problem

$$m_\mu^\varepsilon u_{tt}'' + s_\mu^\varepsilon u_t' - div\, A_\sigma^\varepsilon(\nabla u + b^\varepsilon g) = r_\mu^\varepsilon f \quad in \quad \Omega\times(0,T), \qquad [3.2]$$

$$u\,|_{t=0} = \rho^\varepsilon u_0, \; u_t'\,|_{t=0} = \varphi^\varepsilon u_1 \quad in \quad \Omega, \quad u = 0 \quad on \quad \partial\Omega\times(0,T),$$

where the oscillating multipliers $\rho^\varepsilon = \rho(x/\varepsilon)$, $\varphi^\varepsilon = \varphi(x/\varepsilon)$ were introduced for further context, assuming that $\rho(y)$ and $\varphi(y)$ are 1-periodic smooth functions. Problem [3.2] models wave processes with the dumping coefficient s_μ^ε in periodic composite media. For fixed parameters μ, σ and ε, problem [3.2] has a unique solution, for example, according to Duvaut and Lions (1972). Therefore, it is natural to find asymptotic expansions of this solution.

3.3. Regular homogenized asymptotic expansions of solutions

The equation of problems [3.2] is hyperbolic for fixed parameters μ, σ and ε. However, if μ or σ is small, this hyperbolicity may degenerate on the set Ω_0^ε, which models a low-density and weakly conducting material. Accordingly, let us first consider the simpler case $\sigma = 1$, which is in some sense regular.

Further, it will be convenient to use the Laplace transform, which is an isomorphism of the corresponding spaces (Agranovich and Vishik 1964). For completeness, the precise definitions will be given here. For a fixed positive ω, the space $L^2_\omega(0,\infty;L^2(\Omega))$ is defined as the set of functions from $L^2_{loc}(0,\infty;L^2(\Omega))$ for which $\|u\|_{L^2_\omega(0,\infty;L^2(\Omega))}=\left\|e^{-\omega t}u\right\|_{L^2(0,\infty;L^2(\Omega))}$ is a finite value. The last equality defines the norm of the space $L^2_\omega(0,\infty;L^2(\Omega))$ relative to which the space is complete. Denote $E_\omega(L^2(\Omega))=E_\omega(\mathbb{C}_\omega;L^2(\Omega))$ the set of functions $U(z)=U(a+ib)$ with values in $L^2(\Omega)$, which are continuous and holomorphic on the complex half-plane $\mathbb{C}_\omega=\{z\in\mathbb{C}:z=a+ib,\ a>\omega\}$. The norm:

$$\|U\|^2_{E_\omega(L^2(\Omega))}=\int_{-\infty}^{\infty}\|U(\omega+ib)\|^2_{L^2(\Omega)}\,db$$

is a finite value, where ω is a fixed real and the integral is understood as the limit in the average of the integral from $-K$ to K when $K\to\infty$.

It is proven in Agranovich and Vishik (1964) that the Laplace transform

$$L(u(t))=\int_0^\infty e^{-zt}u(t)\,dt=U(z)\qquad for\quad z\in\mathbb{C}_\omega$$

is an isometric isomorphism from $L^2_\omega(0,\infty;L^2(\Omega))$ to $E_\omega(L^2(\Omega))$ for fixed ω. The spaces $E_\omega(H^1_0(\Omega))$ and $E_\omega(H^{-1}(\Omega))$ are defined similarly, for which analogous isomorphisms also hold. In addition, the Laplace transform commutes with differentiation with respect to spatial variables; the time convolution operator is translated by the Laplace transform into the operator of pointwise multiplication with respect to $z\in\mathbb{C}$, and for derivatives, we have the equalities

$$L(u''_{tt}(t))=z^2U(z)-zu|_{t=0}-u'_t|_{t=0},\quad L(u'_t(t))=zU(z)-u|_{t=0}\,.$$

Let us introduce the notations $U=Lu$, $F=Lf$ and $G=Lg$ for the Laplace transforms of the solution and data of problem [3.2], where $x\in\Omega$ is considered as a parameter and the complex parameter z, for example, is in $\mathbb{C}_0$. Thus, applying the Laplace transform to problem [3.2], we obtain for $U \in H_0^1(\Omega)$ an equivalent problem with a complex parameter of the following form problem:

$$m_\mu^\varepsilon z^2 U + s_\mu^\varepsilon z U - div A_\sigma^\varepsilon \left(\nabla U + b^\varepsilon G\right) \tag{3.3}$$

$$= m_\mu^\varepsilon (z \rho^\varepsilon u_0 + \varphi^\varepsilon u_1) + s_\mu^\varepsilon \rho^\varepsilon u_0 + r_\mu^\varepsilon F \quad in \quad \Omega \times \mathbb{C}_0.$$

The problem is a boundary value problem for an elliptic equation with complex coefficients in the lower terms for a fixed z. So, the problem is solvable for all $z \in \mathbb{C}$, except, possibly, a countable set according to Agranovich and Vishik (1964).

Let us fix $z \in \mathbb{C}_0$ further. Following Bensoussan et al. (1978), the asymptotic approximation to the solution of problem [3.3] can be chosen in the form:

$$U_a = V(x) + \varepsilon V_1(x, x/\varepsilon) + \varepsilon^2 V_2(x, x/\varepsilon) + \varepsilon^3 V_3(x, x/\varepsilon), \tag{3.4}$$

where the functions $V_1(x,y)$, $V_2(x,y)$, $V_3(x,y)$ are 1-periodic in the variable $y \in \mathbb{R}^n$. To define the terms of this asymptotic expansion, the two formulas:

$$\nabla\left(W(x, x/\varepsilon)\right) = \left(\varepsilon^{-1} \nabla_y W(x,y) + \nabla_x W(x,y)\right)\Big|_{y=x/\varepsilon}, \tag{3.5}$$

$$div\left(W(x, x/\varepsilon)\right) = \left(\varepsilon^{-1} div_y \, W(x,y) + div_x \, W(x,y)\right)\Big|_{y=x/\varepsilon}$$

for functions of the form $W = W(x, x/\varepsilon)$, depending on ε in a special way, will often be used in what follows. We substitute U_a instead of U into [3.3] and take into account [3.5]. Then, keeping only the terms essential for what follows, we get:

$$m_\mu^1 z^2 V + s_\mu^1 z V - \varepsilon^{-1} div_y (A_1^1 \nabla_y V_1) - \varepsilon^{-1} div_y A_1^1 (\nabla_x V + b^1 G)$$

$$- div_y (A_1^1 \nabla_y V_2) - div_x A_1^1 (\nabla_x V + b^1 G) - div_x (A_1^1 \nabla_y V_1) \tag{3.6}$$

$$- div_y (A_1^1 \nabla_x V_1) - \varepsilon \, div_x (A_1^1 \nabla_x V_1) - \varepsilon \, div_x (A_1^1 \nabla_y V_2)$$

$$- \varepsilon \, div_y (A_1^1 \nabla_x V_2) = m_\mu^1 (z \rho u_0 + \varphi u_1) + s_\mu^1 \rho u_0 + r_\mu^1 F$$

for $y = x/\varepsilon$ and $x \in \Omega$, where, for example, by definition $m_{\mu}^{1} = m_1(y)$ for $y \in Y_1$, $m_{\mu}^{1} = m_0(y)$ for $y \in Y_0$ and $\rho = \rho(y)$. Recall that the case $\sigma = 1$ is considered.

The main idea of constructing homogenizing asymptotic expansions is the observation that the variables x and y in [3.6] can be considered independent. In the sense that if the left side is equal to zero for these variables, then this is also true for $y = x/\varepsilon$. Equating the coefficients at ε^{-1} in the asymptotic equality, we have:

$$-\,div_y(A_1^1(y)\nabla_y V_1(x,y)) = div_y(A_1^1(y)\nabla_x V(x) + A_1^1(y)b^1(y)G(x)). \qquad [3.7]$$

The equality will be fulfilled if we choose $V_1 = N_a(y)\nabla_x V(x) + N_b(y)G(x)$ and get the vector function $N_a(y)$ and $N_b(y)$ as 1-periodic solutions of the problems:

$$-\,div_y(A_1^1\nabla_y N_a) = div_y A_1^1, \quad -\,div_y(A_1^1\nabla_y N_b) = div_y A_1^1 b^1. \qquad [3.8]$$

We know (Bensoussan et al. 1978) that solutions to the problems exist and are defined up to constants, which will be fixed as $\int_Y N_a(y)dy = 0$ and $\int_Y N_b(y)dy = 0$. Thus, in equality [3.7], it was possible to separate the variables x and y and the equality are fulfilled for all x and y, in particular for $y = x/\varepsilon$. With this choice, the coefficients at ε^{-1} in the asymptotic equality [3.6] are equal to zero.

Equating the coefficients at ε^{0} in the asymptotic equality (5), we obtain:

$$\begin{aligned} -\,div_y(A_1^1\nabla_y V_2) &= div_y\left(A_1^1 N_a\nabla_x^2 V + A_1^1 N_b\nabla_x G\right) \\ &+ div_x\left((A_1^1 + A_1^1\nabla_y N_a)\nabla_x V + (A_1^1 b^1 + A_1^1\nabla_y N_b)G\right) \qquad [3.9] \\ &+ m_{\mu}^{1}\rho z u_0 + m_{\mu}^{1}\varphi u_1 + s_{\mu}^{1}\rho u_0 + r_{\mu}^{1}F - m_{\mu}^{1}z^2 V - s_{\mu}^{1}zV, \end{aligned}$$

where it was taken into account in natural notations that $V_1 = N_a\nabla_x V + N_b G$. By analogy with [3.7], we consider [3.9] as a 1-periodic problem for $V_2 = V_2(x,y)$ with fixed $x \in \Omega$. We know (Bensoussan et al. 1978) that this problem is correct if and only if the right-hand side is orthogonal to 1 in the space $L^2(\Omega)$ and so we have:

$$mz^2 V + szV - div_x A(\nabla_x V) = div_x BG + m_\rho z u_0 + m_\varphi u_1 + m_s u_0 + rF, \quad [3.10]$$

where constants and constant matrices are defined by the following equalities:

$$m = \int_Y m_\mu^1(y)dy = \mu \int_{Y_0} m_0(y)dy + \int_{Y_1} m_1(y)dy, \quad s = \int_Y s_\mu^1(y)dy,$$

$$r = \int_Y r_\mu^1(y)dy, \ m_\rho = \int_Y m_\mu^1 \rho dy, \ m_\varphi = \int_Y m_\mu^1 \varphi dy, \ m_s = \int_Y s_\mu^1(y)\rho(y) dy, \quad [3.11]$$

$$A = \int_Y \left(A_1^1(y) + A_1^1(y)\nabla_y N_a(y) \right) dy, \ B = \int_Y \left(A_1^1(y) b^1(y) + A_1^1(y)\nabla_y N_b(y) \right) dy.$$

Equality [3.10] defines a homogenized problem with constant coefficients for problem [3.2]. Indeed, applying the inverse Laplace transform to this equality, we have for $v = v(t,x)$ the following homogenized initial-boundary value problem:

$$m v_{tt}'' + s v_t' - div A(\nabla v) = r f + div B g \quad in \quad \Omega \times (0,T), \quad [3.12]$$

$$v|_{t=0} = m^\rho u_0, \ v_t'|_{t=0} = m^\varphi u_1 + m^{s\rho} u_0 \quad in \quad \Omega, \ v = 0 \quad on \quad \partial\Omega \times (0,T)$$

with the constants $m^\rho = m^{-1} m_\rho$, $m^\varphi = m^{-1} m_\varphi$, $m^{s\rho} = m^{-1} m_s - s m^{-2} m_\rho$ (note that $m^{s\rho} = 0$ for $\rho = 1$ or $s = 0$), since the Laplace transform of $m v_{tt}''$ coincides with $m(z^2 V - z v|_{t=0} - v_t'|_{t=0})$ and equality [3.10] can be rewritten in the form:

$$m\left(z^2 V - m^{-1} m_\rho z u_0 - m^{-1} m_\varphi u_1\right) + s\left(z V - m^{-1} m_\rho u_0\right)$$

$$- div_x A(\nabla_x V) = div_x BG + m\left(m^{-1} m_s - s m^{-2} m_\rho\right) u_0 + r F.$$

We know (Bensoussan et al. 1978) that the matrix A is elliptic and symmetric and therefore there is a unique regular solution to problem [3.12]. Consequently, two terms in the asymptotic approximation [3.4] are defined. To determine the third term

$V_2 = V_2(x,y)$, we subtract from [3.9] equality [3.10] for the Laplace transform of the solution $L(v) = V(x)$ to problem [3.12]. Then, in natural notation, we obtain:

$$-div_y(A_1^1 \nabla_y V_2) = m_\rho^\circ z\, u_0 + m_\varphi^\circ u_1 + m_s^\circ u_0 + r^\circ F - s^\circ z V \qquad [3.13]$$

$$- m^\circ z^2 V + (div_y A_1^1 N_a + A^\circ)\nabla_x^2 V + (div_y A_1^1 N_b + B^\circ)\nabla_x G,$$

where the following 1-periodic matrix functions and 1-periodic functions:

$$A^\circ = A_1^1 + A_1^1 \nabla_y N_a - A, \quad B^\circ = A_1^1 b^1 + A_1^1 \nabla_y N_b - B,$$

$$m_\rho^\circ = m_\mu^1 \rho - m_\rho, \; m_\varphi^\circ = m_\mu^1 \varphi - m_\varphi, \; m_s^\circ = s_\mu^1 \rho - m_s,$$

$$r^\circ = r_\mu^1 - r, \; s^\circ = s_\mu^1 - s, \; m^\circ = m_\mu^1 - m$$

are such that their components and the functions are orthogonal to 1 in $L^2(Y)$. Thus, equality [3.13] and therefore equality [3.9] will be satisfied if we choose:

$$V_2 = N_a^1(y)\nabla_x^2 V(x) + N_b^1(y)\nabla_x G(x) + N_r(y)F(x)$$

$$+ N_\rho(y)\, z\, u_0(x) + N_\varphi(y) u_1(x) + N_{ss}(y) u_0(x) \qquad [3.14]$$

$$- N_s(y)\, z\, V(x) - N_m(y) z^2 V(x),$$

and define, for example, the functions $N_s = N_s(y)$ and $N_m = N_m(y)$ as 1-periodic solutions of the following problems:

$$div_y(A_1^1 \nabla_y N_s) = s^\circ, \quad div_y(A_1^1 \nabla_y N_m) = m^\circ. \qquad [3.15]$$

We know (Bensoussan et al. 1978) that solutions to the problems exist and are defined up to constants, which will be fixed as $\int_Y N_s(y)dy = 0$ and $\int_Y N_m(y)dy = 0$. Similarly, we can define 1-periodic functions $N_\rho(y)$, $N_\varphi(y)$, $N_{ss}(y)$, $N_r(y)$ and matrix functions $N_a^1(y)$, $N_b^1(y)$ in equality [3.14]. With such a choice of $V_2 = V_2(x,y)$, all coefficients at ε^0 in asymptotic equality [3.6] are equal to zero. Thus, the asymptotic equality [3.6] is satisfied up to $O(\varepsilon)$. Hence, the asymptotic expansion of $U_a = U_a(x, x/\varepsilon)$ also approximates the equation of problem [3.3] up to $O(\varepsilon)$. Following Sandrakov (1997b, 1998a, 1998b, 2007) and using the inverse Laplace transform of $U_a = U_a(x, x/\varepsilon)$, we can verify by energy methods that the last statement implies, for example, the accuracy estimate

$$\| u - v \|^2_{C^0([0,T];L^2(\Omega_1^\varepsilon))} + \mu \| u - v \|^2_{C^0([0,T];L^2(\Omega_0^\varepsilon))} \leq C\varepsilon, \qquad [3.16]$$

where u is the solution of problem [3.2], v is the solution of the homogenized problem [3.12] and the constant C does not depend on μ and ε when $0 < \mu \leq 1$ and $0 < \varepsilon \leq \varepsilon_0$ for relevant ε_0. In fact, in Sandrakov (1997b, 1998a, 1998b, 2007), more precise estimates for the energy norm are proved, but the formulations of the estimates are more complicated, since when calculating the derivatives we must use the fact that the asymptotic expansion also depends on the fast variables x/ε. In addition, for the case of small μ, we can derive the accuracy estimate

$$\| u - v_\circ \|^2_{C^0([0,T];L^2(\Omega_1^\varepsilon))} + \mu \| u - v_\circ \|^2_{C^0([0,T];L^2(\Omega_0^\varepsilon))} \leq C(\mu^2 + \varepsilon), \qquad [3.17]$$

where u is the solution of problem [3.2], $v_\circ$ is the solution of the homogenized problem [3.12] for $\mu = 0$ and the constant C does not depend on μ and ε when $0 < \mu \leq \mu_0$ and $0 < \varepsilon \leq \varepsilon_0$ for relevant μ_0 and ε_0.

It is possible to define $V_3 = V_3(x, x/\varepsilon)$ in the asymptotic approximation [3.4] so that it approximates the equation of problem [3.3] up to $O(\varepsilon^2)$. However, this does not lead to an improvement in the exponent of ε in the accuracy estimates [3.16] and [3.17], since the term $V_1 = N_a \nabla_x V + N_b G$ in the expansion [3.4] may not satisfy the homogeneous boundary conditions on the boundary $\partial\Omega$, which indicates possible boundary layer effects. In fact, to check the presented accuracy estimates, it is sufficient to restrict ourselves to the case of $V_2(x, x/\varepsilon) = 0$ and $V_3(x, x/\varepsilon) = 0$ in definition [3.4]. But, without understanding the structure of the function $V_2 = V_2(x, x/\varepsilon)$, it would be impossible to find out the exact form of the homogenized equation [3.12], which is the basis of the method here. Moreover, using ordinary asymptotic expansions, it is easy to improve the exponent of the small parameter μ in estimate [3.17], which means that this parameter is non-singular. Observe, furthermore, that it is not convenient to use $V_2(x, x/\varepsilon)$ directly from [3.14] in checking the accuracy estimates, since after applying the inverse transformation to [3.14], the delta functions $M^{\rho}\delta_t'(t)u_0$ and $M^{\varphi}\delta(t)u_1 + M^{s\rho}\delta(t)u_0$ arise in this equality, where $M_{\varepsilon}^{\rho} = N_{\rho}^{\varepsilon} - m^{\rho}N_m^{\varepsilon}$, $M_{\varepsilon}^{\varphi} = N_{\varphi}^{\varepsilon} - m^{\varphi}N_m^{\varepsilon}$, $M_{\varepsilon}^{s\rho} = N_{ss}^{\varepsilon} - m^{s\rho}N_m^{\varepsilon} - m^{\rho}N_s^{\varepsilon}$ and, for example, $N_m^{\varepsilon} = N_m(x/\varepsilon)$. Which again points to boundary layer effects near the boundary $t=0$. To eliminate the delta functions from [3.14], we can add to the asymptotic approximation the term $\varepsilon^2 w_{tt}''$, where w satisfies the initial conditions:

$$w|_{t=0} = -M_{\varepsilon}^{\rho}u_0, \qquad w_t'|_{t=0} = -M_{\varepsilon}^{\varphi}u_1 - M_{\varepsilon}^{s\rho}u_0$$

and the Laplace transform of w_{tt}'' satisfies the equation of problem [3.3] up to $O(\varepsilon)$. w is easy to define and $w=0$ only if $\rho=\varphi=1$. However, such a refinement does not improve the accuracy estimates [3.16] and [3.17] through the influence of boundary layer effects near the boundary $\partial\Omega$. But, the refinement is actual if harmonic waves with frequency ϖ are considered, since in this case Ω can be replaced by a suitable torus, which is a manifold without boundary.

Thus, according to accuracy estimate [3.16] and [3.17], when modeling wave processes in composite media, instead of solving problem [3.2] (e.g. numerically), it is possible to solve the homogenized problem [3.12] with guaranteed accuracy for small ε (and μ, respectively). Naturally, the numerical solution of problem [3.2] is significantly more complicated than the solution of problem [3.12] with constant coefficients, since a very fine computational grid is required to take into account the geometry of a very large number of material cells.

3.4. Singular homogenized asymptotic expansions of solutions

For very small σ, the considered accuracy estimates are not entirely correct, which is associated with the loss of hyperbolicity in [3.2] on the domain Ω_0^{ε}, corresponding to a weakly conducting material. This indicates the singularity of the parameter and leads to the need to replace 1-periodic problems [3.8] and [3.15] (which are usually called *cell problems*) with Neumann problems on the cell part Y_1.

In this connection, we recall the statement on the solvability of the problems. This statement is well known in the case of bounded coefficients and smooth boundary $\Gamma = \partial Y_1$ (as a submanifold with boundary of the n-dimensional torus). For Lipschitz coefficients and boundaries, this was proved in Mitrea and Taylor (1999). A survey of solvability results for problems with Lipschitz coefficients and boundaries is given in Agranovich (2002). Finally, for the bounded coefficients and the Lipschitz boundary, this statement was proved in Kim and Sakellaris (2024). More precisely, it was proved there for domains with the Lipschitz boundary, but the proofs are local and hence apply to manifolds.

THEOREM 3.1.– *Let* $V \in L^2(Y_1)$, $W \in H_{per}^{-1/2}(\Gamma)$ *and* $D \in L^2(Y_1)^n$ *be such that:*

$$\int_{Y_1} V(y)\,dy = \int_{\Gamma} W(s)\,ds,$$

where Y_1 *is considered as a connected submanifold of the* n *-torus. Then, there is a unique solution* $U \in H^1_{per}(Y_1)/\mathbb{R}$ *of the following 1-periodic problem:*

$$-\,div_y\left(A_1 \nabla_y U\right) = V + div_y D \;\; in \;\; Y_1, \quad -\Upsilon\cdot\left(A_1 \nabla_y U\right) = W + \Upsilon\cdot D \;\; on \;\; \Gamma,$$

where Υ *is the outward normal to* Γ *and* $\Upsilon\cdot D$ *is the inner product of* Υ *and D.*

According to the assumptions, the domain Ω is composed of two domains Ω_0^ε and Ω_1^ε, which are separated by a common boundary $\Gamma_\varepsilon = \partial\Omega_1^\varepsilon \setminus \partial\Omega$. Hence (Bakhvalov and Panasenko 1989), problem [3.2] can be rewritten in an equivalent form as the initial-boundary value problem with conjugation conditions at the interface of these domains. Thus, problem [3.2] is equivalent to:

$$\mu m_0^\varepsilon u_{tt}'' + \mu s_0^\varepsilon u_t' - \sigma div A_0^\varepsilon(\nabla u + b_0^\varepsilon g) = \mu r_0^\varepsilon f \;\; in \;\; \Omega_0^\varepsilon \times (0,T),$$

$$m_1^\varepsilon u_{tt}'' + s_1^\varepsilon u_t' - div A_1^\varepsilon(\nabla u + b_1^\varepsilon g) = r_1^\varepsilon f \;\; in \;\; \Omega_1^\varepsilon \times (0,T), \qquad [3.18]$$

$$u|_{t=0} = \rho^\varepsilon u_0, \;\; u_t'|_{t=0} = \varphi^\varepsilon u_1 \;\; in \;\; \Omega, \quad u = 0 \;\; on \;\; \partial\Omega \times (0,T),$$

$$\Upsilon_\varepsilon\cdot(A_1^\varepsilon \nabla u + A_1^\varepsilon b_1^\varepsilon g) = \sigma \Upsilon_\varepsilon\cdot(A_0^\varepsilon \nabla u + A_0^\varepsilon b_0^\varepsilon g) \;\; on \;\; \Gamma_\varepsilon \times (0,T)$$

and $u \in H_0^1(\Omega)$ for almost all $t \in (0,T)$. The last equality means the continuity of the flows on the internal boundary (the contact surface of the considered media Ω_0^ε and Ω_1^ε). In this equality, for example, the trace of $\Upsilon_\varepsilon\cdot(A_1^\varepsilon \nabla u + A_1^\varepsilon b_1^\varepsilon g)$ at a point

$\gamma \in \Gamma_\varepsilon$ is calculated as the limit value of this function for $x \to \gamma$, where $x \in \Omega_1^\varepsilon$. Here and below, Υ_ε denotes the external normal to the locally Lipschitz boundary Γ_ε and, for example, the notation $m_0^\varepsilon = m_0(x/\varepsilon)$, $r_1^\varepsilon = r_1(x/\varepsilon)$ is introduced.

This equivalent form for problem [3.2] clearly explains that we are considering media with small density and conductivity coefficients on the domain Ω_0^ε for small μ and σ. It is clear that we can multiply the equations in [3.18] by σ^{-1} and consider the case of large density and conductivity coefficients on the domain Ω_1^ε for small σ and $\mu = \sigma$. Accordingly, multiplying these equations by $\sigma^{-1/2}$, we can consider the case of small coefficients on Ω_0^ε and large ones on Ω_1^ε for small σ and $\mu = \sigma$. Other variants are also possible when multiplying by μ^{-c}. Of course, the multiplications do not affect the final form of the solution asymptotic expansions for problem [3.2] and the homogenized problem defining the expansions, which are mainly based on the assumption that F_1^ε is connected as ε -periodic set.

Applying the Laplace transform to problem [3.18] in the notation of [3.3], we get the following equivalent problem for $U \in H_0^1(\Omega)$ with a complex parameter:

$$\mu\, m_0^\varepsilon z^2 U + \mu\, s_0^\varepsilon z U - \sigma\, div\, A_0^\varepsilon (\nabla U + b_0^\varepsilon G)$$

$$= \mu\, m_0^\varepsilon (z \rho^\varepsilon u_0 + \varphi^\varepsilon u_1) + \mu\, s_0^\varepsilon \rho^\varepsilon u_0 + \mu\, r_0^\varepsilon F \ \ in\ \ \Omega_0^\varepsilon,$$

$$m_1^\varepsilon z^2 U + s_1^\varepsilon z U - div\, A_1^\varepsilon (\nabla U + b_1^\varepsilon G) \qquad [3.19]$$

$$= m_1^\varepsilon (z \rho^\varepsilon u_0 + \varphi^\varepsilon u_1) + s_0^\varepsilon \rho^\varepsilon u_0 + r_1^\varepsilon F \ \ in\ \ \Omega_1^\varepsilon$$

$$\Upsilon_\varepsilon \cdot (A_1^\varepsilon \nabla U + A_1^\varepsilon b_1^\varepsilon G) = \sigma \Upsilon_\varepsilon \cdot (A_0^\varepsilon \nabla U + A_0^\varepsilon\, b_0^\varepsilon G) \ \ on\ \ \Gamma_\varepsilon.$$

Following Sandrakov (1991), to construct the solution asymptotics for problem [3.19], we should describe the kernel of the highest-order operator on 1-periodic functions for $\varepsilon = 1$. For the first equation of [3.19], the kernel is trivial, for example, when $\sigma = \varepsilon^2$ (and real part of z^2 is positive for a fixed $z \in \mathbb{C}_0$). For the second equation, the kernel is formed by functions that are multiples of 1. Therefore, we will choose the asymptotics on the domain Ω_1^ε in the form [3.4] with $V_3 = 0$, and to determine the asymptotics on the set Ω_0^ε, we add [3.4] to the function $V_0 = V_0(x, x/\varepsilon)$, which is equal to zero on Ω_1^ε. Thus, we will choose:

$$U_a = V(x) + \varepsilon V_1(x, x/\varepsilon) + \varepsilon^2 V_2(x, x/\varepsilon) + V_0(x, x/\varepsilon). \qquad [3.20]$$

Note that the choice of the first term form, depending only on x, is precisely related to the fact that the kernel is formed by multiples of 1.

Let us substitute asymptotics [3.20] instead of U into the second equation from [3.19] and take into account [3.5]. Then, keeping only the terms essential for what follows, we have exactly the asymptotic equality [3.6], considered for $y = x/\varepsilon$ and $x \in \Omega_1^\varepsilon$. Similarly, for the conjugation condition from [3.19], we obtain:

$$\Upsilon \cdot (A_1\nabla_y V_1 + A_1\nabla_x V + A_1 b_1 G + \varepsilon A_1\nabla_y V_2 + \varepsilon A_1 \nabla_x V_1) \qquad [3.21]$$

$$= \varepsilon \Upsilon \cdot (\vartheta A_0 \nabla_y V_0 + \varepsilon\vartheta A_0 \nabla_x V_0 + \varepsilon\vartheta A_0 \nabla_x V + \varepsilon\vartheta A_0 b_0 G + \varepsilon\vartheta A_0 \nabla_y V_1)$$

for $y = x/\varepsilon$ and $x \in \Gamma_\varepsilon$, where $\vartheta = \sigma/\varepsilon^2$ is considered as a parameter. As a model example, we can imagine that $\sigma = \varepsilon^2$ and therefore $\vartheta = 1$.

Equating the coefficients at ε^{-1} in the asymptotic equality [3.6], we have equation [3.7]. Accordingly, the coefficient at ε^0 in [3.21] leads to this relation:

$$-\Upsilon \cdot (A_1 \nabla_y V_1) = \Upsilon \cdot (A_1 \nabla_x V + A_1 b_1 G) \quad \textit{for} \quad y = x/\varepsilon \ \textit{and} \ x \in \Gamma_\varepsilon.$$

Thus, we can choose $V_1 = N_a^1(y)\nabla_x V(x) + N_b^1(y) G(x)$ and define the vector function $N_a^1(y)$ and $N_b^1(y)$ as 1-periodic solutions of the Neumann problems:

$$-div_y(A_1 \nabla_y N_a^1) = div_y A_1 \ \textit{in} \ Y_1, \ -\Upsilon \cdot A_1 \nabla_y N_a^1 = \Upsilon \cdot A_1 \ \textit{on} \ \Gamma, \qquad [3.22]$$

$$-div_y(A_1 \nabla_y N_b^1) = div_y A_1 b_1 \ \textit{in} \ Y_1, \ -\Upsilon \cdot A_1 \nabla_y N_b^1 = \Upsilon \cdot A_1 b_1 \ \textit{on} \ \Gamma.$$

By virtue of Theorem 1, solutions to problems [3.22] exist and are defined up to constants, which will fix with the conditions $\int_{Y_1} N_a^1(y)dy = 0$ and $\int_{Y_1} N_b^1(y)dy = 0$.

Let us extend such function vectors to Y_1 as 1-periodic solutions of the problems:

$$-div_y(A_0\nabla_y N_a^1) = 0 \ \ in \ \ Y_0, \quad N_a^1 = N_a^1 \ \ on \ \ \Gamma \cup Y_1, \qquad [3.23]$$

$$-div_y(A_0\nabla_y N_b^1) = 0 \ \ in \ \ Y_0, \quad N_b^1 = N_b^1 \ \ on \ \ \Gamma \cup Y_1.$$

We know (Duvaut and Lions 1972) that such solutions are uniquely determined.

Equating the coefficients at ε^0 in the asymptotic equality [3.6], we have equation [3.9]. Accordingly, the coefficient at ε^1 in [3.21] leads to this relation:

$$-\Upsilon\cdot(A_1\nabla_y V_2) = -\Upsilon\cdot(\vartheta A_0\nabla_y V_0) + \Upsilon\cdot(A_1 N_a^1\nabla_x^2 V + A_1 N_b^1\nabla_x G)$$

for $y = x/\varepsilon$ and $x \in \Gamma_\varepsilon$. By analogy with [3.7], we consider equation [3.9] with such boundary conditions as a 1-periodic problem for $V_2 = V_2(x,y)$ with fixed $x \in \Omega$. By virtue of Theorem 1, such a problem is correct if and only if:

$$m^1 z^2 V + s^1 z V - div_x A^1(\nabla_x V) = div_x B^1 G \qquad [3.24]$$

$$+ m_\rho^1 z\, u_0 + m_\varphi^1 u_1 + m_s^1 u_0 + r^1 F + \vartheta \int_\Gamma \Upsilon\cdot A_0\, \nabla_y V_0\, ds.$$

where the constants m^1, s^1, m_ρ^1, m_φ^1, m_s^1, r^1 and the matrices A^1, B^1 with constant components are defined by the equalities [3.11], where the integrals over Y are replaced by integrals over Y_1. In comparison with [3.10], here appeared the term of interaction (mutual resonance) between continuous media Ω_0^ε and Ω_1^ε. This interaction is relevant when σ is of the order ε^2 and is zeroed in the appropriate sense for small and large ϑ in accordance with Sandrakov (1998a, 1998b, 2000). This interaction is much more complex when the F_1^ε has several connected mutually separated components, which leads to multiphase homogenized models for systems of equations (Sandrakov 1998b, 2004a, 2004b, 2004c, 2007).

Let us substitute asymptotics [3.20] instead of U to the first equation from [3.19] and take into account [3.5]. Then, keeping only the terms essential for what follows, we have the following equation:

$$m_0 z^2 V_0 + s_0 z V_0 - \theta\, div_y(A_0\nabla_y V_0)$$

$$= r_0 F + m_0 z \rho u_0 + m_0 \varphi u_1 + s_0 \rho u_0 - s_0 z V - m_0 z^2 V \qquad [3.25]$$

for $y = x/\varepsilon$ and $x \in \Omega_0^\varepsilon$, where $\theta = \vartheta/\mu = \sigma/\mu\varepsilon^2$ is considered as a parameter. This equation will be satisfied if we choose $V_0 = V_0(z,x,y)$ in the form:

$$V_0 = P_r(z,y)F(z,x) + P_\rho(z,y)\, z u_0(x) + P_\varphi(z,y) u_1(x) \tag{3.26}$$

$$+ P_{ss}(z,y)u_0(x) - P_s(z,y)\, zV(z,x) \; - P_m(z,y)\, z^2\, V(z,x),$$

where P_r, P_ρ, P_φ, P_{ss}, P_s, P_m are defined as 1-periodic solutions of the problems:

$$m_0(z^2 P_r - m_0^{-1} r_0) + s_0 z P_r - \theta div_y(A_0 \nabla_y P_r) = 0 \;\; in \;\; Y_0,$$

$$m_0(z^2 P_\rho - \rho) + s_0 z P_\rho - \theta\, div_y(A_0 \nabla_y P_\rho) = 0 \;\; in \;\; Y_0,$$

$$m_0(z^2 P_\varphi - \varphi) + s_0 z P_\varphi - \theta\, div_y(A_0 \nabla_y P_\varphi) = 0 \;\; in \;\; Y_0, \tag{3.27}$$

$$m_0(z^2 P_{ss} - m_0^{-1} s_0 \rho) + s_0 z P_{ss} - \theta\, div_y(A_0 \nabla_y P_{ss}) = 0 \;\; in \;\; Y_0,$$

$$m_0(z^2 P_s - m_0^{-1} s_0) + s_0 z P_s - \theta\, div_y(A_0 \nabla_y P_s) = 0 \;\; in \;\; Y_0,$$

$$m_0(z^2 P_m - 1) + s_0 z P_m - \theta\, div_y(A_0 \nabla_y P_m) = 0 \;\; in \;\; Y_0,$$

and $P_r = 0$, $P_\rho = 0$, $P_\varphi = 0$, $P_{ss} = 0$, $P_s = 0$, $P_m = 0$ on $\Gamma \cup Y_1$.

The problems are uniquely solvable for a fixed $z \in \mathbb{C}_0$ when the real part of z^2 is positive. In fact, this follows directly from the initial-boundary value problems for the inverse Laplace transform of the problem solutions. Indeed, applying the inverse transform to [3.27], we get equivalent unsteady 1-periodic problems of the form:

$$m_0 (p_r)''_{tt} + s_0 (p_r)'_t - \theta\, div_y(A_0 \nabla_y p_r) = 0 \;\; in \;\; Y_0 \times (0,\infty),$$

$$p_r \,|_{t=0} = 0, \; (p_r)'_t \,|_{t=0} = m_0^{-1} r_0 \;\; in \;\; Y_0, \; p_r = 0 \;\; on \;\; \Gamma \times (0,\infty),$$

$$m_0(p_\rho)''_{tt} + s_0(p_\rho)'_t - \theta\, div_y(A_0 \nabla_y p_\rho) = 0 \;\; in \;\; Y_0 \times (0,\infty), \tag{3.28}$$

$$p_\rho \,|_{t=0} = 0, \;\; (p_\rho)'_t \,|_{t=0} = \rho(y) \;\; in \;\; Y_0, \;\; p_\rho = 0 \;\; on \;\; \Gamma \times (0,\infty),$$

$$m_0(p_\varphi)''_{tt} + s_0(p_\varphi)'_t - \theta\, div_y(A_0 \nabla_y p_\varphi) = 0 \;\; in \;\; Y_0 \times (0,\infty),$$

$$p_\varphi \,|_{t=0} = 0, \;\; (p_\varphi)'_t \,|_{t=0} = \varphi(y) \;\; in \;\; Y_0, \;\; p_\varphi = 0 \;\; on \;\; \Gamma \times (0,\infty),$$

and p_{ss}, p_s, p_m are solutions of similar problems, where $Lp_r = P_r$, $Lp_\rho = P_\rho$, $Lp_\varphi = P_\varphi$, $Lp_{ss} = P_{ss}$, $Lp_s = P_s$, $Lp_m = P_m$ are the Laplace transforms of the problem solutions, extended by zero to Y_1. It is well known (Duvaut and Lions 1972) that suitable solutions to the problems [3.28] exist and are uniquely defined. So, the function $V_0 = V_0(z,x,y)$ is defined and the equality [3.25] takes the form:

$$\theta\, div_y(A_0 \nabla_y V_0) = m_0(z^2 P_r - m_0^{-1} r_0) F - m_0(z^2 P_m - 1)\, z^2 V$$

$$- m_0(z^2 P_s - m_0^{-1} s_0) z V + m_0(z^2 P_\rho - \rho) z u_0 + m_0(z^2 P_\varphi - \varphi) u_1$$

$$+ m_0(z^2 P_{ss} - m_0^{-1} s_0 \rho)\, u_0 + s_0 z P_r F + s_0 z P_\rho z u_0$$

$$+ s_0 z P_\varphi u_1 + s_0 z P_{ss} u_0 - s_0 z P_s\, z V - s_0 z P_m\, z^2\, V.$$

Integrating this equality over Y_0 and by parts, we obtain the relation:

$$-\theta \int_\Gamma \Upsilon \cdot A_0 \nabla_y V_0 \, ds = Q_r F - Q_m z^2 V - Q_s\, z V + Q_\rho\, z u_0 + Q_\varphi u_1 + Q_{ss} u_0,$$

or, using the structure of the initial conditions in [3.24], the equivalent equality:

$$-\theta \int_\Gamma \Upsilon \cdot A_0 \nabla_y V_0 \, ds = Q_r F - Q_s (z V - m_1^\rho u_0)$$

$$- Q_m (z^2 V - m_1^\rho z\, u_0 - m_1^\varphi u_1 - m_1^{s\rho} u_0) \tag{3.29}$$

$$+ (Q_\rho - m_1^\rho Q_m) z u_0 + (Q_\varphi - m_1^\varphi Q_m) u_1$$

$$+ (Q_{ss} - m_1^{s\rho} Q_m - m_1^\rho Q_s)\, u_0,$$

where $m_1^\rho = m_\rho^1 / m^1$, $m_1^\varphi = m_\varphi^1 / m^1$, $m_1^{s\rho} = (m_s^1 - s^1 m_\rho^1) / m^1$ as in problem [3.12] and we use the following notation:

$$Q_r = \int_{Y_0} m_0 (z^2 P_r - m_0^{-1} r_0) + s_0 z P_r \, dy, \quad Q_s = \int_{Y_0} m_0 (z^2 P_s - m_0^{-1} s_0) + s_0 z P_s \, dy,$$

$$Q_\rho = \int_{Y_0} m_0 (z^2 P_\rho - \rho) + s_0 \, z \, P_\rho \, dy, \quad Q_\varphi = \int_{Y_0} m_0 (z^2 P_\varphi - \varphi) + s_0 \, z \, P_\varphi \, dy,$$

$$Q_{ss} = \int_{Y_0} m_0 (z^2 P_{ss} - m_0^{-1} s_0 \rho) + s_0 \, z \, P_{ss} \, dy, \quad Q_m = \int_{Y_0} m_0 (z^2 P_m - 1) + s_0 \, z \, P_m \, dy.$$

The inverse Laplace transform of these functions (depending only on z, since, for example, $P_m = P_m(z, y)$ by definition) are formed by the equalities

$$q_r = \int_{Y_0} m_0 (p_r)_{tt}'' dy + \int_{Y_0} s_0 (p_r)_t' dy, \quad q_s = \int_{Y_0} m_0 (p_s)_{tt}'' dy + \int_{Y_0} s_0 (p_s)_t' dy,$$

$$q_\rho = \int_{Y_0} m_0 (p_\rho)_{tt}'' dy + \int_{Y_0} s_0 (p_\rho)_t' dy, \quad q_\varphi = \int_{Y_0} m_0 (p_\varphi)_{tt}'' dy + \int_{Y_0} s_0 (p_\varphi)_t' dy,$$

$$q_{ss} = \int_{Y_0} m_0 (p_{ss})_{tt}'' dy + \int_{Y_0} s_0 (p_{ss})_t' dy, \quad q_m = \int_{Y_0} m_0 (p_m)_{tt}'' dy + \int_{Y_0} s_0 (p_m)_t' dy,$$

which depend only on $t \in [0, \infty)$ and are determined by solutions to problems [3.28].

Equality [3.24], taking into account [3.29] and $\theta = \vartheta / \mu$, defines a homogenized problem for problem [3.2]. Indeed, applying the inverse Laplace transform to this equality, we have the following homogenized initial-boundary value problem:

$$m^1 v_{tt}'' + s^1 v_t' - \mu q_m * (v_{tt}'') - \mu q_s * (v_t') - \operatorname{div} A^1 (\nabla v) = r^1 f - \mu q_r * f$$

$$+ \operatorname{div} B^1 g + \mu (q_\rho^1)_t' u_0 + \mu (q_\varphi^1) u_1 + \mu (q_{s\rho}^1) u_0 \quad in \ \Omega \times (0, T), \tag{3.30}$$

$$v|_{t=0} = m_1^\rho u_0, \ v_t'|_{t=0} = m_1^\varphi u_1 + m_1^{s\rho} u_0 \ in \ \Omega, \ v = 0 \ on \ \partial\Omega \times (0, T),$$

where $q_\rho^1 = q_\rho - m_1^\rho q_m$, $q_\varphi^1 = q_\varphi - m_1^\varphi q_m$, $q_{s\rho}^1 = q_{ss} - m_1^{s\rho} q_m - m_1^\rho q_s$ and the initial conditions were defined as in [3.12] using [3.24] and [3.29]. In addition, in this problem, $*$ denotes the convolution operator with respect to $t \in [0, \infty)$, for example; by definition, we have the equality relation:

$$q_m * (v_{tt}'') = \int_0^t q_m(t-\tau)(v_{\tau\tau}''(\tau, x))\, d\tau.$$

We know (Bensoussan et al. 1978) that the matrix A^1 is elliptic. For homogeneous initial conditions, it was proven in Sandrakov (1997b, 1998a, 1998b) that a unique solution to this problem exists for a fixed μ, and is regular. For the non-homogeneous case, we can prove this result (without regularity) directly for the Laplace image following Sandrakov and Semenov (2022) and Sandrakov et al. (2023), where a first-order equation with respect to t is considered (in fact, for the case $m_\mu^\varepsilon = 0$). The proof of regularity for this problem is more complicated, as explained in Sandrakov et al. (2023).

Note also that in the case $m_\mu^\varepsilon = 0$, problem [3.2] models diffusion and filtration processes in the considered composite medium and all the presented constructions are preserved in this case through zeroing all densities and eliminating the second initial condition. Accuracy estimates in this case are discussed in Sandrakov (2004a, 2004b, 2004c, 2007) and Sandrakov et al. (2023).

Thus, three terms are defined in [3.20]. For the fourth term $V_2 = V_2(z, x, y)$, we can obtain a representation that is similar to [3.14]. We can also determine more generally $V_0 = V_0^0(z, x, y) + \varepsilon V_1^0(z, x, y)$ in [3.20]. However, the relevant equality for $V_1^0 = V_1^0(z, x, y)$, similar to [3.26], will contain 20 terms and will be very cumbersome. Note that, in accordance with [3.26], the inverse Laplace transform from the approximation [3.20] (for $V_2 = 0$) has the following representation:

$$u_a = v + \varepsilon N_a^{1\varepsilon} \nabla v + \varepsilon N_b^{1\varepsilon} g + p_r^\varepsilon * f \qquad [3.31]$$

$$+ (M_\rho^\varepsilon)_t' u_0 + (M_\varphi^\varepsilon) u_1 + (M_{s\rho}^\varepsilon) u_0 - p_s^\varepsilon * (v_t') - p_m^\varepsilon * (v_{tt}''),$$

where $N_a^{1\varepsilon} = N_a^1(x/\varepsilon)$, $N_b^{1\varepsilon} = N_b^1(x/\varepsilon)$, $p_r^\varepsilon = p_r^\varepsilon(t,x/\varepsilon)$, $p_s^\varepsilon = p_s^\varepsilon(t,x/\varepsilon)$, $p_m^\varepsilon = p_m^\varepsilon(t,x/\varepsilon)$ are determined by solutions to the cell problems [3.22], [3.23] and [3.28]. The same solutions are used in the following notations:

$$M_\rho^\varepsilon = p_\rho^\varepsilon - m_1^\rho p_m^\varepsilon, \quad M_\varphi^\varepsilon = p_\varphi^\varepsilon - m_1^\varphi p_m^\varepsilon, \quad M_{s\rho}^\varepsilon = p_{ss}^\varepsilon - m_1^{s\rho} p_m^\varepsilon - m_1^\rho p_s^\varepsilon.$$

Note that these oscillating functions, such as the functions p_r^ε, p_s^ε, p_m^ε, p_ρ^ε, p_φ^ε and p_{ss}^ε, are all equal to zero at Ω_1^ε, which reflects rather complex mutual resonances on the domain Ω_0^ε, which corresponds to a weakly conducting material.

For homogeneous initial conditions in [3.2], this follows from the results of Sandrakov (1997b, 1998a, 1998b, 2007), for example, that:

$$\| u-u_a \|^2_{C^0([0,T];L^2(\Omega_1^\varepsilon))} + \mu \| u-u_a \|^2_{C^0([0,T];L^2(\Omega_0^\varepsilon))} \le C(\varepsilon+\sigma), \tag{3.32}$$

where u is the solution of problem [3.2], u_a is defined by [3.31] through the solution of the homogenized problem [3.30] (and cell problems) and the constant C does not depend on μ, σ and ε for $0<\mu\le 1$, $0<\sigma\le\sigma_0$, $0<\varepsilon\le\varepsilon_0$ for suitable σ_0 and ε_0. This estimate can be refined for sufficiently small μ, which significantly simplifies the homogenized problem [3.30], since for $\mu=0$ this problem essentially coincides with problem [3.12], but the oscillating terms in [3.31] are still essential in the accuracy estimate. In fact, analogues of some of these terms arose as boundary layer terms near the boundary $t=0$ in section 3.3.

In the case of non-uniform initial conditions in [3.2], the situation with the accuracy of estimates is more complicated due to the solution regularity for problems [3.28], since these solutions are not sufficiently regular in the general case.

Nevertheless, using smooth approximations, it can be shown that the left-hand side of [3.32] tends to zero as $\sigma\to 0$ and $\varepsilon\to 0$ uniformly with respect to μ as $0<\mu\le 1$. Similarly, this approach leads to the following accuracy estimate:

$$\| u - v \|^2_{C^0([0,T];L^2(\Omega_1^\varepsilon))} \le C(\varepsilon+\sigma),$$

where u is the solution of problem [3.2], v is the solution of the homogenized problem [3.30] and the constant C does not depend on μ, σ and ε for $0<\mu\leq 1$, $0<\sigma\leq\sigma_0$, $0<\varepsilon\leq\varepsilon_0$ for suitable σ_0 and ε_0.

Thus, when modeling wave processes in composite materials with contrasting properties, instead of solving problem [3.2] (e.g. numerically), it is possible to solve the homogenized problem [3.30] with guaranteed accuracy for small σ and ε. In addition, the presented accuracy estimates reflect the appearance of strongly oscillating terms in the solution asymptotics for problem [3.2], which characterizes significant energy absorption in the domain with low conductivity and can be used for designing wave-absorbing composite materials. Compared to the homogenized problem [3.12], the homogenized problem [3.30] may contain convolutions and nontrivial external forces even if $f=0$ and $g=0$ (when there are no external forces in [3.2]). In terms of elasticity, the latter means that the presence of microstresses in a composite material (even in the absence of external forces) may lead to the emergence of an effect on it that is analogous to external forces, the so-called residual stress effect. According to [3.12], this effect can only be observed for composite materials with contrasting properties for the models considered.

3.5. Computational aspects of modeling by homogenization

It is useful to note the advantages of the presented approach to studying problem [3.2]. If the assumptions of section 3.3 are fulfilled, then it is possible to solve problem [3.12] for the equation with constant coefficients instead of problem [3.2] computationally (with guaranteed accuracy by [3.16]), which is much simpler. Also, by separating the variables, it is possible to represent the exact solution of problem [3.12] as a series in eigenfunctions of the problem operator. For some domains, such eigenfunctions are well known; otherwise, they can be found computationally.

In turn, if the assumptions of section 3.4 are met, then the computational situation is much more complicated. In fact, for very small σ, we cannot find a computational solution to problem [3.2] with guaranteed accuracy, since standard accuracy estimates contain σ^{-1} as a factor and it is necessary to choose a grid step to solve problem [3.2] significantly smaller than σ. In addition, this grid step must be significantly smaller than ε in order to take into account the geometry of the sets Ω_0^ε and Ω_1^ε. Consequently, even by choosing a very small grid step, we cannot guarantee the accuracy of the numerical calculations, since accuracy estimates strongly depend on the geometry of these sets. This is actually confirmed by expansion [3.31], which is a good approximation to the solution of problem [3.2] due to estimate [3.32], and is quite complexly oscillating on the set Ω_0^ε, which indicates the significant absorption of energy on the set. Following Sandrakov (2007) and Sandrakov et al. (2023), the magnitude of such energy absorption can be calculated, which may be relevant for applications.

We now address the computational aspects for problem [3.30], the solution of which approximates the solution of problem [3.2] due to [3.32]. Note that a similar estimate was obtained for the energy norm, but the formulation of such an estimate is more cumbersome. Details for related problems can be found in Sandrakov (1997b, 1998a, 1998b, 2007) and Sandrakov et al. (2023). Next, we remark that problem [3.30] has constant coefficients and the only unusual thing is the presence of convolutions with respect to the time variable. We know (Sandrakov 1998a, 1998b, 2007; Sandrakov et al. 2023). that these convolutions are correctly defined and are expressed through solutions to problems [3.28]. Computational solution of problems [3.28] is standard. Moreover, these solutions can be represented as series through the eigenfunctions of the problem operator. Hence, it is possible to solve problem [3.30] computationally in parallel with problems [3.28] or using series to represent the solutions. The solution is standard and has guaranteed accuracy, at least for a fixed $\theta = \vartheta / \mu$ in [3.28]. Naturally, in a specific problem, the parameter θ is fixed. Moreover, if in the problem the parameter μ is small, then it is easy to check that the influence of convolutions is of the order μ and therefore these convolutions can be considered equal to zero in [3.30]. Consequently, problem [3.30] for small μ, such as problem [3.12], has constant coefficients and can be solved by standard computational methods with guaranteed accuracy.

In conclusion, we note that the results presented for problem [3.2] are new. Typically (Sandrakov 1996, 1998a, 1998b, 2007; Sandrakov et al. 2023), this problem was studied for external forces with homogeneous initial conditions and when $s_\mu^\varepsilon = 0$. But, for such initial conditions without external forces, the solution to problem [3.2] is trivial. Whereas, as noted for problem [3.30], inhomogeneous initial conditions can induce external forces in the homogenized problem, but the converse is not true, how easy it is to check.

Note also that when applied to photonic crystals, it is necessary to simulate the passage of waves through them. Such waves, of course, do not satisfy homogeneous initial conditions. Thus, the cases considered here are of greatest interest for such wave propagation modeling.

3.6. Spectral aspects of modeling by homogenization

It is useful to clarify some spectral properties for problem [3.24] (taking into account [3.29]) isometrically isomorphic to the homogenized problem [3.30] containing time convolutions. In particular, the influence of these convolutions will be clarified in comparison with the homogenized problem [3.12]. For simplicity of presentation, we assume that $m_\mu^\varepsilon = r_\mu^\varepsilon$, $s_\mu^\varepsilon = 0$, $u_0 = u_1 = 0$ and $g = 0$ in problem [3.2]. In this case, among problems [3.28], only the following:

$$m_0 (p)''_{tt} - \theta \, div_y (A_0 \nabla_y p) = 0 \ \ in \ \ Y_0 \times (0,\infty), \tag{3.33}$$

$$p|_{t=0} = 0, \ (p)'_t|_{t=0} = 1 \ \ in \ \ Y_0, \quad p - 0 \ \ on \ \ \Gamma \times (0,\infty)$$

is actual. The solution of the problem admits the known (and directly provable) form:

$$p = \sum_{l=1}^{\infty} (\alpha\omega_l)^{-1} c_l \, \varphi_l(y) \sin \alpha\omega_l t, \tag{3.34}$$

where $\alpha^2 = \theta$, $\omega_l^2 = \lambda_l$, $c_l = \int_{Y_0} m_0(y) \varphi_l(y) \, dy$ and the spectral problem:

$$-div_y (A_0 \nabla_y \varphi_l) = \lambda_l m_0 \varphi_l \ \ in \ \ Y_0, \quad \varphi_l = 0 \ \ on \ \ \Gamma \cup Y_1, \tag{3.35}$$

for 1-periodic eigenfunctions $\varphi_l = \varphi_l(y)$ with $\int_{Y_0} m_0 \varphi_l \varphi_h \, dy = \delta_{lh}$ is under consideration when $l,h = 1,2,\ldots$. It is well known that the λ_l are real positive eigenvalues of finite multiplicity with a limit point at infinity, and it may be assumed that $0 < \lambda_1 \le \lambda_2 \le \ldots$ (counting multiplicities). Remark that Y_0 may have several connected components, for example, Y_0^1 and Y_0^2. Then, $\lambda_{l_1}^1$, $\varphi_{l_1}^1$ and $\lambda_{l_2}^2$, $\varphi_{l_2}^2$ are well defined for each component. Extending $\varphi_{l_1}^1$ (respectively, $\varphi_{l_2}^2$) by zero to Y_0^2 (respectively, Y_0^1), we can assume that equations [3.35] hold for $\lambda_l \in \{\lambda_{l_1}^1\} \cup \{\lambda_{l_2}^2\}$ and the corresponding eigenfunction $\varphi_l = \varphi_l(y)$ (taking into account multiplicities).

Differentiating the representation [3.34] twice with respect to t (formally for now), multiplying by m_0 and integrating over Y_0, we obtain:

$$q(t) = q_m(t) = \sum_{l=1}^{\infty} \alpha\omega_l \, c_l^2 \sin \alpha\omega_l t, \tag{3.36}$$

for the convolution kernel in [3.30], which has the well-known Laplace transform:

$$Q(z) = -\sum_{l=1}^{\infty} \frac{c_l^2 \theta \lambda_l}{z^2 + \theta \lambda_l} = -\frac{1}{2}\sum_{l=1}^{\infty} \frac{c_l^2 i\alpha\omega_l}{z + i\alpha\omega_l} + \frac{1}{2}\sum_{l=1}^{\infty} \frac{c_l^2 i\alpha\omega_l}{z - i\alpha\omega_l}. \qquad [3.37]$$

Let us introduce the notation:

$$\Theta(\lambda) = m^1 - \mu \sum_{l=1}^{\infty} c_l^2 \theta \lambda_l (\lambda + \theta \lambda_l)^{-1}$$

and denote $A^1 = D$ for simplicity. Then, problem [3.24], which is the Laplace transform of problem [3.30] in the case under consideration, has the form:

$$\lambda \Theta(\lambda) V - div_x D(\nabla_x V) = \Theta(\lambda) F \ in\ \Omega,\ V = 0\ on\ \partial\Omega \qquad [3.38]$$

for $\lambda = z^2 \in \mathbb{C}_0$. This is proven in Sandrakov (2006b, 2007) that the function $\Theta(\lambda)$ admits a meromorphic (analytical and single-valued) extension to the complex plane $\mathbb{C}$ with first-order poles at the points of the set $\Lambda_\theta^e = \{-\theta\lambda_l : c_l \neq 0 \ for\ l = 1,2,\dots\}$, which is an infinite countable set (for fixed θ) and can be represented as:

$$\Lambda_\theta^e = \{-z_h^e \ \ for\ \ h = 1,2,\dots\} \ \ with\ \ 0 < z_1^e < z_2^e < \dots,$$

where $z_h^e = \theta\lambda_l$ for a suitable $l = l_h$. Note that this statement follows from the convergence of the series [3.36] in $L^1(0,\infty)$, first proved in Sandrakov (1996, 1997a), from which follows the convergence of the series in the definition of $\Theta(\lambda)$ for each $\lambda \in \mathbb{C} \backslash \Lambda_\theta^e$, and as a consequence the convergence of the series [3.37], where it is taken into account that $\lambda = z^2$. It is also easy to check that this series [3.36] does not converge in $L^p(0,\infty)$ for $3/2 \leq p$.

Fortunately, $p = 1$ is sufficient for the convergence of series [3.36], [3.37] and the correctness of the problems [3.24], [3.30] and [3.38]. In addition, it is verified by examples that there are eigenfunctions φ_l for which $c_l = 0$. It follows from [3.35] that:

$$c_l = \lambda_l^{-1} \int_\Gamma \Upsilon \cdot A_0 \nabla_y \varphi_l \, ds.$$

Therefore, an eigenvalue λ_l with $l = 1, 2, ...$ is a pole of $\Theta(\lambda)$ if (and only if) the eigenfunction φ_l has a non-zero average flow across the boundary Γ. In particular, as is well known, this is the case for φ_1. It is proven in Sandrakov (2007) that Λ_θ^e is countable and infinite. This is natural, since the only difference between [3.10] and [3.24] is precisely the boundary integral, the appearance of which leads to convolutions in the homogenized problem [3.30].

Interestingly, the real parts of $\Theta(\lambda)$ and $\lambda\Theta(\lambda)$ are bounded away from zero and $|\Theta(\lambda)|$ is bounded on $\mathbb{C}_0$ (Sandrakov 1997b, 2007; Sandrakov et al. 2023), characterizing the smearing of the homogenized density on the complex half-plane in the Laplace image. So, for example, problem [3.38] is solvable on $\mathbb{C}_0$.

Let $z_0^e = 0$. According to Sandrakov (2006b, 2007) on each of the intervals:

$$... (-z_{h+1}^e, -z_h^e), ..., (-z_3^e, -z_2^e), (-z_2^e, -z_1^e), (-z_1^e, -z_0^e) \qquad [3.39]$$

of the negative real half-line, with the exception of the first, there is a unique value $-z_{h+1}^0$ such that $\Theta(-z_{h+1}^0) = 0$. The set:

$$\Lambda_\theta^0 = \{-z_{h+1}^0 : \Theta(-z_{h+1}^0) = 0 \ for \ h = 1, 2, ...\}$$

is countable and the function $\Theta(\lambda)$ is invertible for $\lambda \in \mathbb{C} \backslash (\Lambda_\theta^e \cup \Lambda_\theta^0)$. So, problem [3.38] is associated with the operator family $B_\lambda : H^2(\Omega) \cap H_0^1(\Omega) \to L^2(\Omega)$, where

$$B_\lambda = \lambda - \Theta(\lambda)^{-1} div_x D(\nabla_x V). \qquad [3.40]$$

If the operator B_λ^{-1} is defined for $\lambda \in \mathbb{C}$, then $V = B_\lambda^{-1} F$ is a solution to [3.38] (hereafter, the operators are linear and continuous). Hence, the spectrum of problem [3.38] is naturally understood as the set of characteristic numbers of the operator $B_{-\eta}$; details may be found in Gohberg and Sigal (1971) and Vainberg (1989). The minus appearing in the notation $B_{-\eta}$ is explained by the tradition of sign choice in the Laplace transform and the resolvent in notation of Kato (1966). It should be stressed that the spectrum of [3.38] and so B_λ (where λ is treated as a spectral parameter) is not defined in the classical sense, since the dependence on λ is not linear.

For $\lambda \in \Lambda_\theta^e$, problem [3.38] is not defined. For $\lambda \in \Lambda_\theta^0$, the family B_λ is not defined exactly (it has a pole at this point), and problem [3.38] has a unique trivial solution. We know (Sandrakov 1997b, 2004b, 2007) that B_λ is invertible for $\lambda \in \mathbb{C}_0$. The following statement was proven in Sandrakov (2007).

THEOREM 3.2.– *For* $\lambda \in \mathbb{C} \backslash (\Lambda_\theta^e \cup \Lambda_\theta^0)$, *operators* $B_\lambda^{-1} : H^2(\Omega) \cap H_0^1(\Omega) \to L^2(\Omega)$ *form a finite meromorphic Fredholm family with a countable set of poles* Λ_p^θ.

Recall that operators $S_\lambda : H_1 \to H_2$ (e.g. between Hilbert spaces H_1 and H_2) form a finite meromorphic Fredholm family for λ in an open subset $\mathrm{O} \subset \mathbb{C}$ if the following representation:

$$S_\lambda = \sum_{k=1}^{K} S_{-k}^{fd} (\lambda - \lambda^\circ)^{-k} + S_0^F + \sum_{h=1}^{\infty} S_h^c (\lambda - \lambda^\circ)^h \quad for \quad \lambda \in \Phi \backslash \{\lambda^\circ\}$$

holds for some neighborhood $\Phi \subset \mathrm{O}$ of every point λ° in O, where $0 \leq K < \infty$ is fixed, S_{-k}^{fd} is a finite-dimensional operator, S_0^F is a Fredholm operator, S_h^c is an operator and the series converges in the operator norm. Terminology from Gohberg and Sigal (1971) and Vainberg (1989, Ch. 4) is used here. See also Reed and Simon (1978, Theorem XIII.13), where different terminology and sources are used. In addition, $K = 1$ for families of self-adjoint operators according to Kato (1966).

Remark that a fairly complete invertibility theory is available for the finite meromorphic Fredholm families according to Vainberg (1989) and Reed and Simon (1978). Moreover, it is well established that the poles of B_λ^{-1} (for $\lambda \in \mathbb{C} \backslash (\Lambda_\theta^e \cup \Lambda_\theta^0)$) are the characteristic numbers of B_λ from [3.40] and vice versa. In fact, this is a generalization of a known theorem from complex analysis, according to which the poles of a meromorphic function are the zeros of the inverse function and vice versa.

In order to define the set structure of poles Λ_p^θ specified in Theorem 3.2, we denote by λ_j° and Φ_j° the eigenvalues and functions of the Dirichlet problem:

$$-\, div_x D(\nabla_x \Phi_j^\circ) = \lambda_j^\circ \Phi_j^\circ \ in \ \Omega, \ \ \Phi_j^\circ = 0 \ on \ \partial\Omega \ for \ j = 1, 2, ... \qquad [3.41]$$

where λ_j° are real eigenvalues of finite multiplicity that have a limit point at infinity, and it may be assumed that $0 < \lambda_1^\circ < \lambda_2^\circ < \ldots$ (without counting the multiplicity). Consider the meromorphic functions $\Theta(\lambda)_j^\circ = \lambda\,\Theta(\lambda) + \lambda_j^\circ$ on $\mathbb{C}$ for $j = 1, 2, \ldots$.

According to Sandrakov (2006b, 2007) for fixed $j = 1, 2, \ldots$ on each of intervals [3.39], there is a unique value $-z_h^j$ such that $\Theta(-z_h^j)_j^\circ = 0$. The set of poles:

$$\Lambda_p^\theta = \cup_{j=1}^{\infty} \Lambda_\theta^j = \cup_{j=1}^{\infty} \{ -z_h^j : \Theta(-z_h^j)_j^\circ = 0 \ for \ h = 1, 2, \ldots \}$$

is countable and on each of the intervals [3.39] for $h = 1, 2, \ldots$, we have:

$$-z_{h+1}^e < \ldots < -z_{h+1}^j < -z_{h+1}^{j-1} < \ldots < -z_{h+1}^1 < -z_{h+1}^0 < -z_h^e, \qquad [3.42]$$

where $j = 1, 2, \ldots$, $-z_h^e \leftarrow -z_h^j$ for $j \to \infty$ and $-z_1^e < \ldots < -z_1^j < \ldots < -z_1^1 < -z_0^e$.

Returning to composite media, we recall that $\lambda = z^2$ in problem [3.38], which is equivalent to the homogenized problem [3.24]. Naturally, we may denote by κ_h^e and κ_h^j the square roots of z_h^e and z_h^j introduced in [3.42]. Thus, we have:

$$\kappa_h^e = \sqrt{z_h^e}, \quad \kappa_h^j = \sqrt{z_h^j} \quad for \ h = 1, 2, \ldots, \quad j = 0, 1, \ldots, \qquad [3.43]$$

where $\pm i\kappa_h^e = \pm i\alpha\omega_h$ is a first-order pole of the function $Q(z)$ in notations [3.37]. Accordingly, $\pm i\,\kappa_h^j$, $j = 1, 2 \ldots$ are the first-order poles of the resolving operator of problem [3.38], and outside the indicated poles, this problem is uniquely solvable.

Let us consider the half-space $\mathbb{R}_-^2 = \{ (x_1, x_2) : x_1 < 0 \}$ filled with a homogeneous material, for example, a homogenized material modeled by problem [3.12] for $m = 1$, $A = a^2 E$, $s = 0$, $g = 0$, $m^\rho = 1$ and $m^\varphi = 1$. In order to consider harmonic oscillations, we consider a rectangular domain $\Omega^- = (-l_1, 0) \times (0, l_2)$ in this half-space and will continue the functions on the domain in a periodic manner with the cell Ω^-. Now, we take a similar domain $\Omega = (0, L) \times (0, l_2)$ in a composite material in contact with the material under consideration and modeled by the homogenized problem [3.30] (with the same assumptions as for [3.12] and $D = b^2 E$). Due to the presented accuracy estimates, such problems approximate the solutions of the original problem [3.2] under relevant assumptions. A schematic view of the contact composite continuous medium is shown in Figure 3.2.

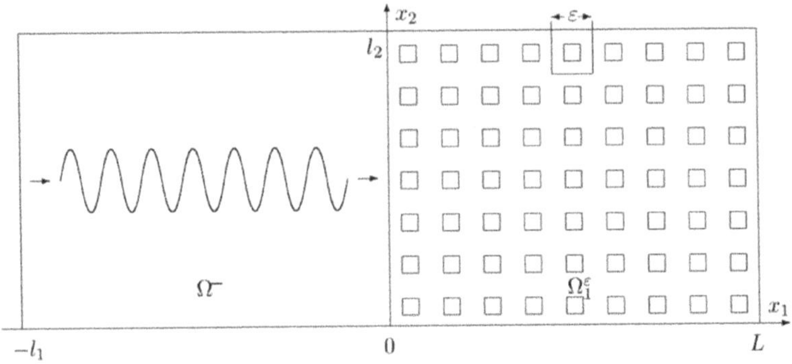

Figure 3.2. *Model of contact composite medium*

Thus, for the domain $\Omega^- = (-l_1, 0)\times(0, l_2)$, we consider the following problem:

$$v''_{tt} - a^2\Delta v = 0 \quad in \quad \Omega^-\times(0,\infty), \qquad [3.44]$$

$$v\,|_{t=0} = u_0,\; v'_t\,|_{t=0} = u_1 \;\; in \;\; \Omega^-, \;\; v \text{ is } \Omega^-\text{-periodic.}$$

Let us also consider a spectral Ω^--periodic problem similar to [3.41] with the following eigenfunctions and values:

$$\Phi_{l_1 l_2}^{n_1 n_2} = e^{i\,2\pi(l_1^{-1} n_1 x_1 + l_2^{-1} n_2 x_2)}, \;\; \lambda_{l_1 l_2}^{n_1 n_2} = (2\pi a)^2 (l_1^{-2} n_1^2 + l_2^{-2} n_2^2), \qquad [3.45]$$

with $(n_1, n_2) \in \mathbb{Z}^2$ and $-a^2\Delta\Phi_{l_1 l_2}^{n_1 n_2} = \lambda_{l_1 l_2}^{n_1 n_2}\Phi_{l_1 l_2}^{n_1 n_2}$. For a fixed (n_1, n_2), the function

$$e_{l_1 l_2 \pm}^{n_1 n_2} = e^{\pm i\omega t}\Phi_{l_1 l_2}^{n_1 n_2} = e^{\pm i\omega t} e^{i\,2\pi(l_1^{-1} n_1 x_1 + l_2^{-1} n_2 x_2)} \qquad [3.46]$$

with $\omega = \omega_{l_1 l_2}^{n_1 n_2} = (\lambda_{l_1 l_2}^{n_1 n_2})^{1/2}$ meets [3.44] (for obvious u_0 and u_1) and determines the incoming and outgoing waves in the half-space. In particular, for $(n_1, n_2) = (1, 0)$ these are the simplest incoming and outgoing plane waves, which satisfy [3.44]. By varying l_1 in the definition of these waves, we can obtain plane waves with an arbitrary non-zero frequency. From a mathematical point of view, this means that the Laplacian spectrum is continuous and coincides with the semi-axis.

Returning to the main domain $\Omega=(0,L)\times(0,l_2)$ containing weakly conducting inclusions, of course, care should be taken that the inclusions do not cross the boundary $x_1=0$, since significant oscillations will arise at this boundary in accordance with representation [3.31], which will complicate the future. For a similar reason, it is convenient to assume that L and l_2 are multiples of ε, for example, $L=m_1\varepsilon$ and $l_2=m_2\varepsilon$ for positive integers m_1 and m_2.

For $F=F(x_1,x_2)$, problem [3.38] with Ω-periodic boundary conditions is isometrically isomorphic to the following initial-boundary value problem:

$$w_t'-\mu\eta*(w_t')-b^2\Delta w=0 \quad in \quad \Omega\times(0,\infty), \qquad [3.47]$$

$$w|_{t=0}=F \ \ in \ \ \Omega, \ \ w \text{ is } \Omega\text{-periodic},$$

where the Laplace transform of $\eta=\eta(t)$ is $\mathrm{N}(\lambda)=\sum_{l=1}^{\infty}c_l^2\theta\lambda_l(\lambda+\theta\lambda_l)^{-1}$ and so $\Theta(\lambda)=1-\mu N(\lambda)$. Let us add the equality $\int_\Omega F\,dx=0$ to the conditions of this problem in order to get rid of trivial solutions and be able to use the Poincaré inequality. We can prove that there is a unique solution to this problem such that:

$$\|w\|_{L^2(0,\infty;H^1_{per}(\Omega))}+\|w_t'\|_{L^2(0,\infty;H^{-1}_{per}(\Omega))}\leq C\|F\|_{L^2(\Omega)}, \qquad [3.48]$$

for every $F\in L^2(\Omega)$ according to Sandrakov (1996, 1997b) and Sandrakov et al. (2023), where the case of homogeneous Dirichlet conditions was considered. Here, for example, $H^1_{per}(\Omega)$ denotes the Sobolev space of Ω-periodic functions.

In particular, the unique solution $w=w^{n_1n_2}_{Ll_2}$ to problem [3.47] is determined for $F=\Phi^{n_1n_2}_{Ll_2}$ from [3.45], where a,l_1 are replaced by b,L. It is natural to call such a solution a homogenized wave, since for $\mu=0$ this solution has the form [3.46] for $i\omega=\lambda^{n_1n_2}_{Ll_2}$, although increasing solutions are not usually considered.

Further, we assume that these eigenfunctions and values Φ_L^j and λ_L^j are renumbered by the index $j = 1, 2, ...$ as for problem [3.41], emphasizing the dependence on L. Then, the Laplace transform of solution $w = w_L^j$ to problem [3.47] has the form:

$$Lw_L^j = (\lambda + \Theta(\lambda)^{-1}\lambda_L^j)^{-1}\Phi_L^j = (\lambda\Theta(\lambda) + \lambda_L^j)^{-1}\Theta(\lambda)\Phi_L^j \qquad [3.49]$$

according to Sandrakov (1996, 1997b, 2007). For fixed L, h and j on the interval $(-z_h^e, -z_h^0)$, there is only one point $-z_h^j$, where Lw_L^j has a pole by [3.42]. Here, $z_1^0 = z_0^e = 0$ by definition. These $-z_h^j$ are also points of the spectrum of problem [3.38]. In particular, for $(n_1, n_2) = (1, 0)$, we can consider the function Φ_L^1 and by varying L obtain a spectral point $.z\frac{1}{1}.$ arbitrarily close to $-z_1^0$ according to Sandrakov (2007). But some care is required with the variation of L, since by choosing a very large, for example, $L = 1/\varepsilon$, we can obtain a rapidly oscillating function with period ε in [3.45] and the answer will be more complicated as explained for problem [3.30]. Hence, it is possible to approximate each point of the interval $(-z_1^e, -z_1^0)$ by a spectral point as well as for the intervals $(-z_{h+1}^e, -z_{h+1}^0)$ for $h = 1, 2, ...$. However, this is impossible for the points of the interval $(-z_{h+1}^0, -z_h^e)$ for $h = 1, 2, ...$, since there are no zeros of the function $\lambda\Theta(\lambda) + \lambda_L^j$ there, according to Sandrakov (2006b, 2007) where this function is studied in detail.

Now, it follows from the Laplace transform inversion theorem for meromorphic functions, given and proved in Romanovsky (1961), for fixed j that:

$$w_L^j = \Phi_L^j \sum_{h=1}^{\infty} s_h^j \, e^{-z_h^j t} \qquad [3.50]$$

where the series converges in norms from [3.48] (by this a priori estimate) and the constants s_h^j can be calculated through residues at the poles $-z_h^j$. It is essential here that all poles from each of intervals in [3.39] are included in the homogenized exponential. Perhaps, this is natural since all frequencies $\alpha\omega_l$ (with $c_l \neq 0$), which are determined by the eigenvalues and functions from [3.35], participate in determining the convolution kernel of the homogenized problem [3.30].

Finally, using the notation [3.43], we can present the solution $v = v_L^j$ of the homogenized problem [3.30] (isometrically isomorphic to problem [3.38] in the case

$\lambda = z^2$ and $F = \Phi_L^j$) whose equation is of the form [3.47], where the first derivative is replaced by the second and the initial conditions are replaced by $v|_{t=0} = 0$ and $v'_t|_{t=0} = \Phi_L^j$. The solution is the following form:

$$v_L^j = \Phi_L^j \sum_{h=1}^{\infty} s_h^j \left(e^{i\kappa_h^j t} - e^{-i\kappa_h^j t}\right)(2i\kappa_h^j)^{-1} \qquad [3.51]$$

by representation [3.37], since $\sum_{h=1}^{\infty} s_h^j = 1$ by the initial conditions in [3.38] and [3.50]. It is natural to call this solution a homogenized fundamental wave, while the other fundamental wave will be thc following function:

$$u_L^j = \Phi_L^j \sum_{h=1}^{\infty} 2^{-1} s_h^j \left(e^{i\kappa_h^j t} + e^{-i\kappa_h^j t}\right) \qquad [3.52]$$

for which the initial conditions $u|_{t=0} = \Phi_L^j$ and $u'_t|_{t=0} = 0$ are satisfied. Here ,the series converge because for this case a priori estimates similar to [3.48] proved in (Sandrakov 1998a, 1998b, 2000) are also satisfied.

Note that for fixed j the coefficient at the pole $(\lambda + z_h^j)^{-1}$, $h = 1,2,\ldots$ in [3.49] will be s_h^j multiplied by the finite-dimensional projector onto the eigenspace corresponding to the eigenvalue λ_L^j according to Sandrakov (2007). For example, there are two functions forming the eigenspace with $(n_1, n_2) = (1,0),\ (0,1)$, and there are four functions for $(n_1, n_2) = (1,1),\ (-1,1),\ (1,-1),\ (-1,-1)$ in the next eigenspace. This means that we can replace Φ_L^j in [3.49]–[3.52] with any of these functions (with the same λ_L^j) and therefore with their linear combination. It is implicitly assumed here that these functions Φ_L^j are orthonormal.

Thus, even plane waves in composites with weakly conducting inclusions are quite complex and are almost periodic with respect to time in contrast to [3.46]. The interaction of such waves in a contact composite medium as in Figure 3.2 will be considered in the future. Here, it is only immediately clear that if the waves on the left come with a positive frequency ω from interval $(\kappa_h^e, \kappa_{h+1}^0)$ for $h = 1,2,\ldots$, then there will be a complete reflection of such waves, since the fundamental waves on the right are only zero. Here, ω is from [3.46] for incoming waves. Thus, for composite materials with weakly conducting inclusions, the spectrum (and transmission frequencies) has a semiconductor structure with forbidden zones.

Surprisingly, the width of these zones is characterized by the parameter $\theta = \vartheta / \mu = \sigma / \mu \varepsilon^2$ and the eigenvalues of the inclusions according to [3.35] (for frequencies, the root should be extracted). However, the first of these zones already goes to infinity for very small μ (and fixed σ and ε). This is true for both not very small σ (e.g. $\sigma = \varepsilon$) and very small σ (such as $\sigma = \varepsilon^3$). All these cases of problem [3.2] were also considered separately in Sandrakov (1998a, 1998b, 2004c, 2007) and Sandrakov et al. (2023). In the listed cases, forbidden zones do not arise. This information can be extracted, for example, from the asymptotics of the solutions to problem [3.33] for very small and large θ. For a first-order equation in t, such an asymptotics is presented and justified in Sandrakov (2007).

In turn, on complementary intervals (κ_h^0, κ_h^e) for $h = 1, 2, ...$ the interaction of waves will be very complex and not only through the almost periodicity of fundamental waves but also through fast oscillations on inclusions through the asymptotics [3.31], which will undoubtedly absorb energy. Estimates of such absorption for parabolic equations of the first order in t are given in Sandrakov (2007) and Sandrakov et al. (2023). But, as explained here, homogenized equations of the first [3.28] and second orders [3.30] are investigated in the same way. This remark applies to the case of $s = 0$, where the frequencies were found from the equality $\omega^2 + z_h^j = 0$. In general, the equation $\omega^2 + s\,\omega + z_h^j = 0$ should be considered for frequencies and complex roots with a negative real part may appear which will characterize decreasing (for $s > 0$) terms in fundamental waves.

A complete description of the interaction of such waves in a contact composite medium is undoubtedly of interest, especially for Maxwell's equations. But the facts given about forbidden zones may also have applications. Indeed, Figure 3.2 can be considered as a contact fragment with a reactor, where the cells are filled or not. In the latter case, we can consider $\mu = \sigma = 0$ and use the homogenized equation [3.12]. While the filled cells can have characteristics related to the homogenized problem [3.30]. Thus, by the reflection of waves from the contact boundary, we can check the presence of forbidden zones and predict the filling of the cells.

At last, metamaterials (as well as photonic crystals) are usually called composite materials whose properties are defined by the resonant properties of their constituent elements, the so-called metaatoms. Metamaterials can be synthesized by introducing various periodic structures with a variety of geometric forms into some material, which leads to various unusual properties of the materials. It is possible that it is precisely the oscillating terms in the presented asymptotic expansions [3.31], in interaction with the convolutions in them and in the equations, which describe such resonant properties by fundamental waves and their interaction through a significant redistribution of energy, actually to mutual vibrations of inclusions.

3.7. Conclusion

The homogenizations of initial-boundary value problems for wave equations with asymptotically degenerate rapidly oscillating periodic coefficients are considered. Such problems model wave processes in heterogeneous periodic media such as photonic crystals. The homogenized problems have the form of initial-boundary value problems for integro-differential equations in convolutions and with external forces, even if in the original problem the forces were zero. The appearance of problems with convolutions indicates the possible presence of gaps in the spectrum of the relevant model, as is the case with semiconductors. Statements on the approximation of solutions to the original problem by solutions of the homogenized problem in various cases are discussed. Thus, homogenization methods can be useful for modeling wave processes in composites similar to photonic crystals. The methods are especially relevant for composites with very different properties. In fact, it is precisely such composites that are relevant in applications, since it is hardly reasonable to create composites from materials with really similar properties. Undoubtedly, extending these methods to the system of Maxwell's equations is highly relevant and promising for applications to photonic crystals and metamaterials.

3.8. References

Agranovich, M.S. (2002). Spectral problems for second-order strongly elliptic systems in smooth and non-smooth domains. *Math. Surveys*, 57(5), 847–920. doi: 10.1070/rm2002v057n05abeh000552.

Agranovich, M.S. and Vishik, M.I. (1964). Elliptic problems with a parameter and parabolic problems of general type. *Math. Surveys*, 19(3), 53–157. doi: 10.1070/RM1964v019n03ABEH001149.

Allaire, G. (1992). Homogenization and two-scale convergence. *SIAM J. Math. Anal.*, 23(6), 1482–1518. doi: 10.1137/05230.

Bakhvalov, N.S. (1974). Average characteristics of bodies with periodic structure. *Doklady Akademii Nauk SSSR*, 218(5), 1046–1048.

Bakhvalov, N.S. (1975). Averaging of differential equations with partial derivatives having rapidly oscillating coefficients. *Doklady Akad. Nauk SSSR*, 221, 516–519.

Bakhvalov, N.S. and Panasenko, G.P. (1989). *Homogenization: Averaging Processes in Periodic Media.* Kluwer, Dordrecht. doi: 10.1007/978-94-009-2247-1.

Bensoussan, A., Lions, J.L., Papanicolaou, G. (1978). *Asymptotic Analysis for Periodic Structures.* North-Holland, Amsterdam.

Cioranescu, D., Damlamian, A., Griso, G. (2018). *The Periodic Unfolding Method.* Springer, Singapore. doi: 10.1007/978-981-13-3032-2.

Cioranescu, D., Damlamian, A., Griso, G. (2023). The periodic unfolding method for countably many scales. *Adv. Math. Sci. Appl.*, 32(1), 17–58.

Diaz, J.I., Gomez-Castro, D., Shaposhnikova, T.A., Zubova, M.N. (2019a). A nonlocal memory strange term arising in the critical scale homogenization of a diffusion equation with dynamic boundary conditions. *Electron. J. Differ. Equ.*, 77, 1–13.

Diaz, J.I., Gomez-Castro, D., Shaposhnikova, T.A., Zubova, M.N. (2019b). Classification of homogenized limits of diffusion problems with spatially dependent reaction over critical-size particles. *Appl. Anal.*, 98, 232–255. doi: 10.1080/00036811.2018.1441997.

Diaz, J.I., Podolskiy, A.V., Shaposhnikova, T.A. (2022a). On the homogenization of an optimal control problem in a domain perforated by holes of critical size and arbitrary shape. *Doklady Math.*, 105(1), 6–13. doi: 10.1134/S1064562422010033.

Diaz, J.I., Podolskiy, A.V., Shaposhnikova, T.A. (2022b). On the convergence of controls and cost functionals in some optimal control heterogeneous problems when the homogenization process gives rise to some strange terms. *J. Math. Anal. Appl.*, 506(1), 125559. doi: 10.1016/j.jmaa.2021.125559.

Duvaut, G. and Lions, J.L. (1972). *Les inéquations en mécanique et en physique.* Dunod, Paris.

Garttner, S., Frolkovic, P., Knabner, P., Ray, N. (2022). Efficiency of micro-macro models for reactive two-mineral systems. *Multiscale Model. Simul.*, 20(1), 433–461. doi: 10.1137/20M1380648.

Garttner, S., Knabner, P., Ray, N. (2024). Local existence of strong solutions to micro-macro models for reactive transport in evolving porous media. *Eur. J. Appl. Math.*, 35(1), 127–154. doi: 10.1017/S095679252300013X.

Gohberg, I.C. and Sigal, E.I. (1971). An operator generalization of the logarithmic residue theorem and the theorem of Rouche. *Mathe. USSR – Sbornik*, 13(4), 603–625. doi: 10.1070/SM1971v013n04ABEH003702.

Jager, W. and Woukeng, L. (2021). Homogenization of Richards' equations in multiscale porous media with soft inclusions. *J. Differ. Equ.*, 281, 503–549. doi: 10.1016/j.jde.2021.02.012.

Kato, T. (1966). *Perturbation Theory for Linear Operators.* Springer-Verlag, New York.

Kim, S. and Sakellaris, G. (2024). The Neumann Green function and scale-invariant regularity estimates for elliptic equations with Neumann data in Lipschitz domains. *Calc. Var.*, 63, 219. doi: 10.1007/s00526-024-02825-2.

Kinderlehrer, D. and Stampacchia, G. (1980). *An Introduction to Variational Inequalities and their Applications*. Academic Press, New York.

Mascheroni, P., Miller, L., Penta, R. (2025). Homogenization of solute transport in double porosity materials. *Mathematical Methods in the Applied Sciences*. doi: 10.1002/mma.10973.

Mitrea, M. and Taylor, M. (1999). Boundary layer methods for Lipschitz domains in Riemannian manifolds. *J. Funct. Anal.*, 163(2), 181–251. doi: 10.1006/jfan.1998.3383.

Piat, V.C. and Sandrakov, G.V. (2004). Homogenization of some variational inequalities for elasto-plastic torsion problems. *Asymptot. Anal.*, 40(1), 1–23. doi: 10.3233/ASY-2004-637.

Ray, N. and Schulz, R. (2023). Existence and uniqueness of solutions to a flow and transport problem with degenerating coefficients. *Eur. Appl. Math.*, 34(1), 55–76. doi: 10.1017/S0956792522000018.

Reed, M. and Simon, B. (1978). *Methods of Modern Mathematical Physics. IV. Analysis of Operators*. Academic Press, New York.

Romanovsky, P.I. (1961). *Fourier Series, Field Theory and Laplace Transformations (Including Analytic and Special Functions)*. Nauka, Moscow (in Russian).

Sanchez-Palencia, E. (1980). *Nonhomogeneous Media and Vibration Theory*. Springer-Verlag, Berlin. doi: 10.1007/3-540-10000-8.

Sandrakov, G.V. (1991). Averaging principles for equations with rapidly oscillating coefficients. *Math. USSR – Sbornik*, 68(2), 503–553. doi: 10.1070/SM1991v068n02ABEH002111.

Sandrakov, G.V. (1996). Homogenization of the non-stationary stokes flow in a periodic porous medium. *Dokl. Akad. Nauk*, 347(3), 312–315.

Sandrakov, G.V. (1997a). Homogenization of non-stationary Stokes equations with viscosity in a perforated domain. *Izv. Math.*, 61(1), 113–141. doi: 10.1070/IM1997v061n01ABEH000107.

Sandrakov, G.V. (1997b). Homogenization of nonstationary equations with contrasting coefficients. *Dokl. Akad. Nauk*, 355(5), 605–608.

Sandrakov, G.V. (1998a). Homogenization of nonstationary problems of the theory of strong nonhomogeneous elastic media. *Dokl. Akad. Nauk*, 358(3), 308–311.

Sandrakov, G.V. (1998b). Homogenization of dynamic problems of the elasticity theory with very different coefficients. *Dokl. Akad. Nauk*, 362(4), 449–452.

Sandrakov, G.V. (2000). Averaging of dynamic processes in composites periodically reinforced by fibers. *Dokl. Math.*, 61(3), 346–349.

Sandrakov, G.V. (2004a). Multiphase homogenized models for diffusion in highly nonhomogeneous media. *Dokl. Math.*, 70(1), 507–511.

Sandrakov, G.V. (2004b). Multiphase models of nonstationary diffusion arising from homogenization. *Comput. Math. Math. Phys.*, 44(10), 1741–1756.

Sandrakov, G.V. (2004c). Multiphase diffusion models in homogenization. *Dokl. Akad. Nauk*, 394(2), 171–174.

Sandrakov, G.V. (2004d). Homogenization of variational inequalities for problems with regular obstacles. *Dokl. Akad. Nauk*, 397(2), 170–173.

Sandrakov, G.V. (2005). Homogenization of variational inequalities for obstacle problems. *Sb. Math.*, 196(3–4), 541–560. doi: 10.1070/SM2005v196n04ABEH000891.

Sandrakov, G.V. (2006a). Homogenization of nonlinear equations and variational inequalities with obstacles. *Dokl. Math.*, 73(2), 178–181. doi: 10.1134/S1064562406020062.

Sandrakov, G.V. (2006b). Spectral properties of homogenized diffusion models in a strongly inhomogeneous medium. *Dokl. Math.*, 74(3), 815–818. doi: 10.1134/S1064562406060081.

Sandrakov, G.V. (2007). Multiphase homogenized diffusion models for problems with several parameters. *Izv. Math.*, 71(6), 1193–1252. doi: 10.1070/IM2007v071n06ABEH002387.

Sandrakov, G.V. and Semenov, V.V. (2022). Homogenized models with memory effect for heterogeneous periodic media. *J. Optim., Differ. Equ. Appl.*, 30(2), 1–18. doi: 10.15421/142206.

Sandrakov, G.V., Lyashko, S.I., Semenov, V.V. (2023). Simulation of filtration processes for inhomogeneous media and homogenization. *Cybernet. Syst. Anal.*, 59(2), 212–230. doi: 10.1007/s10559-023-00556-4.

Vainberg, V. (1989). *Asymptotic Methods in Equations of Mathematical Physics*. CRC Press, London. doi: 10.1201/9781003580423.

4

Polynomial Operator Interpolation and its Applications

V.L. MAKAROV[1] **and O.F. KASHPUR**[2]

[1] *The Institute of Mathematics of the National Academy of Sciences of Ukraine, Kyiv, Ukraine*

[2] *Taras Shevchenko National University of Kyiv, Ukraine*

4.1. Introduction

Currently, the problem of approximating operators in abstract Hilbert and finite-dimensional Euclidean spaces is a topical problem in applied mathematics.

Solving problems of identifying objects of unknown structure, synthesis of optimal systems (in Wiener's formulation), can be regarded as variants of problems of approximating operators in the corresponding functional spaces. Interpolation is one of the methods of operator approximation.

Quite often, information about the object under study is contained in a set of input–output pairs $(x_i, F(x_i)) \subset X \times Y$, $i = \overline{1,m}$, where $F : X \to Y$ is generally a nonlinear operator. Based on this information, it is necessary to construct a mathematical model of the object. The so-called black box problem arises, the solution of which, on a set of operator polynomials, leads to polynomial operator interpolation.

For a color version of all of the figures in this chapter, see www.iste.co.uk/koroliouk/computational2.zip.

Computational Methods and Mathematical Modeling in Cyberphysics and Engineering Applications 2,
coordinated by Dmitri KOROLIOUK, Sergiy LYASHKO and Nikolaos LIMNIOS. © ISTE Ltd 2026.

The justification for the approximation of a continuous one-variable function by polynomials is the Weierstrass theorem. Later, Stone (1948) generalized this theorem to the class of continuous functions of several variables. For continuous functional, the analogue of the Weierstrass theorem is the Frechet theorem (Fréchet 1910). Prenter (1970) proved a similar theorem for a continuous operator defined on a real separable space. For Banach spaces, such a result was obtained by Istrăţescu (1977) and Daugavet (1994). Many applied problems confirm the importance of research in polynomial interpolation. In practice, nonlinear systems described by operator polynomials are pretty common.

Let us note some features of polynomial operator interpolation. First of all, this is the non-uniqueness of the solution of the interpolation problem. Unlike classical interpolation formulas, the operator interpolant may not preserve the polynomial of the corresponding degree.

The property of preserving polynomials of the corresponding degree is important when obtaining estimates of the accuracy of interpolation formulas and studying the convergence of interpolation processes with an increase in the number of nodes, as well as for constructing quadrature formulas when calculating continual integrals. Interpolation polynomials with this property are appropriate for use in problems of identification, modeling and a wide class of problems that allow us to reduce the study of a nonlinear structure to the study of its polynomial approximation.

The importance of approximating operators by polynomials is confirmed by many applied problems (Porter 1976; Cox 2020). In practice, nonlinear systems described by polynomials often arise. Such systems are called polynomial. The study of such systems is of great importance (Prenter 1971b). Their fairly simple mathematical description explains interest in such systems.

In the theory of operator approximation, interpolation methods for approximating functionals and operators have been developed both in abstract linear and in concrete functional spaces. The main results related to this research area were obtained in the works of Ulm (1967a, 1967b), Ulm and Pohl (1969), Prenter (1971a), Kergin (1980), Micchelli (1980), Porter (1980), Rvachev (1982), Egorov et al. (1993), Egorov et al. (1993), Makarov and Khlobystov (1998), Makarov et al. (2000, 2010), Litvin (2002), Fillipson (2004) and others.

In this chapter, the authors present solutions to some applied problems based on operator interpolation formulas.

4.2. Formulation of Lagrange's operator interpolation problem

Let X, Y be linear spaces, Π_n be a set of operator polynomials $P_n : X \to Y$ of degree n:

$$\Pi_n = \left\{P_n(x) : P_n(x) = L_0 + L_1 x + \cdots + L_n x^n\right\},$$

where $L_0 \in Y$, $L_k x^k = L_k(x, x, \ldots, x)$ is the k-th operator degree obtained from the k-linear symmetric operator $L_k(x_1, x_2, \ldots, x_k) : X^k \to Y$, when $x_1 = x_2 = \cdots = x_k = x$, $k = \overline{1, n}$.

Let us give an example of the operator polynomial $P_n(x)$ from the set Π_n. Let $X = Y = L_2(0,1)$, $L_0 \in Y$. Continuous symmetric k-linear operator forms $L_k(x_1, x_2, \ldots, x_k) : X^k \to Y$, $k = \overline{1, n}$, are defined as follows:

$$L_k(x_1, x_2, \ldots, x_k) = \int_0^1\int_0^1 \cdots \int_0^1 K_k(t_1, t_2, \ldots, t_k) x_1(t_1) x_2(t_2) \cdots x_k(t_k) dt_1 dt_2 \cdots dt_k$$

where $K_k(t_1, t_2, \ldots, t_k)$, $k = \overline{1, n}$, are continuous symmetric functions of the variables $t_1, t_2, \ldots, t_k$. Suppose $x_1(t) = x_2(t) = \cdots = x_k(t) = x(t)$ in the last equality, then

$$L_k x^k = \int_0^1\int_0^1 \cdots \int_0^1 K_k(t_1, t_2, \ldots, t_k) x(t_1) x(t_2) \cdots x(t_k) dt_1 dt_2 \cdots dt_k, k = \overline{1, n}.$$

Operator polynomials are the basis of many mathematical models from mechanics, economics, ecology, biology and other fields (Torokhti and Howlett 1997; Kapustyan et al. 2011; Cox 2020).

There are several formulations of interpolation problems in linear spaces. Let us consider the operator analogue of the classical interpolation problem of functions of several variables. Let the operator $F : X \to Y$, in the general case nonlinear, be defined by its values $F(x_i) \in Y, i = \overline{1, m}$, at the interpolation nodes $x_i \in X, i = \overline{1, m}$. It is necessary to find the polynomial $P_n \in \Pi_n$ that satisfies the Lagrange interpolation relations

$$P_n(x_i) = F(x_i), \ i = \overline{1, m}.$$

4.3. Solution of the Lagrange's operator interpolation problem

The fundamentals of the general theory of operators' interpolation have been established in monographs (Makarov and Klobystov 1998; Makarov et al. 2000, 2010). We present some results of these studies in more detail, since they will be used in the future for solving applied problems.

Let X, Y be the Hilbert spaces, $(\cdot,\cdot)$ be a scalar product in X, $(\cdot,\cdot)_Y$ be a scalar product in Y, μ be some measure, the first moment of which is equal to zero, and the second is bounded, B be a correlation operator of measure μ, $KerB = \varnothing$. The operator $F : X \to Y$ is defined by its values $F(Bx_i), i = \overline{1,m}$, at the interpolation nodes $\{Bx_i\}_{i=1}^{m} \subset X$. In Makarov and Klobystov (1998) and Makarov et al. (2000, 2010), the Lagrange problem was solved on the sequence of nodes $Bx_i, i = \overline{1,m}$:

$$P_n(Bx_i) = F(Bx_i), \; i = \overline{1,m}. \qquad [4.1]$$

Let us present the main results of these works. For this, we introduce the notation: $\vec{F} = \{F(Bx_i)\}_{i=1}^{m}$, $\Gamma = \left\| \sum_{k=0}^{n} (Bx_i, x_j)^k \right\|_{i,j=1}^{m}$, Γ^+ be the Moore-Penrose matrix for Γ, Z be the matrix whose rows are the orthonormal eigenvectors of the matrix Γ with zero eigenvalue. The authors found a condition for the existence of a unique interpolant that satisfies the relations [4.1]. The following result holds.

THEOREM 4.1.– *It is necessary and sufficient for the solvability of the polynomial operator interpolation problem with relations [4.1] that the condition*

$$Z\vec{F} = 0. \qquad [4.2]$$

and condition [4.2] holds.

In the monographs (Makarov and Khlobystov 1998; Makarov et al. 2000, 2010), the whole set of polynomials of interpolation operators of the n-th degree for the operator F, which is given by its values at the nodes $\{Bx_i\}_{i=1}^{m}$, in the case of satisfying the relations [4.1], is constructively described.

THEOREM 4.2.– *Let condition [4.2] be valid. Then, the formula*

$$P_n^I(x) = Q_n(x) + < \vec{F} - \vec{Q}_n, \Gamma^+ \sum_{p=0}^{n} \{(x_i, x)^p\}_{i=1}^m >, \quad [4.3]$$

for all $Q_n(x) \in \Pi_n$, $\vec{Q}_n = \{Q_n(Bx_i)\}_{i=1}^m$ *describes the whole set* $\Pi_n^I(F)$ *of interpolation of the nodes* $\{Bx_i\}_{i=1}^m$. *Furthermore, the self-adjoined positive operator B can be chosen as the unit operator.*

Further, the authors proved theorems about the convergence of the interpolation process to the interpolated operator in the case of an increase in the number of nodes and found estimates of the accuracy of interpolation formulas on sets of polynomial and entire operators in the metric generated by the scalar product defined on the set Π_n:

$$(P_1, P_2)_H =$$

$$\sum_{k=0}^{n} \int_X \cdots \int_X \left(L_k^{(1)}(v_1, v_2, \ldots, v_k), L_k^{(2)}(v_1, v_2, \ldots, v_k) \right)_Y \mu(dv_k) \cdots \mu(dv_2) \mu(dv_1),$$

$$\| P \|_H = (P, P)_H^{\frac{1}{2}},$$

where $L_k^{(1)}, L_k^{(2)}$ – k-linear operator forms of the polynomials $P_1, P_2 \in \Pi_n$, respectively.

Let us present the solution of one of the extreme problems obtained in Makarov and Khlobystov (1998) and Makarov et al. (2000, 2010). Let us denote

$$<< \vec{a}, \vec{b} >> = \sum_{i=1}^{m} (\alpha_i, \beta_i)_Y, \; \vec{a} = \{\alpha_i\}_{i=1}^m, \; \vec{b} = \{\beta_i\}_{i=1}^m, \; \alpha_i, \beta_i \in Y.$$

THEOREM 4.3.– *If the condition [4.2] is valid, then the polynomial* $P_0(x)$, *defined by the formula*

$$P_0(x) = < \vec{F}, \Gamma^+ \sum_{p=0}^{n} \{(x_i, x)^p\}_{i=1}^m >, \quad [4.4]$$

is a solution of the extremum problem

$$\| P_0 \|_H = \inf_{P_n^I(x) \in \Pi_n^I(F)} \| P_n^I \|_H = \left(<< \Gamma^+ \vec{F}, \vec{F} >> \right)^{1/2}, Q(x) \in \Pi_n, \qquad [4.5]$$

and this solution is unique.

This theorem is a generalization of Porter's theorem (1980) on the interpolation polynomial with minimal norm defined on $L_2[a,b]$ to an abstract Hilbert space. Based on Theorem 4.3, a unique solution is determined to the Lagrange-type interpolation problem that has minimal norm among all interpolation polynomials [4.3] satisfying the interpolation relations [4.1].

It is of considerable interest to distinguish from the set the subset of operator polynomials that preserve polynomials of the same degree, so called c-polynomials. In the case of classical one-variable function interpolation with $m = n+1$, Lagrange polynomial preserves a polynomial of the same degree.

DEFINITION 4.1.– *An operator polynomial* $P_n^c(x;F)$ *of the n-th degree from the set* $\Pi_n(F)$ *is called c-polynomial if it satisfies the condition*

$$P_n^c(x;F) = F(x) \;\; \forall F \in \Pi_n, \;\; \forall x \in X.$$

In the monograph (Makarov and Khlobystov 1998), a theorem is proved about the set $\Pi_n^c(F)$ of operator polynomials that preserves the polynomial of the corresponding degree:

THEOREM 4.4.– *Let condition [4.2] be valid,* $\overset{*}{P_n}(x;F)$ *be some c-polynomial, then the whole set* $\Pi_n^c(F)$ *of c-polynomials of the n-th degree is described by the formula*

$$\Pi_n^c(F) = \left\{ P_n^c(x;F) = P_n(x;F) - P_n(x;\overset{*}{P_n}) + \overset{*}{P_n}(x;F) \right\}, \;\; \forall P_n(x;F) \in \Pi_n(F).$$

DEFINITION 4.2.– *An interpolation problem is called invariantly solvable if its solution exists for any values of the operator at the nodes.*

We introduce the notation: $\Gamma_k = \| (Bx_i, x_j)^k \|_{i,j=1}^m$. In terms of the matrices Γ, Γ_0, Γ_1, the condition for invariant solvability of the Lagrange interpolation problem in the Hilbert space is found:

$$rang(\Gamma_0 + \Gamma_1) + n - 1 \geq m.$$

4.4. Solution operator equations by the interpolation method

The above results can be used to find solutions to operator equations in Hilbert spaces. In this section, an algorithm for solving the inverse interpolation problem based on the minimal norm interpolant [4.4] is constructed.

Let X, Y be Hilbert spaces, X be separable, $A: X \to Y$ be a linear operator for which there exists a bounded inverse operator A^{-1}. Consider the equation

$$Ax = -f, \; f \in X, \; u \in Y. \qquad [4.6]$$

From [4.6], we get

$$x = A^{-1}(-f).$$

To solve problem [4.6] using the interpolation formula [4.4], we choose an orthonormal system of elements e_i, $i = \overline{1,m}$ from the space X and construct a set of interpolation nodes $\aleph(m)$:

$$\aleph(m) = \{x_0 = 0, x_i = e_{i_1} + e_{i_2} + \ldots + e_{i_k}, \; k = \overline{1,n}, \; i_j = \overline{1,m}, \; m \geq n\}.$$

The number of elements of the set $\aleph(m)$ is $M = 1 + \sum_{k=1}^{n} C_{m+k-1}^{k}$. With this choice of interpolation nodes $x_i, i = \overline{1,m}$, as shown in Kashpur and Khlobystov (2002), the matrix Γ is nonsingular. Thus, in [4.4] $\Gamma^{+} = \Gamma^{-1}$.

Let us define interpolation relations

$$f_i = -Ax_i, i = \overline{1,m}.$$

Let us construct an approximation of the inverse operator A^{-1} based on the interpolation polynomial of minimal norm [4.4], which satisfies the interpolation relations [4.1]. Note that when using formula [4.4], it is not necessary to specify an initial approximation for the unknown operator A^{-1}. The approximation to $A^{-1}(-f)$ based on formula [4.4] has the form

$$A^{-1}(-f) \approx P_1(-f) =< \vec{x}, \Gamma_m^{-1} \vec{l}(-f) >, \tag{4.7}$$

where $\vec{x} = \{x_i\}_{i=1}^m$, $\vec{l}(-f) = \{(-f, \overline{f}_i)\}_{i=1}^m$, $\Gamma_m = \left\| \sum_{p=0}^{1} (B\overline{f}_i, \overline{f}_j) \right\|_{i,j=1}^m$.

To determine the vector $\vec{l}(-f)$ and the elements of the matrix Γ_m, it is necessary to specify a self-adjoint positive, trace-class operator B contained in the interpolation relations [4.1], in such a way that the equalities hold

$$B\overline{f}_i = -f_i,\ i = \overline{1,m}. \tag{4.8}$$

We will choose the operator B in the simplest way so that the solution of equation [4.8] is not laborious. From [4.8], we find the unknowns $\tilde{f}_i$, $i = \overline{1,m}$.

The matrix Γ_m and the vector $\vec{l}(-f)$ contained in the interpolation formula [4.7] are defined, and therefore, the interpolant of the minimum norm [4.7] for finding the inverse operator A^{-1} is also defined.

If, to construct the interpolant, we choose interpolation nodes from the set $\aleph(m)$ (Kashpur and Khlobystov 2002), then in the metric of the space H interpolation process [4.4], and hence [4.7], converges to a polynomial operator in the case $m \to \infty$.

Let us put in [4.7] $m = 1, 2, \ldots$, then we obtain that the sequence of approximate solutions

$$x_m = P_1(-f) =< \vec{x}, \Gamma_m^{-1} \vec{l}(-f) >,\ m = 1, 2, \ldots$$

will converge to the exact solution of the boundary value problem [4.6] if $m \to \infty$. In this case, we can find the solution of problem [4.6] with a given accuracy of ε.

The advantages of the proposed algorithm for solving operator equation [4.6] based on formula [4.7] are as follows: based on a once-time constructed algorithm, which is associated with the implementation of interpolation process [4.7], we obtain an approximation for the inverse operator A^{-1} with which we can solve a whole class of problems. The proposed algorithm does not require the existence of an initial approximation to find A^{-1}.

Let us illustrate the solution algorithm for the boundary value problem. Let

$$Ax = x''(t) - tx(t) = -f(t),\ t \in (0,1),\ x(0) = x(1) = 0. \qquad [4.9]$$

We define an orthonormal system of elements $e_i(t) = \sqrt{2}\sin(\pi i t),\ i = \overline{1,m}$, system of interpolation nodes $x_i(t),\ i = \overline{1,M}$, chosen from the set $\aleph(m)$. Let us calculate interpolation relations

$$f_i = -Ax_i = x_i'' - tx_i,\ i = \overline{1,M}.$$

We choose the operator B as follows:

$$B\overline{f} = \int_0^1 G(x,\xi)\overline{f}(\xi)d\xi,$$

where

$$G(x,\xi) - \begin{cases} x(1-\xi), & 0 \le x \le \xi \le 1 \\ \xi(1-x), & 0 \le \xi \le x \le 1 \end{cases}.$$

To find the unknown elements $\overline{f}_i,\ i = \overline{1,m}$, we differentiate the equality twice

$$B\overline{f}_i = -f_i,\ i = \overline{1,M}.$$

We get

$$\overline{f}_i = f_i'',\ i = \overline{1,M}.$$

The elements of the matrix Γ_m are calculated based on scalar products

$$(B\overline{f}_i, \overline{f}_j)_{L_2(0,1)} = -(f_i'', f_j)_{L_2(0,1)} = (f_i', f_j')_{L_2(0,1)},$$

and we will get

$$\Gamma_m = \left\| \sum_{p=0}^{1} (f_i', f_j')_{L_2(0,1)} \right\|_{i,j=1}^{m}.$$

All vectors and matrices from [4.7] are found; therefore, a formula has been found for calculating the m-th approximation x_m to the exact solution of the boundary value problem [4.9]

$$x_m = P_1(-f) = < \vec{x}, \Gamma_m^{-1}\vec{l}(-f) >, m = 1,2,\ldots$$

If in the latter formula $m \to \infty$, then based on the theorem from Kashpur and Khlobystov (2002), we can find an approximate solution x_m to the boundary value problem [4.9] with a given accuracy ε.

Note that the solution of the boundary value problem [4.9] can be written based on Green's function, which is determined using the Airy functions $Ai(t)$, $Bi(t)$:

$$x(t) = -\int_0^1 G(t,z) f(z) dz,$$

where

$$G(t,z) = C \begin{cases} (\sqrt{3}Ai(t) - Bi(t))(Ai(z)Bi(1) - Ai(1)Bi(z)), & 0 \le t \le z \le 1 \\ (\sqrt{3}Ai(z) - Bi(z))(Ai(t)Bi(1) - Ai(1)Bi(t)), & 0 \le t \le z \le 1 \end{cases},$$

$$C = -\frac{\pi}{Bi(1) - \sqrt{3}Ai(1)}.$$

In this case, using the results of Khlobystov and Kashpur (1996), for a linear operator, the error in the metric of the operator's value space is written as:

$$\left\| x - P_1^I(-f) \right\|_{C^2[0,1]} \leq \|A\| \varepsilon_m(x),$$

where $\|A\|$ be the norm of the linear operator A, $\varepsilon_m = \left\| x - \sum_{i=1}^{m} (x, e_i) e_i \right\|$.

4.5. Interpolation in Euclidean spaces

Lagrangian interpolation of a one-variable function was generalized to functions of many variables in the work of Kergin (1980). Since this publication, a significant number of publications on Kergin interpolation and its generalizations have appeared (Micchelli 1980; Micchelli and Milman 1980; Bloom 1981; Bos 1983; Anderson and Passare 1991a, 1991b; Calvi and Filipsson 2004). Further, in Fillipson (2004), Kergin interpolation on the set of many-variable functions and in Banach spaces is considered, where the error formula for the approximation of holomorphic mappings is obtained. A convergence theorem is considered, which can be applied to operators on Banach algebras, in particular, on the algebra of square matrices with complex coefficients. Note that the interpolation formulas given in the work, up to equivalent transformations of integrals, coincide with the formulas obtained earlier by other authors in Ulm (1967) and Sobolevsky (1975). It is impossible to obtain classical Newton interpolation polynomials from the obtained formulas, unlike the formulas constructed in Makarov et al. (2009). In the latter work, the results of the interpolation of many-variable functions are generalized to functionals and operators defined on linear topological and functional spaces. From the interpolation formulas given in Makarov et al. (2009), as a special case, we can obtain previously known (Ulm 1967; Sobolevsky 1975; Makarov and Khlobystov 1998) and classical interpolation structures for functions of many variables (Steffensen 2006).

To construct a unique interpolation polynomial in the Euclidean space E_k, it is necessary that the relation $m = \dfrac{(n+k)!}{n!k!}$ holds between the degree of the polynomial n and the number of nodes m. In addition, the construction of an interpolant of degree n in E_k causes certain difficulties. In the problems of object identification based on observation of its reaction to input signals, the case when the initial information about the object under study is insufficient is particularly interesting:

the number of interpolation relations is less than the dimension of the space of polynomials on which the solution of the interpolation problem is sought in finite-dimensional Euclidean space without additional restrictions. We call such a problem underdetermined. The solution of such problems will be considered in this section.

In the Hilbert space *X*, the equality holds (Gikhman and Skorokhod 1971)

$$B(u,v) = \int_X (x,u)(x,v)\mu(dx) = (Bu,v), u,v,x \in X, \qquad [4.10]$$

where μ be the Gaussian measure on *X*, the first moment of which is equal to zero, $B(u,v)$ be a correlation functional, B be the correlation operator of this measure.

Let us consider the solution of the interpolation problem in a finite-dimensional Euclidean space E_k with the Gaussian measure μ. Let us apply the above results to this space. Let the function $f : E_k \to R_1$ be given by its values at the nodes $\gamma_i = (\gamma_1^i, \gamma_2^i, \ldots, \gamma_k^i)$, $i = \overline{1,m}$, $m \le p$, where p is the dimension of the space of polynomials of degree n in E_k, $u = (u_1, u_2, \ldots, u_k)$, $v = (v_1, v_2, \ldots, v_k)$, $\gamma = (x_1, x_2, \ldots, x_k)$.

Let us put in [4.10] $\mu(dx) = g(x)dx$, $g(x) = \frac{1}{\sqrt{2\pi}} e^{-\frac{x^2}{2}} dx$, $x = \gamma$. Then, the equality [4.10] will be written as follows:

$$B(u,v) = \int_{E_k} (\gamma,u)(\gamma,v)\mu(d\gamma)$$

$$= \int_{-\infty}^{+\infty}\int_{-\infty}^{+\infty} \cdots \int_{-\infty}^{+\infty} (u_1x_1 + u_2x_2 + \cdots + u_kx_k)(v_1x_1 + v_2x_2 + \cdots v_kx_k)$$

$$\times g(x_1)g(x_2)\cdots g(x_k)dx_1dx_2\cdots dx_k = (u_1v_1 + u_2v_2 + \cdots + u_kv_k) = (u,v) = (Iu,v).$$

Therefore, in the case $X = E_k$, the unit operator I can be chosen as the operator B, and the vectors $\gamma_i, i = \overline{1,m}$ as the interpolation nodes.

Let us write down an interpolation polynomial that satisfies the relations

$$P_n(\gamma_i) = f(\gamma_i),\ i = \overline{1,p}, \tag{4.11}$$

in the form of an interpolant of the minimum norm:

$$P_n(\gamma) =< \vec{f}, \Gamma^+ \sum_{p=0}^{n} \{(x_1\gamma_1^j + x_2\gamma_2^j + \cdots + x_k\gamma_k^j)^p\}_{i=1}^m >, \tag{4.12}$$

where $\vec{f} = f(\gamma_i)_{i=1}^m$, $\Gamma = \| \sum_{p=0}^{n} (\gamma_1^i\gamma_1^j + \gamma_2^i\gamma_2^j + \cdots + \gamma_k^i\gamma_k^j)^p \|_{i,j=1}^m$.

The whole set of interpolation polynomials has the form

$$P_n(\gamma) = q(\gamma) + < \vec{f} - \vec{q}, \Gamma^+ \sum_{p=0}^{n} \{(x_1\gamma_1^j + x_2\gamma_2^j + \cdots + x_k\gamma_k^j)^p\}_{i=1}^m >,\ q \in \Pi_n.$$

Let Π_{kn} be the space of polynomials of degree n depending on k variables. In Kashpur and Khlobystov (2015), the following theorem was proved.

THEOREM 4.5.– *Let the function $f : E_k \to R_1$ be given its values $f(\gamma_i)$, $i = \overline{1,m}$. If the interpolation nodes are chosen so that the relevant subsystem from vectors*

$$s_i = \left\{ \left(\frac{j!}{j_1! j_2! \cdots j_k!} \right)^{1/2} \left(\gamma_1^i\right)^{j_1} \left(\gamma_2^i\right)^{j_2} \cdots \left(\gamma_k^i\right)^{j_k}, j_1 + j_2 + \cdots + j_k = j, \right\}_{j=0}^{n} \tag{4.13}$$

are linearly independent, then in the space Π_{kn} interpolation problem of many-variable function with condition [4.11] is invariantly solvable and it has a unique solution with minimum norm under the condition $m \le p$, where p is the dimension of the space Π_{kn}.

Let us apply the above theorems to solve one class of applied problems.

4.6. Construction of surfaces

The problem of constructing surfaces arises when solving many applied problems, in particular in problems about the spread of pollution, in geology, in computer graphics, urban planning, in problems of determining the trajectory of an object's motion, in the automotive industry, when the problem of constructing a mathematical representation of a sketch arises (for the first time such an approach was used by Renault) and later interactive interpolation is already used.

4.6.1. *Example*

Let us consider the construction of the interpolation polynomial of the minimum norm $P_2(x, y)$ based on formula [4.12]. Let us choose interpolation as follows:

$$\gamma_1 = (0,0), \gamma_2 = (1,0), \gamma_3 = (2,0),$$

$$\gamma_4 = (0,2), \gamma_5 = (1,2),$$

$$\gamma_6 = (0,3).$$

Let: $\gamma_i = (x_i, y_i)$, $i = \overline{1,m}$. Then, for the Euclidean space E_2, the vectors s_i, $i = \overline{1,m}$ based on formula [4.13] will be written in the form:

$$s_i = (1, x_i, y_i, x_i^2, \sqrt{2}x_i y_i, y_i^2, ..., x_i^n, \sqrt{n}x_i^{n-1} y_i, ..., \sqrt{n}x_i y_i^{n-1}, y_i^n),\ i = \overline{1,m}.$$

We will get

$$s_1 = (1,0,0,0,0,0),\ s_2 = (1,1,0,1,0,0),\ s_3 = (1,2,0,4,0,0),$$

$$s_4 = (1,0,2,0,0,4),\ s_5 = (1,1,2,1,2\sqrt{2},4),\ s_6 = (1,0,3,0,0,9).$$

Since there is no second-order curve passing through the points γ_i, $i = \overline{1,6}$, the vectors s_i, $i = \overline{1,6}$, are linearly independent and the matrix Γ will be nonsingular. We conclude that to construct the interpolant [4.12], we can choose any subsystem of vectors from [4.13], i.e. the interpolation problem will be invariantly solvable and have a unique solution (in the sense of the minimal norm) in the case when $m \le 6$ $(p = 6)$.

Let us choose $m = 3$, a subsystem of vectors s_1, s_3, s_4. For convenience, we denote them as $\overline{s_1} = s_1, \overline{s_2} = s_3, \overline{s_3} = s_4$. Then, the matrix Γ will be nonsingular, and the interpolation polynomial [4.12] ($n = 2$, $m = 3$), which satisfies relations [4.11], takes the form

$$P_2(\gamma) = P_2(x, y) =< \vec{f}, \Gamma^{-1} \sum_{p=0}^{2} \{(x_i x + y_i y)^p\}_{i=1}^{3} >= \sum_{i=1}^{3} l_i(\gamma) f(\gamma_i),$$

where $l_i(\gamma) = l_i(x, y)$ is the fundamental Lagrange polynomials, $l_i(\gamma_j) = \delta_{ij}$, δ_{ij} be the Kronecker symbol, $i, j = 1,2,3$,

$$l_1(x, y) = 1 - 0.1(x + y + 2x^2 + 2y^2),$$

$$l_2(x, y) = 0.1(x + 2x^2),$$

$$l_3(x, y) = 0.1(y + 2y^2).$$

4.6.2. *Example*

Let the function $f(x, y) = x + 2y - x^2 - 4y^2$ be given by its values at the nodes γ_i, $i = \overline{1, M}$ belonging to the set $\aleph(m)$. Let $m = 2$, $e_0 = (0,0)$, $e_1 = (1,0)$, $e_2 = (0,0)$, then $M = 6$.

In this case, the function $f(x, y)$ and the interpolant of the minimum norm [4.12], which is constructed from the given values, coincide:

Let us present a graphical interpretation of the obtained results for $M = 2$, $M = 4$. In this case, there is a unique interpolation polynomial of minimal norm satisfying the relations [4.11], the error $\Delta_2 = 2.000100200$, $\Delta_4 = 1.507500000$ in the domain $\{(x, y) : -1 \le x \le 1, -1 \le y \le 1\}$.

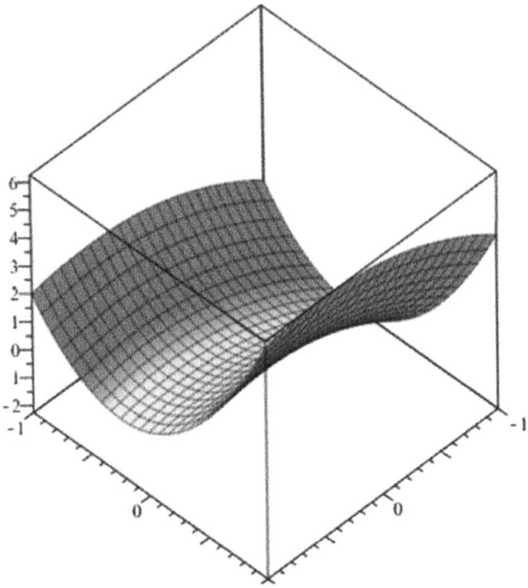

Figure 4.1. *Function* $f(x, y)$

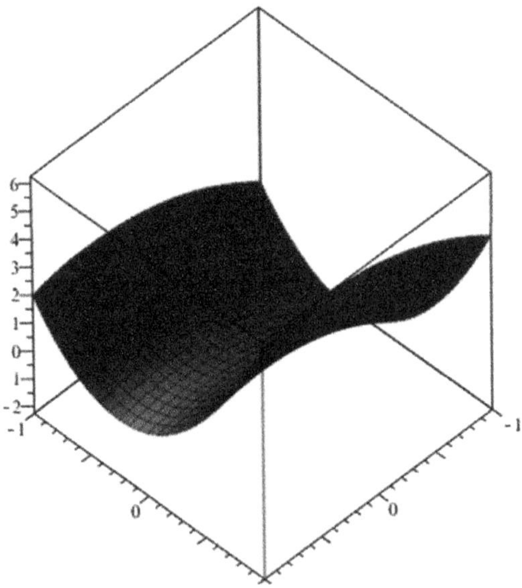

Figure 4.2. *Interpolant* $P_2(x, y)$, $M = 6$

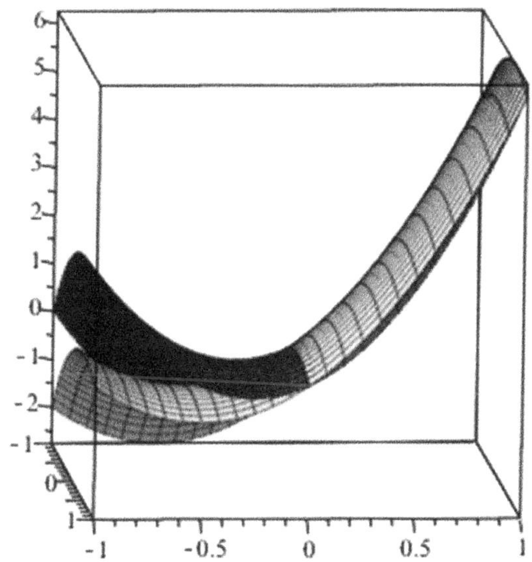

Figure 4.3. *Interpolant* $P_2(x, y)$, $M = 2$

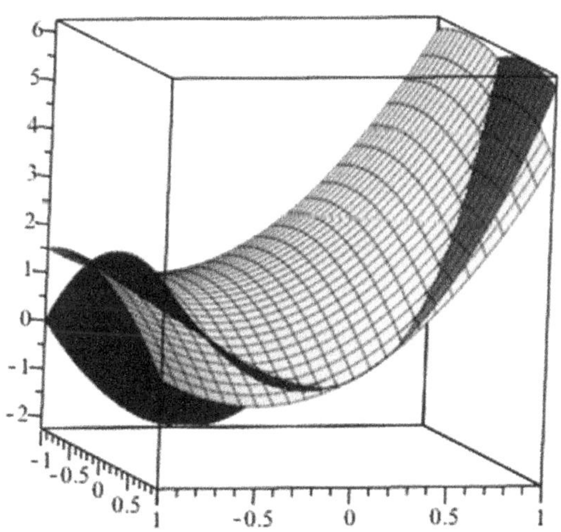

Figure 4.4. *Interpolant* $P_2(x, y)$, $M = 4$

4.6.3. *Example*

The construction of surfaces can be used to solve the problem of determining the trajectory of an object's motion: based on Stone's theorem (Stone 1948), which is a generalization of Weierstrass's theorem to the class of continuous functions in R_n, the trajectory of motion can be represented as the intersection of two surfaces of order n:

$$\begin{cases} P_n^{(1)}(x,y,z) = \sum\limits_{i,j,k=0,i+j+k\le n}^{n} \alpha_{ijk} x^i y^j z^k = 0, \\ P_n^{(2)}(x,y,z) = \sum\limits_{i,j,k=0,i+j+k\le n}^{n} \beta_{ijk} x^i y^j z^k = 0. \end{cases} \qquad [4.14]$$

We will construct the trajectory of an object's motion as a curve that is the intersection of two surfaces; i.e. using polynomial operator interpolation, we will find the unknown coefficients from [4.14].

Consider the interpolant $P_n^I(x)$ from the set [4.3] in the Hilbert space X, when $\vec{F} = \vec{0}$:

$$P_n^I(x) = Q_n(x) - < \vec{Q}_n, \Gamma^+ \sum_{p=0}^{n} \{(x_i, x)^p\}_{i=1}^m >, \; Q_n \in \Pi_n. \qquad [4.15]$$

The operator polynomial [4.15] interpolates the operator F, which at the nodes $v_i, \overline{1,m}$, is equal to $0 \in Y$. Such an interpolant always exists, since the conditions of Theorem 4.1 are satisfied:

$$\Gamma \vec{F} = \vec{0}.$$

Let $X = R_3, Y = R_2$ $v_i = (x_i, y_j, z_i)$, $i = \overline{1,m}$ be known coordinates of the object at time $t_i, i = \overline{1,m}$. Let us denote $\vec{Q}_{ni}(x) = \left\{ Q_{ni}(v_j) \right\}_{j=1}^{m}$, $i = 1,2$. Based on

[4.15], the equation of the object's motion trajectory as a curve of intersection of two surfaces of the n-th order can be written as

$$\begin{cases} P_{n1}^{I}(x) = Q_{n1}(x) - < \vec{Q}_{n1}, \Gamma^{+} \sum_{p=0}^{n} \{(x_i, x)^p\}_{i=1}^{m} >, \\ P_{n2}^{I}(x) = Q_{n2}(x) - < \vec{Q}_{n2}, \Gamma^{+} \sum_{p=0}^{n} \{(x_i, x)^p\}_{i=1}^{m} >, \end{cases} \quad [4.16]$$

$$Q_{n1} \neq Q_{n2},\ Q_{ni} \not\in KerP_{ni}^{I}(Q), i = 1, 2.$$

Thus, the polynomial operator interpolation method provides an explicit analytical representation of the equation of the object's trajectory in the form [4.16] as a curve of intersection of two surfaces of order n. In this case, it is not necessary to solve the corresponding system of linear algebraic equations. The condition $Q_{ni} \not\in KerP_{ni}^{I}(Q),\ i = 1, 2$ is essential. If it is not satisfied, then the left-hand sides of [4.16] will become zero, and the intersection curve of the two surfaces cannot be determined.

NOTE.– In (Makarov et. al. 2021), new conditions for the solvability of systems of linear equations and inequalities were obtained, which are analogues of the Kronecker–Capelli theorem (Strang 2023) and Chernikov (1944), respectively. This chapter provides examples of solving such systems.

4.7. Conclusion

This chapter is devoted to the problem of Lagrange operator interpolation. The application of polynomial operator interpolation for solving operator equations and approximating functions of many variables under conditions of uncertainty is considered and justified.

Examples of the application of polynomial operator interpolation to some classes of applied problems are given, particularly for determining the trajectory of an object's motion.

4.8. References

Anderson, M. and Passare, M. (1991a). Complex kergin interpolation. *Journal of Approximation Theory*, 64, 214–225.

Anderson, M. and Passare, M. (1991b). Complex kergin interpolation and the Fantappiè transform. *Mathematische Zeitschrift*, 208(2), 257–271.

Bloom, T. (1981). Kergin interpolation of entire functions. *Duke Mathematical Journal*, 48, 69–83.

Bos, L. (1983). On kergin interpolation in the disk. *Journal of Approximation Theory*, 37, 251–261.

Calvi, J.-P. and Filipsson, L. (2004). The polynomial projectors that preserve homogeneous differential relations: A new characterization of kergin interpolation. *East Journal on Approximations*, 10(4), 451–454.

Chernikov, S.M. (1944) A generalization of the Kronecker–Capelli theorem on systems of linear equations. *Matematicheskii Sbornik*, 15(57), 3, 437–448.

Cox, D.A. (2020). *Applications of Polynomial Systems*. American Mathematical Society, Providence, RI.

Daugavet, I.K. (1994). On polynomial approximation of operators. *Vestnik of St. Petersburg University. Mathematics*, 3(15), 23–26 (in Russian).

Egorov, A.D., Sobolevskii, P.I., Yanovich, L.A. (1993). *Functional Integrals: Approximate Evaluation and Applications*. Kluwer Academic Publishers, Dordrecht.

Fillipson, L. (2004). Kergin interpolation in Banach spaces. *Journal of Approximation Theory*, 127(1), 108–123.

Fréchet, M. (1910). Sur les fonctionelles continues. *Annales scientifiques de l'École Normale Supérieure*, 27, 193–216.

Gikhman, I.I. and Skorokhod, A.V. (1971). *The Theory of Stochastic Processes I*. Nauka, Moscow (in Russian).

Istrăţescu, V.I. (1977). A Weierstrass theorem for real Banach spaces. *Journal of Approximation Theory*, 19(2), 118–122.

Kapustyan, O.V., Kapustyan, O.A., Sukretna, A.V. (2011). Approximate stabilization for a nonlinear parabolic boundary-value problem. *Ukrainian Mathematical Journal*, 63(5), 759–767.

Kashpur, O.F. and Khlobystov, V.V. (2002). On the accuracy of polynomial interpolation in hilbert space with disturbed nodal values of the operator. *Cybernetics and Systems Analysis*, 38(1), 143–147.

Kashpur, O.F. and Khlobystov, V.V. (2015). Invariance and uniqueness of solutions to polynomial interpolation problems in Euclidean space. *Journal of Computational and Applied Mathematics*, 2(119), 8–14.

Kergin, P. (1980). A natural interpolation of C^k-functions. *Journal of Approximation Theory*, 29(4), 278–293.

Khlobystov, V.V. and Kashpur, O.F. (1996). On interpolation of polynomial operators. *Cybernetics and Systems Analysis*, 32, 398–403.

Litvin, A.N. (2002). *Interlineation of Functions and Some of its Applications*. Osnova, Kharkiv (in Ukrainian).

Makarov, V.L. and Khlobystov, V.V. (1998). *Fundamentals of the Theory of Operator Polynomial Interpolation*. Institute of Mathematics of the National Academy of Sciences of Ukraine, Kyiv (in Russian).

Makarov, V.L., Khlobystov, V.V., Yanovich, L.A. (2000). *Interpolation of Operators*. Naukova Dumka, Kyiv (in Russian).

Makarov, V.L., Khlobystov, V.V., Demkiv, I.I. (2009). Interpolation of functionals of many variables. *Reports of the National Academy of Sciences of Ukraine*, 5, 29–35 (in Ukrainian).

Makarov, V.L., Khlobystov, V.V., Yanovich, L.A. (2010). *Methods of Operator Interpolation*. Institute of Mathematics of the National Academy of Sciences of Ukraine, Kyiv.

Makarov, V.L., Khlobystov, V.V., Kashpur O.F. (2021). Operator interpolation and systems of linear equations and inequalities in Euclidean spaces. *Ukrainian Mathematical Journal*, 72(11), 1758–1770.

Micchelli, C.A. (1980). A constructive approach to Kergin interpolation. *Rocky Mountain Journal of Mathematics*, 10, 485–497.

Micchelli, C.A. and Milman, P. (1980). A formula for Kergin interpolation in $\mathcal{R}^n$. *Journal of Approximation Theory*, 29(4), 294–296.

Porter, W.A. (1976). An overview of polynomic system theory. *Proceedings of the IEEE*, 64(1), 18–26.

Porter, W.A. (1980). Synthesis of polynomic system. *SIAM Journal on Mathematical Analysis*, 11(2), 308–315.

Prenter, P.M. (1970). A Weierstrass theorem for real separable Hilbert spaces. *Journal of Approximation Theory*, 3, 341–351.

Prenter, P.M. (1971a). Lagrange and hermite interpolation in Banach space. *Journal of Approximation Theory*, 4(4), 419–432.

Prenter, P.M. (1971b). On polynomial operators and equations. In *Nonlinear Functional Analysis and Applications*, Rall, L.B. (ed.). Academic Press, New York.

Rvachev, V.A. (1982). *Theory of R-functions and Some its Applications*. Naukova Dumka, Kyiv (in Russian).

Sobolevsky, P.I. (1975). Interpolation of functionals and some approximate formulas for integrals with respect to Gaussian measure. *Proceedings of the Academy of Sciences of the Byelorussian SSR. Series: Physics and Mathematics*, 2, 5–12.

Steffensen, J.F. (2006). *Interpolation*, 2nd edition. Dover Books, Garden City, New York.

Stone, M.H. (1948). The generalized Weierstrass approximation theorem. *Mathematics Magazine*, 21, 167–183.

Strang, G. (2023). *Introduction to Linear Algebra*, 6th edition. Wellesley Cambridge Press, Wellesley, MA.

Torokhti, A. and Howlett, P.G. (1997). On the constructive approximation of non-linear operators in the modelling of dynamical systems. *The Journal of the Australian Mathematical Society. Series B. Applied Mathematics*, 39, 1–27.

Ulm, S. (1967). On generalized divided differences. *Proceedings of the Academy of Sciences of the Estonian SSR. Series: Physics and Mathematics*, 16(1), 13–26.

Ulm, S. and Pol, V. (1969). On the construction of generalized divided differences. *Proceedings of the Academy of Sciences of the Estonian SSR. Series: Physics and Mathematics*, 18(1), 100–102.

Yanovich, L.A. (1976). *Approximate Computation of Continual Integrals with Respect to Gaussian Measures.* Nauka i Tekhnika, Minsk (in Russian).

5

New Fractional Differential Analogues of the Biparabolic Evolution Equation and Some Boundary Value Problems

V.M. BULAVATSKY[1] **and S. LYASHKO**[2]

[1] *V.M. Glushkov Institute of Cybernetics of the NAS of Ukraine, Kyiv, Ukraine*

[2] *Taras Shevchenko National University of Kyiv, Ukraine*

In this chapter, new, non-local in terms of time and geometric variables, fractional-differential analogues of the biparabolic (Fushchich et al. 1990) differential equation with partial derivatives of the fourth order (known in the mathematical theory of heat conduction) are introduced. For these model fractional differential equations, statements are made and closed form solutions of some boundary value problems are obtained. Conditions of the existence of regular solutions to the considered problems are given.

5.1. Introduction

Systemic research (on the basis of mathematical and computer modeling) in the field of studying the fundamental laws of dynamics and control of heat and mass transfer processes is a relevant direction for a number of modern sciences, in particular mathematics and computer science, which is developing intensively (Fushchich et al. 1990; Lyashko et.al. 1995, 2000, 2001, 2003; Bulavatsky 1997; Bulavatsky et al. 2005).

Computational Methods and Mathematical Modeling in Cyberphysics and Engineering Applications 2,
coordinated by Dmitri KOROLIOUK, Sergiy LYASHKO and Nikolaos LIMNIOS.

We know that the classic mathematical model of the dynamics of transfer processes, which is based on a linear equation of the parabolic type, assumes an infinite speed of propagation of disturbances, which leads to a number of well-known paradoxes. In this regard, Fushchich et al. (1990) proposed a generalization of the classical Fourier heat conduction (diffusion) equation in the form:

$$Lu \equiv \mu_1 L_1 u + \mu_2 L_2 u = f, \qquad [5.1]$$

where $L_2 = L_1 L_1$, L_1 is the thermal conductivity (diffusion) operator; $L_1 = \frac{\partial}{\partial t} - \kappa \Delta$, Δ is the Laplace operator in terms of geometric variables; $u(x,t)$ is a field function (temperatures, concentrations, pressures, etc.); f is a source function; μ_1, μ_2 are real parameters; and $\kappa = const > 0$.

Equation [5.1] is called the biparabolic equation of heat conduction. It enables us to describe transport modes with a finite speed of propagation of disturbances (Fushchich et al. 1990) and under the conditions $\mu_1 \neq 0, \mu_2 = 0$ coincides with the classical heat conduction equation. It was proved by Fushchich et al. (1990) that equation [5.1] is invariant with respect to the Galileo group and therefore can be used to describe thermal and diffusion processes, independent of the inertial systems in which they are observed. Equation [5.1] was obtained in Fushchich et al. (1990) from the law of conservation of energy, if the energy e and flow $\vec{q}$ set by ratios:

$$e = e_0 + c_v(u - u_0) + \tau_r c_v L_1 u,$$

$$\vec{q} = -\lambda \nabla u - \tau_r c_v \kappa \nabla \left(L_1 u \right),$$

where $u = u(x,t)$ is the temperature, c_v is the heat capacity, λ is the coefficient of thermal conductivity, $\kappa = \lambda / c_v$, τ_r is a real parameter, and ∇ is the Hamiltonian operator.

The biparabolic equation [5.1] has been repeatedly used to model the dynamics of various evolutionary processes in natural science, in particular to describe the dynamics of deformable water-saturated soil environments in the process of their filtration consolidation under the action of applied external loads (Bulavatsky 1997;

Bulavatsky et al. 2005). Currently, the study of the peculiarities of anomalous transfer processes based on the ideas of fractional integro-differentiation (Kilbas et al. 2006; Sandev et al. 2019) has developed. For example, in works (Bulavatsky 2016, 2019; Bulavatsky et al. 2020), a fractional-differential analogue of the biparabolic evolution equation was introduced and some model boundary value problems were solved taking into account the phenomenon of temporal non-locality, in particular with regard to geofiltration processes in a fractured-porous medium and filtration-consolidation processes in water-saturated soils. At the same time, taking into account the non-locality of the time variable, the operator L_1 is taken in the form $L_1 = {}^C D_{0t}^{\alpha} - \kappa \Delta$ (here ${}^C D_{0t}^{\alpha}$ ${}^C D_{0t}^{\alpha}$ is the Caputo fractional derivative operator order $\alpha \, (0 < \alpha \le 1)$ by variable t (Kilbas et al. 2006) and the corresponding fractional-differential analogue of the biparabolic equation [5.1] is written as follows (Bulavatsky 2016):

$$\mu_1 L_1 u(x,t) + \mu_2 L_1 L_2 u(x,t) = f, \qquad [5.2]$$

where:

$$L_1 := {}^C D_{0t}^{\alpha_1} - \kappa_1 \Delta, \quad L_2 := {}^C D_{0t}^{\alpha_2} - \kappa_2 \Delta, \qquad [5.3]$$

${}^C D_{0t}^{\alpha_1}$, ${}^C D_{0t}^{\alpha_2}$ are operators of the Caputo fractional derivative by variable t of orders α_1, α_2 in accordance $(0 < \alpha_1, \alpha_2 \le 1)$, μ_1, μ_2 are known constants.

Note that equations of the form [5.2] (subject to conditions [5.3]) describe, in particular, some processes of fractional-differential filtration dynamics taking into account memory effects in layered geoporous media. As an example, let us consider a one-dimensional on geometrically variable problem of mathematical modeling of fractional-differential pressure dynamics during geofiltration in a three-layer geoporous medium. Let us assume the presence of a filtering scheme according to which the medium consists of three layers, two of which (upper and lower) have thickness μ_1, μ_2 and are sufficiently permeable with filtering coefficients k_1, k_2 k_1, k_2 in accordance. The specified layers of the geoporous medium are separated by a weakly permeable layer of thickness m_0 m_0 with the filter coefficient k_0 k_0. Within the framework of the classical mathematical filtration model, the problem of analyzing the dynamics of pressure $h_i(x,t) \, (i = 1,2)$, in the upper and lower layers of the geomaterial, respectively, is reduced to the solution of the non-stationary boundary value problem for the following system of equations (Polubarinova-Kochina et al. 1969; Oleinik 1981):

$$\frac{\partial h_1}{\partial t} - a_1 \frac{\partial^2 h_1}{\partial x^2} = -b_1 (h_1 - h_2),$$

$$\frac{\partial h_2}{\partial t} - a_2 \frac{\partial^2 h_2}{\partial x^2} = b_2 (h_1 - h_2),$$

where $a_i = \dfrac{k_i m_i}{\mu_i}$ $(i = 1,2)$ are the piezoconductivity coefficients for each layer of the environment, $b_i = \dfrac{k_0}{m_0 \mu_i}$ $(i = 1,2)$ are the cross-flow parameters, $k_i, \mu_i (i = 1,2)$ are the coefficients of filtration and water discharge of the upper and lower layers.

In the case of taking into account non-local time-varying effects, the corresponding system of filtering equations will be rewritten in the form

$$L_1 h_1 (x,t) = -b_1 (h_1 - h_2),$$

$$L_2 h_2 (x,t) = b_2 (h_1 - h_2),$$

or excluding any of the functions from here h_i $(i = 1,2)$, we receive

$$b_1 L_2 h(x,t) + b_2 L_1 h(x,t) + L_1 L_2 h(x,t) = 0, \qquad [5.4]$$

where $h(x,t)$ is the pressure function. L_1, L_2 are determined according to the ratios:

$$L_1 := {}^C D_{0t}^{\alpha_1} - a_1 \frac{\partial^2}{\partial x^2}, \quad L_2 := {}^C D_{0t}^{\alpha_2} - a_2 \frac{\partial^2}{\partial x^2}. \qquad [5.5]$$

The pressure equation [5.4] under the conditions [5.5] generalizes the accepted biparabolic equation [5.1] and is called a generalized biparabolic equation (see Bulavatsky (2016) where the solutions of some filtration boundary value problems related to this equation are given). Note that in the case of small overflow parameters (e.g. $b_1 \ll 1$) from the model equation [5.4], we obtain a fractional-differential analogue of the classical biparabolic equation [5.1] in the form [5.2], [5.3].

In contrast to the results of Bulavatsky (2016, 2019) and Bulavatsky et al. (2020), this study considers non-local fractional-differential analogues of the classical biparabolic equation [5.1] both in time and space. Extending equations [5.2], [5.3] in the case of taking into account not only memory effects, but also taking into account the abnormality of the process dynamics according to the spatial variable x x, we obtain the following model equation:

$$L_1 u(x,t) + \tau_r L_1 L_2 u(x,t) = f, \qquad [5.6]$$

where indicated:

$$L_1 = {}^C D_{0t}^{\alpha_1} - \kappa_1 \, {}^C D_{0x}^{\gamma}, \; L_2 = {}^C D_{0t}^{\alpha_2} - \kappa_2 \, {}^C D_{0x}^{\gamma} \; (0 < \alpha_1, \alpha_2 \le 1, \; 1 < \gamma \le 2) \qquad [5.7]$$

${}^C D_{0t}^{\alpha}, {}^C D_{0x}^{\gamma}$ are the Caputo fractional derivative operators with respect to the variables t, x, respectively, $\kappa_1, \kappa_2, \tau_r = const > 0$.

Equation [5.6] (subject to conditions [5.7]) is a non-local model equation both in terms of time and spatial variable, which generalizes the classical biparabolic equation [5.1] and is significantly different from the previous fractional-differential analogues of the biparabolic evolutionary equation due to taking into account not only memory effects.

Below, in relation to [5.6], [5.7], statements are made and classical solutions to some model boundary value problems of anomalous dynamics are obtained, both in direct and inverse statements. The conditions for the existence of these solutions have been established, and other fractional-differential analogues of the biparabolic equation have been introduced on the basis of the Hilfer and Caputo–Fabrizio derivatives.

5.2. Some boundary value problems for the fractional–differential analogue of the biparabolic equation with non-locality in time and space

Consider a non-stationary boundary value problem for a finite interval of unit length by a geometric variable x with respect to the model, which is determined according to relations [5.6], [5.7]. The problem is reduced to finding in the area $\Omega = \{(x,t) : 0 < x < 1, 0 < t < T < \infty\}$ the solution of the equation:

$$\left({}^{C}D_{0t}^{\alpha_1} - \kappa_1 {}^{C}D_{0x}^{\gamma}\right)u(x,t) + \tau_r\left({}^{C}D_{0t}^{\alpha_1} - \kappa_1 {}^{C}D_{0x}^{\gamma}\right)$$
$$\left({}^{C}D_{0t}^{\alpha_2} - \kappa_2 {}^{C}D_{0x}^{\gamma}\right)u(x,t) = f(x) \qquad [5.8]$$

under the following boundary conditions:

$$u(0,t) = u(1,t) = 0,\ {}^{C}D_{0x}^{\gamma}u(x,t)\Big|_{\substack{x=0,\\ x=1}} = 0\ (t \geq 0), \qquad [5.9]$$

and initial conditions:

$$u(x,0) = \varphi(x), \qquad {}^{C}D_{0t}^{\alpha_2}u(x,t)\Big|_{t=0} = 0 \qquad (0 \leq x \leq 1), \qquad [5.10]$$

where $\varphi(x)$ is the given initial distribution function, $f(x)$ is a source function, $\tau_r = const$, and $u(x,t)$ is a required function.

Let us consider a new unknown function:

$$\upsilon(x,t) = \tau_r L_2 u(x,t) + u(x,t), \qquad [5.11]$$

where L_2 is determined according to [5.7]. Then, problem [5.8]–[5.10] is reduced to the solution of the following pair of boundary value problems with respect to unknown functions $\upsilon(x,t), u(x,t)$:

$$\left({}^{C}D_{0t}^{\alpha_1} - \kappa_1 {}^{C}D_{0x}^{\gamma}\right)\upsilon(x,t) = f(x), \qquad [5.12]$$

$$\upsilon(0,t) = \upsilon(1,t) = 0 \qquad (t \geq 0), \qquad [5.13]$$

$$\upsilon(x,0) = \psi(x) \qquad (0 \leq x \leq 1), \qquad [5.14]$$

$$\tau_r\left({}^{C}D_{0t}^{\alpha_2} - \kappa_2 {}^{C}D_{0x}^{\gamma}\right)u(x,t) + u(x,t) = \upsilon(x,t), \qquad [5.15]$$

$$u(0,t) = u(1,t) = 0 \qquad (t \geq 0), \qquad [5.16]$$

$$\upsilon(x,0) = \varphi(x) \qquad (0 \leq x \leq 1), \qquad [5.17]$$

where $\psi(x) = \varphi(x) - \tau_r\kappa_2 {}^{C}D_{0x}^{\gamma}\varphi(x)$ is an known function.

We will look for the solution of the boundary value problem [5.12]–[5.14] in the form of a series:

$$\upsilon(x,t) = \sum_{n=1}^{\infty} \upsilon_n(t) X_n(x) \quad (x,t) \in \Omega , \tag{5.18}$$

where $X_n(x) = x^{\gamma-1} E_{\gamma,\gamma}(\lambda_n x^{\gamma})$ are eigenfunctions of the Sturm–Liouville problem of the form:

$${}^{C}D_x^{\gamma} X(x) = \lambda X(x), \quad X(0) = X(1) = 0, \qquad x \in (0,1), \tag{5.19}$$

$E_{\gamma,\gamma}(z)$ is the two-parameter Mittag–Leffler function (Gorenflo et al. 2014).

Problem [5.19] was studied, in particular, in works (Samko et al. 1987; Djrbashian 1993; Aleroev et al. 2013a, 2013b; Khasambiev et al. 2014), in which it was shown that only for the eigenvalues λ_n, which are zeros of the two-parameter Mittag–Leffler function $E_{\gamma,\gamma}(\lambda)$, there are the above eigenfunctions $X_n(x)$, which have the form $\{X_n(x)\}_{n=1}^{\infty} = \left\{x^{\gamma-1} E_{\gamma,\gamma}\left(\lambda_n (1-x)^{\gamma}\right)\right\}_{n=1}^{\infty}$. This system of functions forms a non-orthogonal basis (Aleroev et al. 2013a; Khasambiev et al. 2014).

The system of eigenfunctions of the problem, which conjugate to problem [5.19] is defined as follows (Aleroev et al. 2013a, 2013b; Khasambiev et al. 2014):

$$\{Y_n(x)\}_{n=1}^{\infty} = \left\{(1-x)^{\gamma-1} E_{\gamma,\gamma}\left(\lambda_n (1-x)^{\gamma}\right)\right\}_{n=1}^{\infty}.$$

At the same time, systems of functions $\{X_n(x)\}_{n=1}^{\infty}, \{Y_n(x)\}_{n=1}^{\infty}$ form a biorthogonal system of functions (Aleroev et al. 2013a; Aleroev et al. 2013b). Writing down the expansion of the initial condition function and the source function $f(x)$ in a series according to the eigenfunctions of problem [5.19] in the form:

$$\psi(x) = \sum_{n=1}^{\infty} \psi_n X_n(x), \; f(x) = \sum_{n=1}^{\infty} f_n X_n(x), \tag{5.20}$$

where $\psi_n = \left(\psi(x), Y_n(x)\right)_{L^2(0,1)}$, $f_n = \left(f(x), Y_n(x)\right)_{L^2(0,1)}$, we obtain from [5.12]–[5.14] for finding unknown functions $\upsilon_n(t)$ $(n \in N)$ the following sequence of Cauchy problems:

$$ {}^C D_t^{\alpha_1} \upsilon_n(t) - \kappa_1 \lambda_n \upsilon_n(t) = f_n \qquad (n \in N), \qquad [5.21] $$

$$ \upsilon_n(0) = \psi_n \qquad (n \in N), \qquad [5.22] $$

where indicated:

$$ \psi_n = \varphi_n \left(1 - \tau_r \kappa_2 \lambda_n\right), \ \varphi_n = \left(\varphi(x), Y_n(x)\right)_{L^2(0,1)} \qquad (n \in N). \qquad [5.23] $$

The solution of problems [5.21], [5.22] based on Kilbas et al. (2006) is written in the form:

$$ \upsilon_n(t) = \left(\psi_n + \frac{f_n}{\kappa_1 \lambda_n}\right) E_{\alpha_1}\left(\kappa_1 \lambda_n t^{\alpha_1}\right) - \frac{f_n}{\kappa_1 \lambda_n}, \ (n \in N), \qquad [5.24] $$

where $E_\alpha(z)$ is the one-parameter Mittag–Leffler function (Gorenflo et al. 2014).

Similarly, we seek a solution $u(x,t)$ to problems [5.15]–[5.17] in the form of an expansion in the eigenfunctions $X_n(x)$ of problems [5.19], namely:

$$ u(x,t) = \sum_{n=1}^{\infty} u_n(t) X_n(x) \qquad (x,t) \in \Omega \qquad [5.25] $$

based on [5.15]–[5.17], we obtain the following sequence of Cauchy problems:

$$ {}^C D_{0t}^{\alpha_2} u_n(t) + \omega_n u_n(t) = \frac{1}{\tau_r} \upsilon_n(t) \ \ (n \in N), \qquad [5.26] $$

$$ u_n(0) = \varphi_n \ \ (n \in N), \qquad [5.27] $$

where $\omega_n = \frac{1}{\tau_r} - \kappa_2 \lambda_n$ $(n \in N)$, φ_n are given by relations [5.23], and functions $\upsilon_n(t)$ are determined according to relations [5.24].

The solution of problems [5.26], [5.27] taking into account Kilbas et al. (2006) is written in the form:

$$u_n(t) = \varphi_n E_{\alpha_2}(-\omega_n t^{\alpha_2}) + \frac{1}{\tau_r}\int_0^t (t-\tau)^{\alpha_2 - 1} E_{\alpha_2,\alpha_2}\left(-\omega_n (t-\tau)^{\alpha_2}\right) \upsilon_n(\tau) d\tau$$

$$= \left(\varphi_n + \frac{f_n}{\kappa_1 \tau_r \omega_n \lambda_n}\right) E_{\alpha_2}(-\omega_n t^{\alpha_2}) - \frac{f_n}{\kappa_1 \tau_r \omega_n \lambda_n} + \frac{1}{\tau_r}\left(\psi_n + \frac{f_n}{\kappa_1 \lambda_n}\right) J_n(t), \quad [5.28]$$

where the notation is entered:

$$J_n(t) = \int_0^t (t-\tau)^{\alpha_2 - 1} E_{\alpha_2,\alpha_2}\left(-\omega_n (t-\tau)^{\alpha_2}\right) E_{\alpha_1}(\kappa_1 \lambda_n \tau^{\alpha_1}) d\tau \quad (n \in N). \quad [5.29]$$

Thus, the formal solution of the considered problem [5.8]–[5.10] is determined by relations [5.25], [5.28], [5.29]. It can be shown that if some additional conditions are met, this solution is regular. To do this, assume that the function of the initial condition $\varphi(x)$ and the function of the sources $f(x)$ satisfy the following constraints:

$$\varphi(x), f(x) \in C^2[0,1], \ \varphi(0) = \varphi(1) = \varphi'(0) = 0, \ f(0) = f(1) = f'(0) = 0.$$

Then, as was established by Ali et al. (2021) and Bulavatsky et al. (2022), the following estimates take place:

$$|\varphi_n| \le \frac{C_1}{|\lambda_n|^2}, \ |f_n| \le \frac{C_2}{|\lambda_n|^2} \qquad (C_1, C_2 > 0, n \in N). \quad [5.30]$$

Taking into account the well-known property of the two-parameter Mittag–Leffler function (Gorenflo et al. 2014)

$$\left|E_{\eta,\xi}(z)\right| \le \frac{M_1}{1+|z|} \qquad (z \in \mathbb{C}), \quad [5.31]$$

$$\left(0 < \eta < 2, \mu \le |\arg(z)| \le \pi, \mu \in \left(\frac{\pi\eta}{2}, \min\{\pi, \pi\eta\}\right), \xi \in \mathbb{R}, M_1 = const > 0\right)$$

and taking into account [5.30] together with the estimate of the eigenfunctions $X_n(x)$ $X_n(x)$ in the form (Aleroev et al. 2013a, 2013b; Khasambiev et al. 2014):

$$|X_n(x)| \le \frac{C}{|\lambda_n| x} \qquad (C > 0,\ x > 0,\ n \in N),$$

we obtain the following estimate for the members of the series [5.25]:

$$|u_n(t) X_n(x)| \le \frac{L}{|\lambda_n|^3 x} \qquad (L = const > 0, t > 0, x > 0, n \in N). \qquad [5.32]$$

Since the eigenvalue λ_n in problem [5.19] with $\text{Im}\,(\lambda_n) > 0$ have the following properties (Aleroev et al. 2013a, 2013b; Khasambiev et al. 2014):

a) $|\lambda_k| < |\lambda_{k+1}|$ for $k \ge 1$ $k \ge 1$,

b) for sufficiently large n n and $\arg(\lambda_n) > \frac{\alpha\pi}{2}$, we have:

$$\left|e^{\lambda_n t}\right| < 1, \quad |\lambda_n| \sim O(n^{\gamma}) \quad (\le \gamma \le 2),$$

then taking into account [5.32], we find that the major for the series [5.25] is a convergent generalized harmonic series

$$\sum_{n=1}^{\infty} n^{-3\sigma} \qquad (1 < \sigma < 2).$$

Based on the Weierstrass M-test, the series [5.25] is uniformly convergent on the set $\bar{\Omega}$ and its sum is a continuous function on this set: $u(x,t) \in C\left(\bar{\Omega}\right)$. Similarly, it is possible to establish the convergence of series formed from the corresponding derivatives of partial solutions, for example, for a series of derivatives with respect to a geometric variable x:

$$\sum_{n=1}^{\infty} u_n(t)\, {}^C D_{0x}^{\gamma} X_n(x) = \sum_{n=1}^{\infty} \lambda_n u_n(t) X_n(x), \qquad [5.33]$$

according to [5.32], we have the estimate:

$$\left|\lambda_n u_n(t) X_n(x)\right| \le \frac{M_1}{\left|\lambda_n\right|^2} \qquad (M_1 = const > 0, t > 0, x > 0, n \in N).$$

It follows that series [5.33] is uniformly convergent, since the major for it is a convergent generalized harmonic series of the form:

$$\sum_{n=1}^{\infty} n^{-2\sigma} \qquad (1 < \sigma < 2).$$

Similarly, the convergence of all other series from the corresponding derived partial solutions is proved. Thus, there is a regular solution to the problem under consideration and it is not difficult to establish its unity. Indeed, let $u(x,t)$, $w(x,t)$ be two solutions to this problem and $\bar{u}(x,t) = u(x,t) - w(x,t)$. Then, $\bar{u}(x,t)$ satisfies the equation:

$$(L_1 + \tau_r L_1 L_2)\bar{u}(x,t) = 0, \qquad [5.34]$$

boundary conditions:

$$\bar{u}(0,t) = \bar{u}(1,t) = 0, \ {}^C D_{0x}^{\gamma} \bar{u}(x,t)\Big|_{\substack{x=0,\\ x=1}} = 0 \ (t \ge 0) \qquad [5.35]$$

and initial conditions:

$$\bar{u}(x,0) = 0, \ {}^C D_{0t}^{\alpha_2} \bar{u}\Big|_{t=0} = 0 \ (0 \le x \le 1). \qquad [5.36]$$

Therefore, taking into account relations [5.28] and [5.34]–[5.36], we have:

$$\bar{u}_n(t) = \left(\bar{u}(x,t), Y_n(x)\right)_{L^2(0,1)} = 0 \ (t \in [0,T], \ 0 \le x \le 1, \ n \in N).$$

Given the completeness of the system of functions $\{Y_n(x)\}_{n=1}^{\infty}$ in space $L^2(0,1)$, we obtain $\bar{u}(x,t) = 0$, or $u = w$.

In a separate case $\alpha_1 = \alpha_2 = \alpha$, the obtained solution of the direct problem [5.8]–[5.10] is significantly simplified and takes the form [5.25], where $u_n(t)$ are determined according to the following ratios, which do not contain quadratures:

$$u_n(t) = \left(\varphi_n + \frac{f_n}{\kappa_1 \tau_r \omega_n \lambda_n}\right) E_\alpha(-\omega_n t^\alpha) - \frac{f_n}{\kappa_1 \tau_r \omega_n \lambda_n} + \frac{1}{\tau_r(\omega_n + \kappa_1 \lambda_n)}\left(\psi_n + \frac{f_n}{\kappa_1 \lambda_n}\right) \times$$
$$\times \left[E_\alpha(\kappa_1 \lambda_n t^\alpha) - E_\alpha(-\omega_n t^\alpha)\right] \quad (n \in N) \qquad [5.37]$$

In addition, provided $\kappa_1 = \kappa_2 = \kappa$ in the case of non-locality of the process only by the time variable (corresponding to the value $\gamma \to 2$) and lack of sources ($f = 0$), we obtain the following relations from [5.37] (Bulavatsky 2016):

$$u_n(t) = \varphi_n \left((1 - \tau_r \kappa \lambda_n) E_\alpha(\kappa \lambda_n t^\alpha) + \tau_r \kappa \lambda_n E_\alpha\left(-\left(\frac{1}{\tau_r} - \kappa \lambda_n\right) t^\alpha\right)\right), (n \in N), \quad [5.38]$$

where $\lambda_n = -(n\pi)^2$. In this case, the solution of the problem is determined according to [5.25], [5.38].

If, in particular, we put in relations [5.38] $\alpha \to 1, \gamma \to 2$, then we will have a solution of the corresponding boundary value problem [5.8]–[5.10] within the standard (differential) biparabolic model in the form (Bulavatsky 2016)

$$u(x,t) = 2\sum_{n=1}^{\infty} \varphi_n \left(1 + \tau_r \kappa \tilde{\lambda}_n^2 \left(1 - e^{-\frac{t}{\tau_r}}\right)\right) e^{-\kappa \tilde{\lambda}_n^2 t} \sin(\tilde{\lambda}_n x),$$

where $\varphi_n = \int_0^1 \varphi(x) \sin(\tilde{\lambda}_n x) dx, \quad \tilde{\lambda}_n = n\pi \quad (n \in N).$

In a separate case $\varphi(x) = u_0 = const$, and this solution is taking shape (Bulavatsky et al. 2020):

$$u(x,t) = u_c(x,t) + 4\tau_r \kappa \left(1 - e^{-\frac{t}{\tau_r}}\right) \sum_{n=1,3,5,\ldots}^{\infty} \tilde{\lambda}_n^2 e^{-\kappa \tilde{\lambda}_n^2 t} \sin(\tilde{\lambda}_n x),$$

where $u_c(x,t)$ is the solution of the corresponding boundary value problem with respect to the classical linear diffusion equation:

$$u_c(x,t) = 4 \sum_{n=1,3,5,...}^{\infty} \frac{1}{\tilde{\lambda}_n} e^{-\kappa \tilde{\lambda}_n t} \sin(\tilde{\lambda}_n x).$$

It should be noted that one of the well-known (Bulavatsky 1997; Bulavatsky et al. 2005) interpretations of the biparabolic differential equation defines it as a model equation in the theory of filtration consolidation of water-saturated soil environments (here $u(x,t)$ is the function of excess pressures). The performed numerical experiments show (Bulavatsky et al. 2005, 2022) that the biparabolic mathematical model of the filtration-consolidation process is in some sense intermediate between two well-known models (the classical model by K. Tertsagi and the model taking into account the creep of the soil skeleton by Florin (1961)), since the biparabolic model causes lower pore pressure values than in the Florin model, but higher than in the classical Terzaghi model.

Next, we briefly consider the simplest inverse boundary value problem corresponding to problem [5.8]–[5.10]. The problem is to find a function that depends only on the geometric variable $f(x)$ along with the solution $u(x,t)$ to problems [5.8]–[5.10] in the presence of an additional redefinition condition:

$$u(x,T) = h(x) \qquad x \in (0,1) \tag{5.39}$$

and the following conditions: u is continuous in $\bar{\Omega}$ together with the corresponding derivatives and $f(x) \in C[0,1]$.

We will show in the following that under certain restrictions, there is a unique regular solution of the inverse problem in this formulation.

Suppose that there is an expansion of the function $h(x)$ in a series according to the eigenfunctions of the Sturm–Liouville problem [5.19]:

$$h(x) = \sum_{n=1}^{\infty} h_n X_n(x) \qquad \left(x \in [0,1]\right),$$

where:

$$h_n = \left(h(x), Y_n(x)\right)_{L^2(0,1)} \qquad (n \in N).$$

Then, from condition [5.39], we obtain the relation $u_n(T) = h_n \quad (n \in N)$, from which, taking into account [5.28], [5.29], we have:

$$f_n = \frac{\kappa_1 \tau_r \lambda_n \left(h_n - \tau_r^{-1} \psi_n J_n(T) - \varphi_n E_{\alpha_2}\left(-\omega_n T^{\alpha_2}\right) \right)}{J_n(T) - T^{\alpha_2} E_{\alpha_2,\alpha_2+1}\left(-\omega_n T^{\alpha_2}\right)} \quad (n \in N), \qquad [5.40]$$

where $J_n(T)$ is determined according to [5.29].

In this way, the source function is sought $f(x)$ is determined according to relations [5.20], [5.40] and, if we subordinate the function $h(x)$ with additional restrictions similar to those previously accepted for the function $\varphi(x)$, then taking into account asymptotics:

$$h_n = O\left(\frac{1}{\left|\lambda_n\right|^2} \right) (n \to \infty),$$

and estimates [5.31], we obtain:

$$\left|f_n\right| \le \frac{M_2}{\left|\lambda_n\right|} (M_2 > 0, n \in N).$$

So, for the members of the second row in relations [5.20], we have the estimate:

$$\left|f_n X_n(x)\right| \le \frac{M_*}{\left|\lambda_n\right|^2 x} \qquad (M_* = const > 0, x > 0, n \in N).$$

On the basis of the Weierstrass M-test, by comparison with the convergent generalized harmonic series $\sum_{n=1}^{\infty} n^{-2\sigma}$ $(1 < \sigma < 2)$, hence the absolute and uniform convergence of the series [5.20] for $f(x)$. So $f(x) \in C[0,1]$. Since in the present case:

$$\left|u_n(t) X_n(x)\right| \le \frac{M}{\left|\lambda_n\right|^2} \qquad (M = const > 0, \ n \in N),$$

then, based on the Weierstrass M-test, we have $u(x,t) \in C\left(\bar{\Omega}\right)$. This established the existence of a regular solution of the inverse problem.

5.3. Generalization of the model equation based on Hilfer-type fractional derivatives

An increase in the degree of generality of the non-classical mathematical model of transfer, which is based on relations [5.6], [5.7], is achieved by using in the equations of the model, instead of Caputo derivatives, their well-known generalization-fractional Hilfer derivatives (Hilfer 2000). Note that if the conditions are met $\alpha \in (0,1)$, $\beta \in [0,1]$, $u(t) \in C^1[0,T]$, $\gamma = \alpha + \beta - \alpha\beta$, the Hilfer fractional derivative of the order of the function type is defined as follows (Hilfer 2000; Sandev et al. 2019):

$$D_{0+}^{\alpha,\beta} u(t) = I_{0+}^{\beta(1-\alpha)} \frac{d}{dt} I_{0+}^{1-\gamma} u(t) = I_{0+}^{\beta(1-\alpha)} D_{0+}^{\gamma} u(t), \ \forall t \in [0,T].$$

Here, $D_{0+}^{\gamma} u(t)$ is the fractional Riemann–Liouville derivative of order γ from the function $u(t)$, $I_{0+}^{\nu} u(t)$ is the fractional Riemann–Liouville integral of order $\nu > 0$ from the function $u(t) \in C^1[0,T]$, which is defined by the ratio (Kilbas et al. 2006):

$$I_{0+}^{\nu} u(t) := \frac{1}{\Gamma(\nu)} \int_0^t \frac{u(\tau)}{(t-\tau)^{1-\nu}} d\tau, \quad t > 0,$$

where $\Gamma(\cdot)$ is the Euler gamma function (Abramovitz et al. 1965).

Taking into account the ratios $D_{0+}^{\alpha,0} = D_{0+}^{\alpha}$, $D_{0+}^{\alpha,1} = I_{0+}^{1-\alpha} \frac{d}{dt} = {}^{C}D_{0+}^{\alpha}$, we come to the conclusion that the Hilfer fractional derivative can be considered as an interpolant on the parameter $\beta \in [0,1]$ with respect to Caputo and Riemann–Liouville derivatives of the same order (Hilfer 2000; Sandev et al. 2019).

Note that the solution of a Cauchy-type problem for a linear equation with the Hilfer derivative is of the form:

$$D_{0+}^{\alpha,\beta} u(t) + \lambda u(t) = f(t),$$

$$\lim_{t \to 0+} I_{0+}^{(1-\alpha)(1-\beta)} u(t) = \varsigma_0$$

($\varsigma_0 = const,\ \lambda > 0,\ u, f \in C^1([0,T],\mathbb{R})$), according to works Sandev et al. (2019) and Hilfer (2000) is written as follows:

$$u(t) = \varsigma_0 t^{(1-\alpha)(\beta-1)} E_{\alpha,\alpha+\beta(1-\alpha)}(-\lambda t^{\alpha}) + \int_0^t (t-\tau)^{\alpha-1} E_{\alpha,\alpha}(-\lambda(t-\tau)^{\alpha}) f(\tau) d\tau, \quad [5.41]$$

where $E_{\alpha,\beta}(\cdot)$ is the two-parameter Mittag–Leffler function (Gorenflo et al. 2014).

Let us define the operators L_1, L_2 in model equation [5.6] as:

$$L_1 = D_{0t}^{\alpha_1,\beta_1} - \kappa_1 \, {}^C D_{0x}^{\gamma},\ L_2 = D_{0t}^{\alpha_2,\beta_2} - \kappa_2 \, {}^C D_{0x}^{\gamma}, \quad [5.42]$$

where $D_{0t}^{\alpha_1,\beta_1}$ and $D_{0t}^{\alpha_2,\beta_2}$ are operators of Hilfer fractional derivatives with respect to a variable t, ${}^C D_{0x}^{\gamma}$ is the Caputo fractional derivative operator with respect to x, and $0 < \alpha_1, \alpha_2 \le 1$, $0 \le \beta_1, \beta_2 \le 1$, $1 < \gamma \le 2$.

Then, relations [5.6], [5.42] define a new fractional-differential generalization of the biparabolic evolution equation [5.1]). Indeed, at $\beta_1 = \beta_2 = 1$ from [5.6], [5.42] we obtain the fractional-differential analogue of the biparabolic equation in Caputo derivatives, and with the additional assumption $\alpha_1, \alpha_2 \to 1$, $\gamma \to 2$ from relations [5.6], [5.42], we have the classical biparabolic equation [5.1].

Within the framework of the mathematical model, taking into account the time–space non-locality of the process, which is based on relations [5.6], [5.42], consider the one-dimensional boundary value problem for a finite interval $0 < x < 1$ by the geometric variable. The problem is reduced to finding in the area Ω the solution of the equation

$$\left(D_{0t}^{\alpha_1,\beta_1} - \kappa_1 \, {}^C D_{0x}^{\gamma}\right) u(x,t) + \tau_r \left(D_{0t}^{\alpha_1,\beta_1} - \kappa_1 \, {}^C D_{0x}^{\gamma}\right) \left(D_{0t}^{\alpha_2,\beta_2} - \kappa_2 \, {}^C D_{0x}^{\gamma}\right) u(x,t) = f(x) \quad [5.43]$$

under the following conditions:

$$u(0,t) = u(1,t) = 0,\ {}^{C}D_{0x}^{\gamma}u(x,t)\Big|_{\substack{x=0,\\ x=1}} = 0\ (t \geq 0), \tag{5.44}$$

$$\begin{aligned} &I_t^{1-\delta_1}u(x,t)\Big|_{t=0} = \varphi_1(x),\ I_t^{1-\delta_2}u(x,t)\Big|_{t=0} = \varphi_2(x),\\ &I_t^{1-\delta_1}D_{0t}^{\alpha_2,\beta_2}u(x,t)\Big|_{t=0} = 0 \quad (0 \leq x \leq 1), \end{aligned} \tag{5.45}$$

where $\delta_i = \alpha_i + \beta_i(1-\alpha_i)$ $(i = 1,2)$, $\varphi_i(x)$ $(i = 1,2)$ are given functions of the initial conditions, and $u(x,t)$ is a function to be defined.

Let us consider a function $\upsilon(x,t)$ according to relation [5.11], where $L_2u(x,t) = D_{0t}^{\alpha_2,\beta_2}u(x,t) - \kappa_2\, {}^{C}D_{0x}^{\gamma}u(x,t)$. Then, the considered boundary value problem [5.43]–[5.45] is reduced to finding the solutions of the following boundary value problems with respect to unknown functions $\upsilon(x,t), u(x,t)$:

$$L_1\upsilon(x,t) = f(x), \tag{5.46}$$

$$\upsilon(0,t) = \upsilon(1,t) = 0\ (t \geq 0), \tag{5.47}$$

$$I_t^{1-\delta_1}\upsilon(x,t)\Big|_{t=0} = \psi(x),\ (0 \leq x \leq 1), \tag{5.48}$$

$$\tau_r L_2u(x,t) + u(x,t) = \upsilon(x,t), \tag{5.49}$$

$$u(0,t) = u(1,t) = 0\ (t \geq 0), \tag{5.50}$$

$$I_t^{1-\delta_2}u(x,t)\Big|_{t=0} = \varphi_2(x)\ (0 \leq x \leq 1), \tag{5.51}$$

where $\psi(x) = \varphi_1(x) - \tau_r\kappa_2\, {}^{C}D_{0x}^{\gamma}\varphi_1(x)$, and operators L_1, L_2 are determined according to relations [5.42].

Finding a solution to problem [5.46]–[5.48] in the form of expansion [5.18] by the eigenfunctions of problem [5.19], we obtain a sequence of Cauchy-type problems:

$$D_t^{\alpha_1,\beta_1}\upsilon_n(t)-\kappa_1\lambda_n\upsilon_n(t)=f_n,\; I_t^{1-\delta_1}\upsilon_n(t)\Big|_{t=0}=\psi_n\ (n\in N), \qquad [5.52]$$

where $\psi_n=\varphi_n^{(1)}\left(1-\tau_r\kappa_2\lambda_n\right)$, $\varphi_n^{(1)}=\left(\varphi_1(x),Y_n(x)\right)_{L^2(0,1)}$, $(n\in N)$.

Based on [5.41], the solution to problems [5.52] will be written in the form:

$$\upsilon_n(t)=f_n t^{\alpha_1}E_{\alpha_1,\alpha_1+1}(\kappa_1\lambda_n t^{\alpha_1})+\psi_n t^{\delta_1-1}E_{\alpha_1,\delta_1}(\kappa_1\lambda_n t^{\alpha_1})\ (n\in N). \qquad [5.53]$$

Similarly, looking for a solution to problem [5.49]–[5.51] in the form [5.25], we obtain the following sequence of Cauchy-type problems:

$$D_t^{\alpha_2,\beta_2}u_n(t)+\omega_n u_n(t)=\frac{1}{\tau_r}\upsilon_n(t),\; I_t^{1-\delta_2}u_n(t)\Big|_{t=0}=\varphi_n^{(2)}\ (n\in N), \qquad [5.54]$$

where $\omega_n=\dfrac{1}{\tau_r}-\kappa_2\lambda_n$, $\varphi_n^{(2)}=\left(\varphi_2(x),Y_n(x)\right)_{L^2(0,1)}$ $(n\in N)$, and functions $\upsilon_n(t)$ $(n\in N)$ are determined according to [5.53].

Solving problems [5.54], taking into account relation [5.41], we find the solution of the boundary value problem [5.43]–[5.45] in the form of a series:

$$u(x,t)=$$

$$\sum_{n=1}^{\infty}X_n(x)\left[\varphi_n^{(2)}t^{\delta_2-1}E_{\alpha_2,\delta_2}(-\omega_n t^{\alpha_2})+\frac{1}{\tau_r}\int_0^t(t-\tau)^{\alpha_2-1}E_{\alpha_2,\alpha_2}(-\omega_n(t-\tau)^{\alpha_2})\upsilon_n(\tau)d\tau\right]$$

$$=\sum_{n=1}^{\infty}X_n(x)\left[\varphi_n^{(2)}t^{\delta_2-1}E_{\alpha_2,\delta_2}(-\omega_n t^{\alpha_2})+\frac{1}{\tau_r}J_n^*(t)\right],\ (x,t)\in\Omega, \qquad [5.55]$$

where:

$$J_n^*(t) = \int_0^t (t-\tau)^{\alpha_2 - 1} E_{\alpha_2,\alpha_2}(-\omega_n (t-\tau)^{\alpha_2})$$

$$\left[f_n \tau^{\alpha_1} E_{\alpha_1,\alpha_1+1}(\kappa_1 \lambda_n \tau^{\alpha_1}) + \psi_n \tau^{\delta_1 - 1} E_{\alpha_1,\delta_1}(\kappa_1 \lambda_n \tau^{\alpha_1}) \right] d\tau$$

$$= f_n \int_0^t (t-\tau)^{\alpha_2-1} \tau^{\alpha_1} E_{\alpha_2,\alpha_2}(-\omega_n (t-\tau)^{\alpha_2}) E_{\alpha_1,\alpha_1+1}(\kappa_1 \lambda_n \tau^{\alpha_1}) d\tau +$$

$$+\psi_n \int_0^t (t-\tau)^{\alpha_2-1} \tau^{\delta_1-1} E_{\alpha_2,\alpha_2}(-\omega_n (t-\tau)^{\alpha_2}) E_{\alpha_1,\delta_1}(\kappa_1 \lambda_n \tau^{\alpha_1}) d\tau. \qquad [5.56]$$

Thus, the formal solution of problem [5.43]–[5.45] is determined by relations [5.55], [5.56]. It can be shown that, subject to some additional restrictions on the functions of the initial conditions and the source function, the specified solution is regular. To do this, let us assume that the following restrictions are fulfilled: $\varphi_1(x), \varphi_2(x), f(x) \in C^2[0,1]$, $\varphi_i(0) = \varphi_i(1) = \varphi_i'(0) = 0, (i = 1,2)$, $f(0) = f(1) - f'(0) = 0$. Then, fair following estimates (Ali et al. 2021; Bulavatsky et al. 2022) are:

$$\left|\varphi_n^{(i)}\right| \le \frac{C_i}{|\lambda_n|^2} \quad (i = 1,2), \; \left|f_n\right| \le \frac{C_3}{|\lambda_n|^2} (C_1, C_2, C_3 > 0, \; n \in N). \qquad [5.57]$$

Taking into account [5.57] and estimates of the form [5.31], we have on the basis of the Weierstrass M-test that the series [5.55], [5.56] is uniformly convergent on the set $\Omega_\varepsilon = [0,1] \times (\varepsilon, T]$ for any $\varepsilon > 0$, and $u(x,t)$ is a continuous function. Similarly to what was explained earlier, it is possible to establish the convergence of the series formed from the corresponding derivatives of the partial solutions. Therefore, there is a regular solution to the problem under consideration, and it is not difficult to see that it is unique.

5.4. Fractional–differential analogue of the biparabolic evolution equation with Caputo and Caputo–Fabrizio derivatives

The fractional-differential analogues of the biparabolic evolution equation discussed above contained fractional derivative operators with singular kernels.

Below, we will define a fractional-differential analogue of the biparabolic equation [5.1], which simultaneously contains operators of fractional derivatives, both with singular and non-singular kernels (in particular, Caputo derivatives (Kilbas et al. 2006) and Caputo–Fabrizio derivatives (Caputo et al. 2015, 2016)).

The derivative of the Caputo–Fabrizio function is well established:

$$f(t) \in H^1(0, a), \ a > 0, \ f(0) = 0$$

determined by the following relation (Caputo et al. 2015, 2016):

$$^{CF}D_{0t}^{\alpha} f(t) = \frac{1}{1-\alpha} \int_0^t f'(s)\, e^{-\frac{\alpha}{1-\alpha}(t-s)}\, ds,$$

where $^{CF}D_{0t}^{\alpha}$ is the derivative operator, $\alpha \in [0,1]$ is its order.

It should be noted that attention to the Caputo–Fabrizio derivative is due to its application to the modeling of dynamic processes in viscoelastic materials, thermal environments, electromagnetic systems, geohydrodynamics, etc. (Caputo et al. 2015, 2016; Alkahtani 2016; Al-Salti et al. 2016; Bulavatsky 2023).

Below, the solution of the following Cauchy problem for the equation with the Caputo–Fabrizio derivative is used:

$$^{CF}D_{0t}^{\alpha} u(t) - \lambda u(t) = f(t), u(0) = u_0 \ (0 \le t \le T), \qquad [5.58]$$

where $f(t)$ is a given function, $f(t) \in C(0,\infty)$, $f(0) = -\lambda u_0$.

As shown in Al-Salti et al. (2016), a single continuous solution of this problem is determined by the relation:

$$u(t) = \frac{1-\alpha}{1-\lambda(1-\alpha)} f(t) + \frac{\alpha}{[1-\lambda(1-\alpha)]^2} \int_0^t f(s)\, e^{\frac{\alpha\lambda}{1-\lambda(1-\alpha)}(t-s)}\, ds +$$

$$+ \frac{u_0}{1-\lambda(1-\alpha)} e^{\frac{\alpha\lambda}{1-\lambda(1-\alpha)}t} \left(\lambda \neq \frac{1}{1-\alpha} \right). \qquad [5.59]$$

(For $\lambda = \dfrac{1}{1-\alpha}$, the corresponding solution of problem [5.58] is also given in Al-Salti et al.(2016)).

A fractional-differential analogue of the biparabolic equation [5.1], which simultaneously contains Caputo derivatives of the order α_1 and Caputo–Fabrizio order α_2, by the time variable is obtained using the operators in [5.1] in the form of relations:

$$L_1 = {}^{C}D_{0t}^{\alpha_1} - \kappa_1 \, {}^{C}D_{0x}^{\gamma}, \; L_2 = {}^{CF}D_{0t}^{\alpha_2} - \kappa_2 \, {}^{C}D_{0x}^{\gamma} \; (0 < \alpha_1, \alpha_2 \leq 1, \; 1 < \gamma \leq 2). \quad [5.60]$$

Taking [5.60] concludes that the dynamics of evolutionary processes is non-local in space and time based on the fractional-differential analogue biparabolic equation with Caputo and Caputo–Fabrizio derivatives, which is described by the following model equation:

$$\begin{gathered} \left({}^{C}D_{0t}^{\alpha_1} - \kappa_1 \, {}^{C}D_{0x}^{\gamma}\right) u(x,t) + \tau_r \left({}^{C}D_{0t}^{\alpha_1} - \kappa_1 \, {}^{C}D_{0x}^{\gamma}\right) \\ \left({}^{CF}D_{0t}^{\alpha_2} - \kappa_2 \, {}^{C}D_{0x}^{\gamma}\right) u(x,t) = f(x) \end{gathered}, \quad [5.61]$$

where $0 < \alpha_1, \alpha_2 \leq 1, \; 1 < \gamma \leq 2$.

The corresponding boundary conditions for the finite interval of unit length will be taken as:

$$u(0,t) = u(1,t) = 0, \; {}^{C}D_{0x}^{\gamma} u(x,t)\Big|_{\substack{x=0, \\ x=1}} = 0 \; (t \geq 0), \quad [5.62]$$

and the initial conditions are as follows:

$$u(x,0) = \varphi(x), \; {}^{CF}D_{0t}^{\alpha_2} u(x,t)\Big|_{t=0} = 0 \; (0 \leq x \leq 1), \quad [5.63]$$

where $\varphi(x)$ is a known function of the initial conditions, $f(x)$ is a source function, and $u(x,t)$ is the function we seek.

To solve the boundary value problem [5.61]–[5.63], similarly to the above, we introduce a new unknown function $\upsilon(x,t)$ using the relation:

$$\upsilon(x,t) = \tau_r L_2 u(x,t) + u(x,t) = \tau_r \left({}^{CF}D_{0t}^{\alpha_2} - \kappa_2 \, {}^{C}D_{0x}^{\gamma} \right) u(x,t) + u(x,t).$$

Then, problem [5.61]–[5.63] is reduced to the solution of the following boundary value problems with respect to unknown functions $\upsilon(x,t), u(x,t)$: boundary value problem [5.12]–[5.14] with respect to the $\upsilon(x,t)$ and the boundary value problem with respect to the function $u(x, t)$:

$$\tau_r \left({}^{CF}D_{0t}^{\alpha_2} - \kappa_2 \, {}^{C}D_{0x}^{\gamma} \right) u(x,t) + u(x,t) = \upsilon(x,t), \qquad [5.64]$$

$$u(0,t) = u(1,t) = 0, \; (t \ge 0), \qquad [5.65]$$

$$u(x,0) = \varphi(x), \; (0 \le x \le 1). \qquad [5.66]$$

Solution of the problem with respect to the function $\upsilon(x,t)$ written above in the form [5.18], [5.24]. Looking for a solution to problem [5.64]–[5.66] in the form [5.25], we reduce it to the solution of the following problems:

$${}^{CF}D_{0t}^{\alpha_2} u_n(t) + \omega_n u_n(t) = \frac{1}{\tau_r} \upsilon_n(t) \;\; (n \in N), \qquad [5.67]$$

$$u_n(0) = \varphi_n \;\; (n \in N), \qquad [5.68]$$

where $\omega_n = \dfrac{1}{\tau_r} - \kappa_2 \lambda_n$ $(n \in N)$, and values φ_n are determined according to [5.23].

According to relation [5.59], the solution of problems [5.67], [5.68] is written in the form:

$$u_n(t) = \frac{1}{\tau_r \nu_n}\left((1-\alpha_2)\upsilon_n(t) + \frac{\alpha_2}{\nu_n}\int_0^t \upsilon_n(\tau)\, e^{-\frac{\alpha_2\omega_n}{\nu_n}(t-\tau)} d\tau + \tau_r \varphi_n e^{-\frac{\alpha_2\omega_n}{\nu_n}t} \right)$$

$$= \frac{1}{\tau_r \nu_n}\left\{ (1-\alpha_2)\left[\left(\psi_n + \frac{f_n}{\kappa_1\lambda_n}\right)E_{\alpha_1}\left(\kappa_1\lambda_n t^{\alpha_1}\right) - \frac{f_n}{\kappa_1\lambda_n}\right] + \right.$$

$$\left. + \frac{\alpha_2}{\nu_n}\int_0^t \left[\left(\psi_n + \frac{f_n}{\kappa_1\lambda_n}\right)E_{\alpha_1}\left(\kappa_1\lambda_n \tau^{\alpha_1}\right) - \frac{f_n}{\kappa_1\lambda_n}\right] e^{-\frac{\alpha_2\omega_n}{\nu_n}(t-\tau)} d\tau + \tau_r \varphi_n e^{-\frac{\alpha_2\omega_n}{\nu_n}t} \right\}$$

$$= \frac{1}{\tau_r \nu_n}\left\{ \left(\psi_n + \frac{f_n}{\kappa_1\lambda_n}\right)\left[(1-\alpha_2)E_{\alpha_1}\left(\kappa_1\lambda_n t^{\alpha_1}\right) + \frac{\alpha_2}{\nu_n}G_n(t)\right] \right.$$

$$\left. - \frac{f_n}{\kappa_1\lambda_n\omega_n}\left[\nu_n - \exp\left(-\frac{\alpha_2\omega_n}{\nu_n}t\right)\right] + \varphi_n\tau_r \exp\left(-\frac{\alpha_2\omega_n}{\nu_n}t\right)\right\}, \qquad [5.69]$$

where:

$$\nu_n = 1 + (1-\alpha_2)\omega_n,\ G_n(t) = \int_0^t E_{\alpha_1}(\kappa_1\lambda_n\tau^{\alpha_1})\exp\left(-\frac{\alpha_2\omega_n}{\nu_n}(t-\tau)\right)d\tau. \qquad [5.70]$$

(Note that at $\alpha_2 \to 1$ relations [5.69], [5.70] and [5.28], [5.29] coincide).

Thus, the formal solution of the considered problem is determined by relations [5.25], [5.69], [5.70]. In the case of fulfillment of previously imposed restrictions on functions $f(x)$ and $\varphi(x)$, we have the following estimate for the members of the series [5.25], based on [5.69], [5.70]:

$$|u_n(t)X_n(x)| \le \frac{M_{**}}{|\lambda_n|^4 x} \qquad (M_{**} = const > 0,\ t > 0,\ x > 0,\ n \in N).$$

Then, on the basis of the Weierstrass M-test, we conclude that the series [5.25], [5.69], [5.70] converge uniformly on the set $\bar{\Omega}$, regarding the continuity of the solution $u(x,t)$ in problems [5.61]–[5.63] on this set.

5.5. Conclusions

The work is devoted to the introduction of non-local in space–time fractional–differential analogues with respect to the known biparabolic evolutionary equation (Fushchich et al. 1990), which is the basis of the non-classical mathematical model for describing the dynamics of thermal and diffusion processes and allows us to describe these processes more adequately compared to the model based on the classical parabolic type equation. These fractional-differential analogues are based on the use of derivatives such as Caputo, Caputo–Fabrizio and Hilfer.

Statements of some boundary value problems have been made, and closed solutions of these problems have been obtained. The conditions of existence of regular solutions of these problems in relation to the proposed fractional-differential model equations with space–time non-locality, are given.

5.6. References

Abramovitz, M. and Stegun, I.A. (1965). *Handbook of Mathematical Functions*. Dover, New York.

Al-Salti, N., Karimov, E., Kerbal, S. (2016). Boundary-value problem for fractional heat equation involving Caputo-Fabrizio derivative. *NTMSCI*, 4, 79–89. doi: 10.20852/ntmsci.2016422308.

Aleroev, T.S., Kirane, M., Tang, Y.-F. (2013a). Boundary-value problems for differential equations of fractional order, *J. Math. Sci.*, 194(5), 499–512.

Aleroev, T.S., Kirane, M., Malik, S.A. (2013b). Determination of source term for a time fractional diffusion equation with an integral type over-determining condition, *Electronic J. Diff. Eq.*, 270, 1–16.

Ali, M., Aziz, S., Malik, S.A. (2021). On the recoveri of the time dependent diffusion coefficient for a space fractional diffusion equation, *Ànal. Math. Phys.*, 11, 103–123. doi: 10.1007/s13324-021-00537-w.

Alkahtani, B.S.T. (2016). Model of heat with Caputo-Fabrizio derivative with fractional order. *J. Comput. Theor. Nanosci*, 13(5), 2994–2999.

Bulavatsky, V.M. (1997). Biparabolic mathematical model of the filtration consolidation problem. *Dopov. NAN Ukrainy*, 8, 13–17.

Bulavatsky, V.M. (2016). Fractional differential analog of biparabolic evolution equation and some of its applications. *Cybern. Syst. Analysis*, 52(5), 737–747. doi: 10.1007/s10559-016-9875-5.

Bulavatsky, V.M. (2019). Some nonlocal boundary-value problems for the biparabolic evolution equation and its fractional-differential analog. *Cybern. Syst. Analysis*, 55(5), 796–804. doi: 10.1007/s10559-019-00190-z.

Bulavatsky, V.M. (2023). Boundary-value problems of fractional-differential consolidation dynamics for the model with the Caputo-Fabrizio derivative. *Cybern. Syst. Analysis*, 59(4), 651–659. doi: 10.1007/s10559-023-00600-3.

Bulavatsky, V.M. and Bohaienko, V.O. (2020). Some consolidation dynamics problems within the framework of the biparabolic mathematical model and its fractional-differential analog. *Cybern. Syst. Analysis*, 56(5), 770–783. doi: 10.1007/s10559-020-00298-7.

Bulavatsky, V.M. and Bohaienko, V.O. (2022). Boundary-value problems of space-time fractional differential filtration dynamics in fractured-porous media. *Cybern. Syst. Analysis*, 58(3), 358–375. doi: 10.1007/s10559-022-00468-9.

Bulavatsky, V.M., Krivonos, Yu.G., Skopetsky, V.V. (2005). *Nonclassical Mathematical Models of Heat and Mass Transfer Processes*. Naukova Dumka, Kyiv (in Ukrainian).

Caputo, M. and Fabrizio, M. (2015). A new definition of fractional derivative without singular kernel. *Progress Fractional Differ. Appl.*, 1(2), 73–85.

Caputo, M. and Fabrizio, M. (2016). Applications of new time and spatial fractional derivatives with exponential kernels. *Progress Fractional Differ. Appli.*, 2(1), 1–15.

Djrbashian, M.M. (1993). *Harmonic Analysis and Boundary-Value Problems in the Complex Domain*, Springer Basel AG, Basel.

Florin, V.A. (1961). *Fundamentals of Soil Mechanics Volume 2*. Gosstroiizdat, Moscow (in Russian).

Fushchich, V.I., Galitsyn, A.S., Polubinskii, A.S. (1990). A new mathematical model of heat conduction processes, *Ukr. Math. J.*, 42(2), 210–216.

Gorenflo, R., Kilbas, A.A., Mainardi, F., Rogosin, S.V. (2014). *Mittag-Leffler Functions, Related Topics and Applications*. Springer Verlag, Berlin-Heidelberg.

Hilfer, R. (2000). Fractional time evolution. In *Applications of Fractional Calculus in Physics*, Hilfer, R. (ed.). World Scientific, Singapore, 87–130.

Hilfer, R., Luchko, Y., Tomovski, Z. (2009). Operational method for the solution of fractional differential equations with generalized Riemann–Liouville fractional derivatives. *Fraction Calculus Appl. Analysis*, 12(3), 299–318.

Khasambiev M.V. and Aleroev, T.S. (2014). Boundary-value problem for one-dimensional fractional differential advection-diffusion equation, *Vestn. Mos. Gos. Stroit. Univ.*, 6, 71–76.

Kilbas, A.A., Srivastava, H.M., Trujillo, J.J. (2006). *Theory and Applications of Fractional Differential Equations*. Elsevier, Amsterdam.

Lyashko, S.I. (1995). Numerical solution of pseudoparabolic equations. *Cybernet. Systems. Anal.*, 31(5). 718–722. doi: 10.1007/BF02366325.

Lyashko, S.I. and Nomirovskii, D.A. (2003). Generalized solutions and optimal controls in systems describing the dynamics of a viscous stratified fluid. *Differ. Equ.*, 39(5), 90–98. doi: 10.1023/A.

Lyashko, S.I., Klyushin, D.A., Palienko, L.I. (2000). Simulation and generalized optimization in pseudohyperbolical systems. *J. Autom. Inf. Sci.*, 32(5). 108–117. doi: 10.1615/JAutomatInfScien.v32.i5.80.

Lyashko, S.I., Nomirovskii, D.A., Sergienko T.I. (2001). Trajectory and final controllability in hyperbolic and pseudohyperbolic systems with generalized actions. *Cybernet. Systems Anal.*, 37(5), 756–763. doi: 10.1023/A.

Oleinik, A.Ya. (1981). *Drainage Geohydrodynamics*. Naukova Dumka, Kyiv (in Russian).

Polubarinova-Kochina, P.Ya., Pryazhinskaya, V.G., Emikh, V.N. (1969). *Mathematical Methods in Irrigation*. Nauka, Moscow (in Russian).

Samko, S.G., Kilbas, A.A., Marichev, O.I. (1987). *Fractional Integrals and Derivatives and Some of their Applications*. Nauka i Tekhnika, Minsk (in Russian).

Sandev, T. and Tomovsky, Z. (2019). *Fractional Equations and Models. Theory and Applications*. Springer, Cham.

6

Optimal Control for Integro-differential Systems of Hyperbolic Type

A.V. ANIKUSHYN[1], Kh.M. HRANISHAK[1], V.S. LYASHKO[2] and O.S. BONDAR[1]

[1] *Taras Shevchenko National University of Kyiv, Ukraine*

[2] *V.M. Glushkov Institute of Cybernetics of the NAS of Ukraine, Kyiv, Ukraine*

6.1. Introduction

Integro-differential equations of hyperbolic type have long been a standard object of study in pure and applied mathematics and often arise in the investigation of processes in viscoelastic media.

It is well known that the behavior of some viscoelastic materials (polymers, suspensions, emulsions) exhibit memory properties. Therefore, the forces arising within the body depend not only on the current values of deformation and/or velocity gradient but also on the entire temporal history of motion. Typically, memory fades over time: changes that occurred in the recent past have a greater impact on the current stress than those that happened in the distant past.

Based on these arguments, many models for viscoelastic materials lead to equations of motion that have the form of a linear hyperbolic differential equation perturbed by a dissipative integral term of the Volterra type, with a non-negative, decaying convolution kernel (Cannarsa and Sforza 2011).

Computational Methods and Mathematical Modeling in Cyberphysics and Engineering Applications 2,
coordinated by Dmitri KOROLIOUK, Sergiy LYASHKO and Nikolaos LIMNIOS.

Examples of viscoelastic materials include, but are not limited to, amorphous polymers, semi-crystalline polymers, biopolymers, metals at very high temperatures and bituminous materials.

According to Lakes (1998), viscoelastic materials include discs in the human spine, skin tissue, wood, polymer foams used in seating, etc. Materials that behave elastically at room temperature often acquire significant viscoelastic properties when heated.

For example, this happens with metal turbine blades in jet engines. Viscoelasticity also plays a role in the movement and behavior of tectonic plates that float and move independently on the Earth's mantle and are responsible for earthquakes, volcanoes, etc (Kelly 2013).

Detailed physical justifications for such models are described, for instance, in the monographs (Renardy et al. 1987; Fabrizio and Morro 1992).

In this chapter, we consider a model that describes the evolution of a homogeneous isotropic linear viscoelastic solid occupying a (bounded) volume Ω at rest. The unknown function u represents the displacement relative to the initial configuration (Conti et al. 2008).

In the cylindrical domain $(x,t)\in Q=\Omega\times(0,T)$, where $\Omega\subset R^n$ is a bounded simply connected domain with a regular boundary $\partial\Omega$, we consider the initial-boundary value problem

$$Lu=f, \qquad [6.1]$$

for the hyperbolic integro-differential operator

$$Lu:=u_{tt}+Au+Bu, \qquad [6.2]$$

with homogeneous Dirichlet initial and boundary conditions

$$u\,|_{\partial\Omega}=0, \qquad [6.3]$$

$$u\,|_{t=0}=u_t\big|_{t=0}=0. \qquad [6.4]$$

Here, the operator A is independent of the variable t and is defined by the second-order differential expression

$$Au = -\sum_{i,j=1}^{n} \left(a_{ij} u_{x_j}\right)_{x_i} + \sum_{i=1}^{n} a_i u_{x_i} + au, \qquad [6.5]$$

while the Volterra-type integral operator B takes the form

$$Bu = \int_0^t \sum_{i=1}^{n} K_i(t,\tau) u_{x_i x_i}(x,\tau) d\tau. \qquad [6.6]$$

Delay and memory effects widely arise in a variety of areas of science and engineering, in particular, physics, biology, chemistry, ecology, material engineering, mechanism design, etc. Such phenomena are commonly modeled by ordinary and partial delay differential equations. To study these problems mathematically, a new area of analysis and control of delay differential equations has emerged (Anikushyn and Pokojovy 2017, 2019; Demchenko et al. 2019). Over the past decades, numerous problems' formulations for linear and nonlinear integro-differential hyperbolic-type equations and their systems have massively drawn attention in various communities (Anikushyn 2024). In particular, the literature contains numerous studies that are closely related to the formulation of [6.1]–[6.6].

For instance, in the works of Loretti and Sforza (2010), an integro-differential equation of hyperbolic type is considered:

$$u_{tt}(t,x) - u_{xx}(t,x) + \beta \int_0^t e^{-\eta(t-s)} u_{xx}(s,x) ds = 0, \quad t \in (0,T), x \in (0,\pi) \qquad [6.7]$$

with zero initial conditions: $u(0,x) = u_t(0,x) = 0, x \in (0,\pi)$ and boundary conditions:

$$u(t,x) = \begin{cases} 0, & x = 0, \\ g(t), & x = \pi, \end{cases}$$

where $g(t)$ is the control function. The cited work addresses the problem of determining the control g, for which the weak solution of equation [6.7], satisfying the given initial-boundary conditions, meets the conditions at the final point:

$$u(T,x) = u_0(x), u_t(T,x) = u_1(x), x \in (0,\pi).$$

The proof uses harmonic analysis and Hilbert's uniqueness method.

In Diagana (2014), the integro-differential equation is studied:

$$u_{tt} + Bu_t + Au = \int_{-\infty}^{t} C(t-s)u(s)ds + f(t,u),$$

where A, B are the certain linear self-adjoint operators. By employing methods of operator theory and Schauder's fixed point principle, the existence of some generalized solutions is established.

In Engler (1991), the problem

$$u_{tt}(x,t) = \operatorname{div}_x\left(g(\nabla u(x,t)) + \int_0^{\infty} a(s)h(\nabla u(x,t), \nabla u(x,t-s))ds \right) + f(x,t),$$

is studied with initial-boundary conditions

$$u(x,t)\,|_{\partial\Omega} = 0, u(x,0) = u_0, u_t(x,0) = u_1.$$

The author constructs global weak solutions to the specified equation. It is shown that disturbances from the equilibrium states propagate with limited speed, the smoothing effects of the solution operator are identified, and conditions for the asymptotic stability of the equilibrium states are provided.

The work by Gan and Yin (2015) is devoted to the application of the finite element method to the linear integro-differential boundary problem of the second order:

$$u_{tt}(x,t) - \nabla\cdot(A\nabla u) - \int_0^t \nabla\cdot(B\nabla u(s))ds = f(x,t)$$

with initial-boundary conditions

$$u(x,t)\,|_{\partial\Omega} = 0, u(x,0) = u_0, u_t(x,0) = u_1$$

in a two-dimensional spatial domain Ω. Note that the operator B is assumed to be positive definite. Optimal error estimates for the semi-discrete scheme in L_2 and

H_1 norms are obtained. Numerical experiments confirm the efficiency of symmetric schemes and demonstrate that the convergence rate in L_2 norm for fully discrete backward Euler and Crank–Nicholson schemes is $O(h^2+\tau)$ and $O(h^2+\tau^2)$, respectively.

In Karaa et al. (2015), the hp-continuous Galerkin method is applied to a class of linear hyperbolic integro-differential equations of the second order. The problem takes the form:

$$u_{tt}-\nabla\cdot\left(A(x)\nabla u+\int_0^t B(x,t,s)\nabla u(s)ds\right)=f(x,t),$$

$$u(x,t)|_{\partial\Omega}=0,$$

$$u(x,0)=u_0,u_t(x,0)=u_1,x\in\Omega.$$

Based on the analysis of the extended Ritz–Volterra type mixed projection, the authors obtain a priori hp-error estimates in $L^{\infty}(L_2)$ norm for velocity and displacement. Convergence order results are obtained, and a numerical experiment is presented.

Saedpanah (2014) investigates the existence and uniqueness of solutions for the problem

$$\rho u_{tt}(x,t)+Au(x,t)-\int_0^t \beta(t-s)Au(x,s)ds=f(x,t)$$

with mixed homogeneous Dirichlet and non-homogeneous Neumann conditions:

$$u(x,t)|_{\Gamma_1}=0,\sigma(u,x,t)\cdot n|_{\Gamma_2}=g(x,t),$$

$$u(x,0)=u_0,u_t(x,0)=v_0,x\in\Omega.$$

We outline that in this formulation, the differential operators in the integral component and outside it are identical, and the kernel of the integral operator depends only on the difference of the arguments. Existence and uniqueness of the solution are proven using the Galerkin method. Regularity estimates are obtained.

The proposed approach is also used to prove regularity of any order for models with smooth kernels, under appropriate assumptions on the initial conditions.

The goal of the work (Cannarsa and Sforza 2011) was to study the damping effect of memory terms on the asymptotic behavior of solutions to second-order evolutionary equations in Hilbert spaces. The global existence and asymptotic behavior at infinity of solutions to the semilinear integro-differential equation

$$u_{tt}(t)+Au(t)-\int_0^t g(t-s)Au(s)ds=\nabla F(u(t))+f(t)$$

are considered. For kernels that decay exponentially at infinity, under certain assumptions, exponential stability of weak solutions in the energy norm is obtained.

In Shen et al. (2014), the authors studied the optimal control problem for a system of the form:

$$u_{tt}+Au+\int_0^t C(t,s)u(s)ds=f+Bu,$$

$$u(x,t)|_{\partial\Omega}=0,$$

$$u(x,0)=u_0,u_t(x,0)=u_1,x\in\Omega.$$

Here, A is a linear, self-adjoint, strongly elliptic differential operator of the second order, B is a certain linear continuous operator, and C is a differential operator of the second order.

In Conti et al. (2008), the authors established some results regarding the exponential and polynomial decay of the energy associated with the linear Volterra integro-differential equation:

$$u_{tt}-\alpha\Delta u+\beta u_t+\int_0^t \mu(s)\Delta u(t-s)ds=0,$$

providing sufficient conditions for energy decay without using differential inequalities that include the convolution kernel μ.

The issue of well-posedness for abstract hyperbolic Volterra-type equations is also addressed in Dafermos (1970). The work proves the existence and uniqueness of solutions to the equation:

$$(\rho u_t)_t = C(t)u(t) - \int_0^t G(t,s)u(s)ds + f(t)$$

in certain spaces, and their asymptotic behavior is investigated.

In Anikushyn (2024) and Anikushyn et al. (2024), the authors considered the optimal control problem for the system described by [6.1]–[6.6] with different boundary conditions. By using priory inequalities in negative norms, it shows the existence and uniqueness of generalized solution from $L_2(Q)$ for the right-hand sides from $H_0^1(Q)$ and derives the existence of an optimal control. In the study, control is exerted through the right-hand side of the equation using a certain special operator. The work examines some examples of such control operators (illustrating various control mechanisms) and the corresponding function spaces. Different aspects of similar optimal control problem were considered as well in Anikushyn and Hranishak (2024) and Anikushyn et al. (2025).

In Anikushyn and Nomirovskyi (2008), the following optimization problem for a hyperbolic system was considered:

$$L_D u = u_{tt} + Au = f(x,t;h) \qquad [6.8]$$

$$\| u(h) - u^* \|_S \longrightarrow \min. \qquad [6.9]$$

Here, the operator A is given by expression [6.5]. To study the optimization problem [6.8]–[6.9], it is necessary to ensure, in particular, the existence and uniqueness of the solution to equation [6.8] (with some initial and boundary conditions), which belongs to the space S for all control functions $f(h)$, assuming h belongs to the set of admissible control elements U_∂. Thus, the problem of studying certain specific properties of the operator L_D arises. In the cited work, in particular, a priori estimates were obtained in negative norms:

$$\|u\|_{S_0^{k-1}} \le c_1 \|L_D u\|_{(V_T^{2-k})^-} \le c_2 \|L_D u\|_{(H_T^{2-k})^-} \le c_3 \|u\|_{H_0^k},$$

$$\|v\|_{S_T^{k-1}} \le c_1 \|\overset{*}{L}_D v\|_{(V_0^{2-k})^-} \le c_2 \|\overset{*}{L}_D v\|_{(H_0^{2-k})^-} \le c_3 \|v\|_{H_T^k}$$

for an equation of the form [6.8] for certain spaces, based on which the well-posedness of the generalized formulation was shown, and the aforementioned optimization problem was studied.

Afterwards, in Anikushyn (2013), the problem [6.1]–[6.6] was considered, and estimates of the form

$$\|u\|_{S_0^0} \le c_1 \|Lu\|_{(V_T^1)^-} \le c_2 \|Lu\|_{(H_T^1)^-} \le c_3 \|u\|_{H_0^1},$$

were obtained, based on which the well-posendess of the corresponding initial-boundary problem was proven. Thus, the results of Anikushyn and Nomirovskyi (2008), which were obtained in a scale of spaces for arbitrary $k \in Z$, were extended to the integro-differential type of operators for the case $k = 1$. At the same time, the question of the existence and uniqueness of a generalized solution for integro-differential operators for other values of *k*, studied in Anikushyn and Nomirovskyi (2008), remains open.

We note that the results of Anikushyn and Nomirovskyi (2008) and Anikushyn (2013) were proven using the method of a priori inequalities in negative norms. Estimates in negative norms were first obtained in the 1950s during the study of elliptic boundary value problems (Berezanskyi 1968), and later used in the works of Didenko (1972), where such estimates were obtained for a wide class of second-order differential operators and their construction methodology was developed.

The application of the theory of a priori inequalities in negative norms to various optimization problems of systems with distributed parameters is thoroughly described in Lyashko (2002) and, in part, in Anikushyn and Nomirovskyi (2011) and Klyushin et al. (2011). The approach developed by Lyashko and his students proved to be quite effective for the study of various issues of well-posedness, optimal control and controllability of systems with distributed parameters. Notably, several works (Nomirovskyi 1999; Lyashko and Semenov 2001; Lyashko et al. 2001; Lyashko 2002; Lyashko and Nomirovskyi 2003a, 2003b; Nomirovskyi 2004, 2006, 2007; Semenov 2005; Semenov and Denisov 2005; Klyushin et al. 2011; Nomirovskii and Vostrikov 2016; Tymchyshyn and Nomirovskii 2021; Sergienko et al. 2023) explored various formulations for partial differential equations.

It turned out that this approach can also be successfully applied to integro-differential equations with Volterra-type integral components. For example, the problems with integro-differential equations of the hyperbolic type were studied in

Anikushyn (2013, 2024), Anikushyn and Hranishak (2024) and Anikushyn et al. (2024, 2025), of the parabolic and pseudoparabolic types in Hulianytskyi and Semenov (2012), Anikushyn and Hulianytskyi (2014) and Anikushyn and Andaral (2025). Some special types of equations with Volterra-type integral components were also considered in Anikushyn (2014) and Anikushyn and Kosteeva (2018).

In our work, we prove a priori inequalities for the differential part of the operator [6.2] in a new triple of spaces, different from those studied in Anikushyn (2013). Estimates for the integral component [6.6] will be obtained using a method similar to the one in Anikushyn (2013). Based on the obtained a priori inequalities, we present theorems on the existence and uniqueness of the generalized solution as well as on the continuous dependence of the solution on the right-hand side of the equation. Thus, the results of Anikushyn and Nomirovskyi (2008) regarding the well-posedness of the initial-boundary value problem for $k=2$ is also established for the integro-differential equation [6.1]. Moreover, we provide an example of a control problem and prove the corresponding theorem on optimal control existence.

6.2. Main notations and spaces

Consider a linear operator L defined by the equation [6.2]. As before, $u(x,t)$ is the function describing the state of the system in the domain $(x,t)\in Q=\Omega\times(0,T)$, where Ω is a bounded domain in the n-dimensional space with a smooth boundary $\partial\Omega$. The operators A and B have the form [6.5] and [6.6], respectively.

We assume that the kernels $K_i(t,\tau)$ are continuously differentiable on $[0,T]\times[0,T]$. Furthermore, let a be continuous, and the coefficients a_{ij}, a_i are the continuously differentiable functions in $\bar{\Omega}$, $a_{ij}=a_{ji}$, and the operator A is uniformly elliptic in the domain $\bar{\Omega}$, i.e. for arbitrary real numbers λ_i, the inequality

$$\sum_{i,j=1}^{n} a_{ij}(x)\lambda_i\lambda_j \geq \alpha\sum_{i=1}^{n}\lambda_i^2$$

holds, where α is a positive constant that does not depend on $x\in\bar{\Omega}$ and λ_i. The function $u(x,t)$ satisfies the initial conditions [6.4] and the boundary conditions [6.3].

Let $C_0^\infty(Q)$ denote the set of all functions $u \in C^\infty(\bar{Q})$ that satisfy the boundary conditions [6.3] and the initial conditions

$$u\big|_{t=0} = u_t\big|_{t=0} = \ldots = 0,$$

and let $C_T^\infty(Q)$ denote the set of functions that satisfy the conditions [6.3] and

$$v\big|_{t=T} = v_t\big|_{t=T} = \ldots = 0.$$

We also consider the adjoint operator L^*, which can be formally given by

$$L^* u = v_{tt} + A^* u + B^* u,$$

where

$$A^* v = -\sum_{i,j=1}^{n} \left(a_{ij} v_{x_j}\right)_{x_i} - \sum_{i=1}^{n} (a_i v)_{x_i} + av, \quad B^* v = \int_t^T \sum_{i=1}^{n} K_i(\tau, t) v_{x_i x_i}(x, \tau) d\tau.$$

Spaces $C_0^\infty(Q)$ and $C_T^\infty(Q)$ are assumed to be the domains of the operators L and L^*, respectively.

Let H_0^0, H_0^2, V_0^0 and S_0^1 denote the completion of the set $C_0^\infty(Q)$ with respect to the norms

$$\|u\|_{H_0^0}^2 = \int_Q u^2 + \sum_{i=1}^{n} \left(\int_0^t u_{x_i}(\xi) d\xi \right)^2 dQ, \|u\|_{H_0^2}^2 = \int_Q u_{tt}^2 + \sum_{i=1}^{n} u_{tx_i}^2 dQ,$$

$$\|u\|_{V_0^0}^2 = \int_Q u^2 + \sum_{i=1}^{n} \left(\int_0^t u_{x_i}(\xi) d\xi \right)^2 dQ + \int_\Omega \sum_{i=1}^{n} \left(\int_0^T u_{x_i}(\xi) d\xi \right)^2 d\Omega,$$

$$\|u\|_{S_0^1}^2 = \int_Q u_t^2 + \sum_{i=1}^{n} u_{x_i}^2 dQ + \int_\Omega u_t^2\big|_{t=T} + \sum_{i=1}^{n} u_{x_i}^2\big|_{t=T} d\Omega,$$

respectively, and let H_T^0, H_T^2, V_T^0 and S_T^1 denote the completion of the set $C_T^\infty(Q)$ with respect to the norms

$$\|v\|_{H_T^0}^2 = \int_Q v^2 + \sum_{i=1}^n \left(\int_t^T v_{x_i}(\xi)d\xi\right)^2 dQ, \|v\|_{H_T^2}^2 = \int_Q v_{tt}^2 + \sum_{i=1}^n v_{tx_i}^2 dQ,$$

$$\|v\|_{V_T^0}^2 = \int_Q v^2 + \sum_{i=1}^n \left(\int_t^T v_{x_i}(\xi)d\xi\right)^2 dQ + \int_\Omega \sum_{i=1}^n \left(\int_0^T v_{x_i}(\xi)d\xi\right)^2 d\Omega,$$

$$\|v\|_{S_T^1}^2 = \int_Q v_t^2 + \sum_{i=1}^n v_{x_i}^2 dQ + \int_\Omega v_t^2\Big|_{t=0} + \sum_{i=1}^n v_{x_i}^2\Big|_{t=0} d\Omega,$$

respectively.

It is easy to check that the introduced spaces are Hilbert spaces. Thus, we denote the corresponding negative spaces relative to $L_2(Q)$ by $\left(H_0^0\right)^-$, $\left(H_0^2\right)^-$, $\left(V_0^0\right)^-$, $\left(S_0^1\right)^-$, $\left(H_T^0\right)^-$, $\left(H_T^2\right)^-$, $\left(V_T^0\right)^-$, $\left(S_T^1\right)^-$, and let, for example, $\langle\cdot,\cdot\rangle_H$ be the natural bilinear form on $H_0^0 \times (H_0^0)^-$. We will also introduce corresponding bilinear forms on other pairs of spaces. The inner product in the space $L_2(Q)$ is denoted by $(\cdot,\cdot)_{L_2(Q)}$.

6.3. Generalized control problem

Let the right-hand side f of equation [6.1] depend on the independent variables x,t, as well as on the control h from the set of admissible controls $\mathcal{U}$ belonging to the space $\mathcal{H}$. The functional

$$J(h) = \Phi(u(h)) \qquad [6.10]$$

is defined on the solutions of the problems [6.1]–[6.4] and needs to be minimized over the set $\mathcal{U}$. The investigation of the posed problem will be performed in the spaces introduced above. Let us consider the case of generalized (singular) control of the systems [6.1]–[6.4]. In this situation, the right-hand side $f(x,t,h)$ of equation [6.1] may belong to the space $(V_T^0)^{-1}$.

We will consider the solution of the problems [6.1]–[6.4] in the following sense:

DEFINITION 6.1.– *Function $u(x,t) \in S_0^1$ is called a solution of the problems [6.1]–[6.4] with the right-hand side $f \in (V_T^0)^-$ if there exists a sequence of functions $u_i(x,t) \in C_0^\infty(Q), i = 1,2,...$ such that*

$$\|u - u_i\|_{S_0^1} \to 0, \|Lu_i - f\|_{(V_T^0)^-} \to 0, i \to \infty.$$

The results in the following will be based on the following statement, which will be proven below.

LEMMA 6.1.– *For some positive constants c_1, c_2, the following inequalities hold for all $u \in H_0^2$ and $v \in H_T^2$:*

$$\|u\|_{S_0^1} \le c_1 \|Lu\|_{(V_T^0)^-} \le c_2 \|u\|_{H_0^2},$$

$$\|v\|_{S_T^1} \le c_1 \left\|L^* v\right\|_{(V_0^0)^-} \le c_2 \|v\|_{H_T^2}.$$

THEOREM 6.1.– *Assume that the state of the system is determined by the solution of the problems [6.1]–[6.4] and the following conditions hold:*

– The quality criterion $\Phi : S_0^1 \to R$ is a lower semi-continuous functional with respect to the state of the system $u(x,t,h)$ (i.e., if $u_i \xrightarrow{w} u^*$ in S_0^1, then $\liminf_{i\to\infty} \Phi(u_i) \ge \Phi(u^*)$) and bounded from below;

– the set of admissible controls $\mathcal{U} \subset \mathcal{H}$ is weakly compact in $\mathcal{H}$;

– $\mathcal{H}$ is a reflexive Banach space;

– $f(x,t,\cdot)$ acts weakly continuously from $\mathcal{H}$ to $(V_T^0)^-$;

– for the operators L, L^*, the estimates from Lemma 6.1 hold.

Then, there exists an optimal control for the systems [6.1]–[6.4].

PROOF.– We select a minimizing sequence $\{h_k\}_{k=1}^{\infty}, h_k \in \mathcal{U}$, i.e.:

$$J(h_k) \to \inf_{h \in \mathcal{U}} J(h), k \to \infty. \qquad [6.11]$$

Due to the weak compactness of the set $\mathcal{U}$, a subsequence can be extracted from this sequence that converges weakly in $\mathcal{H}$ to some $h^* \in \mathcal{U}$ (we will relabel it again as $\{h_k\}_{k=1}^{\infty}$). Since f is weakly continuous, the corresponding sequence $f(h_k)$ converges weakly and hence is bounded in the space $(V_T^0)^-$. To further prove Theorem 6.1, we will present two auxiliary statements.

LEMMA 6.2.– *For any function $f \in (V_T^0)^-$, there exists a unique generalized solution to the problems [6.1]–[6.4] in the sense of Definition 6.1.*

LEMMA 6.3.– *Let $u(x,t)$ be a generalized solution of the systems [6.1]–[6.4] with the right-hand side $f \in (V_T^0)^-$ in the sense of Definition 6.1. Then, the following inequality holds:*

$$\|u\|_{S_0^1} \le c\|f\|_{(V_T^0)^-}.$$

To prove Lemma 6.2, we first consider the case $f \in (S_T^1)^-$. We introduce the functional

$$l(v) \equiv \langle v, f \rangle_S, v \in C_T^{\infty}(Q),$$

where $\langle \cdot,\cdot \rangle_S$ is a bilinear form constructed by the spaces $S_T^1, (S_T^1)^-$.

Due to the Schwarz inequality and exploiting Lemma 6.1, we have

$$| l(v) |=| \langle v, f \rangle_S | \le \|v\|_{S_T^1} \cdot \|f\|_{(S_T^1)^-} \le c\left\|L^* v\right\|_{(V_0^0)^-}.$$

It follows from the last inequality that the functional $l(v)$ can be considered as a linear continuous function of $L^* v \in (V_0^0)^-$ (i.e., $l(v) = \bar{l}(L^* v)$). By the Hahn–Banach theorem, we extend the functional $\bar{l}(\cdot)$ linearly and continuously over the entire space $(V_0^0)^-$. It follows from the theorem on the general form of a linear continuous functional over $(V_0^0)^-$, that there exists a function $u \in V_0^0$ such that $\bar{l}(w) = \langle u, w \rangle_V$ for all $w \in (V_0^0)^-$. Here, $\langle \cdot, \cdot \rangle_V$ is a bilinear form constructed by the spaces $V_0^0, (V_0^0)^-$.

Applying this functional to the elements $w = L^* v$, $v \in C_T^\infty(Q)$, we obtain

$$\langle u, L^* v \rangle_V = \langle u, w \rangle_V \equiv \bar{l}(w) = \bar{l}(L^* v) = l(v) = \langle v, f \rangle_S.$$

The space $C_0^\infty(Q)$ is dense in H_0^2; thus, there exists a sequence of functions $u_i(x,t) \in C_0^\infty(Q)$ such that $\left\| u_i - u \right\|_{H_0^2} \to 0$, and hence, $\left\| u_i - u_j \right\|_{H_0^2} \to 0, i, j \to \infty$.

Due to the linearity of the operator L and using the right inequality in Lemma 6.1, we have

$$\| Lu_i - Lu_j \|_{(V_T^0)^-} \to 0, i, j \to \infty.$$

Since the space $(V_T^0)^-$ is complete, there exists a functional $\hat{f} \in (V_T^0)^-$ such that

$$\| Lu_i - \hat{f} \|_{(V_T^0)^-} \to 0, i \to \infty.$$

By the Schwarz inequality, for any $v \in C_T^\infty(Q)$,

$$\left| \langle v, Lu_i - \hat{f} \rangle_V \right| \leq \| v \|_{V_T^0} \| Lu_i - \hat{f} \|_{(V_T^0)^-},$$

and hence,

$$\langle v, Lu_i - \hat{f} \rangle_V \to 0, \; i \to \infty. \tag{6.12}$$

Since $u_i \to u$ in H_0^2, then $u_i \to u$ in V_0^0 and it follows that $\langle u_i, L^* v \rangle_V \to \langle u, L^* v \rangle_V$. By the definition of the operator $L^*(\cdot)$ and the proven equality $\langle u, L^* v \rangle_V = \langle v, f \rangle_S$, we have

$$\left|\langle v, Lu_i\rangle_V - \langle v, f\rangle_S\right| \to 0,\ i \to \infty. \tag{6.13}$$

Now, consider the inequality

$$|\langle v, f - \hat{f}\rangle_S| \le |\langle v, Lu_i - f\rangle_S| + |\langle v, Lu_i - \hat{f}\rangle_S|.$$

Using [6.12] and [6.13], as $i \to \infty$, we get

$$\langle v, f - \hat{f}\rangle_V = 0, \text{ for all } v \in C_T^\infty(Q).$$

Due to the density of $C_T^\infty(Q)$ in the space V_T^0, it follows that $\| f - \hat{f} \|_{(V_T^0)^-} = 0$, and therefore, $\| Lu_i - f \|_{(V_T^0)^-} \to 0$ as $i \to \infty$.

Finally, exploiting the left inequalities in Lemma 6.1, we obtain the statement of Lemma 6.2.

To prove Lemma 6.3, we will show that the inequalities from Lemma 6.1 lead to

$$\|u\|_{V_0^0} \le c\|Lu\|_{(S_T^1)^-}, \tag{6.14}$$

$$\|v\|_{V_T^0} \le c\left\|L^* v\right\|_{(S_0^1)^-}. \tag{6.15}$$

Indeed, by virtue of Lemma 6.2, there exists a unique solution $u(x,t)$ for the problems [6.1]–[6.4] with the right-hand side $f \in (V_T^0)^-$.

Hence, for any $f \in (V_T^0)^-$, $v \in V_T^0$ such that $L^* v \in (S_0^1)^-$, we have the equality

$$\langle u, L^* v\rangle_S = \langle v, f\rangle_V. \tag{6.16}$$

From the Schwarz inequality, it follows that

$$\left\langle \frac{v}{\| L^* v \|_{(S_0^1)^-}}, f \right\rangle_V \le \|u\|_{S_0^1}.$$

Thus, the set of functionals $\{v / \| L^* v \|_{(S_0^1)^-}\}$ is bounded in the norm of the space V_T^0, which proves the validity of inequality [6.15]. Inequality [6.14] is proven similarly. Applying the Schwarz inequality to the right-hand side of [6.16] and using inequality [6.15], we obtain

$$\langle u, L^* v \rangle_S = \langle v, f \rangle_V \leq \|v\|_{V_T^0} \|f\|_{(V_T^0)^-} \leq c \left\| L^* v \right\|_{(S_0^1)^-} \|f\|_{(V_T^0)^-} .$$

Thus, the statement of Lemma 6.3 follows.

Now, we continue the proof of Theorem 6.1. Due to the weak continuity of $f(h)$, the sequence $\{f(h_k)\}_{k=1}^{\infty}$ is weakly convergent and hence bounded in the space $(V_T^0)^-$. It follows from Lemma 6.3, that the sequence of solutions $\{u_k\}_{k=1}^{\infty}$ corresponding to the right-hand sides $f(h_k)$, $k = 1, 2, \ldots$, is bounded in the space S_0^1. Since the last space is reflexive, a weakly convergent subsequence can be extracted from $\{u_k\}_{k=1}^{\infty}$ (relabel it again as $\{u_k\}_{k=1}^{\infty}$), which weakly converges to some element $u^* \in S_0^1$.

The following equality holds for any functions $u(x,t) \in C_T^{\infty}(Q)$ such that $L^* v \in (V_T^0)^-$

$$\langle u_k, L^* v \rangle_V = \langle v, f(h_k) \rangle_S . \qquad [6.17]$$

Since $h_k \xrightarrow{w} h^*$ in $\mathcal{H}$, and $f(\cdot)$ is weakly continuous from $\mathcal{H}$ to $(V_T^0)^-$, we have

$$\langle v, f(h_k) \rangle_S \to \langle v, f(h^*) \rangle_S .$$

Furthermore, since $u_k \xrightarrow{w} u^*$, we obtain

$$\langle u_k, L^* v \rangle_V \to \langle u^*, L^* v \rangle_V .$$

Thus, passing to the limit in [6.17] as $k \to \infty$, we conclude that $u^*(x,t)$ is a solution of the problems [6.1]–[6.4] with the right-hand side $f(h^*)$ in the sense of Definition 6.1. Finally, exploiting the weak lower semi continuity of the quality criterion $J(h)$, we obtain

$$J(h^*) \le \liminf_{k\to\infty} J(h_k) = \inf_{h\in\mathcal{U}} J(h),$$

hence, h^* is an optimal control, which concludes the proof of Theorem 6.1.

6.4. A priori inequalities for the differential part of the operator

In this and the subsequent sections, we prove priory inequalities from Lemma 6.1 using a method from Anikushyn et al. (2024). Denote by $L_D u = u_{tt} + Au$ the differential part of the operator [6.2].

LEMMA 6.4.– *There exists a positive constant* c *such that for any function* $u \in C_0^\infty(Q)$, *the following inequality holds*:

$$\|L_D u\|_{(H_T^0)^-} \le c\|u\|_{H_0^2}.$$

PROOF.– Let c_0 be a constant that majorizes the functions $|a|$, $|a_i|$, $|a_{ij}|$ for all $x \in \bar{\Omega}$ (the mentioned functions are bounded since they are continuous on the compact $\bar{\Omega}$). In the following, we define

$$\bar{v}(x,t) = \int_T^t v(x,\xi)\,d\xi.$$

For $u \in C_0^\infty$, we have $L_D u, v \in L_2(Q)$, so we are able to consider the inner product:

$$(L_D u, v)_{L_2(Q)} = (u_{tt}, v)_{L_2(Q)} + \sum_{i,j=1}^{n}\left(-\left(a_{ij}u_{x_j}\right)_{x_i}, v\right)_{L_2(Q)} + \qquad [6.18]$$

$$+\sum_{i=1}^{n}\left(a_i u_{x_i},v\right)_{L_2(Q)}+\left(au,v\right)_{L_2(Q)}.$$

Estimate each term using the Cauchy–Schwarz inequality in integral form:

$$\left|\left(u_{tt},v\right)_{L_2(Q)}\right|=\left|\int_Q u_{tt}v\,dQ\right|\le\left(\int_Q u_{tt}^2dQ\int_Q v^2dQ\right)^{\frac{1}{2}}=\|u_{tt}\|_{L_2(Q)}\|v\|_{L_2(Q)}\le\|u\|_{H_0^2}\|v\|_{H_T^0},$$

$$\left|\sum_{i,j=1}^{n}\left(-\left(a_{ij}u_{x_j}\right)_{x_i},v\right)_{L_2(Q)}\right|=\left|\sum_{i,j=1}^{n}\int_Q a_{ij}u_{x_j}v_{x_i}dQ\right|\le c_0\sum_{i,j=1}^{n}\int_Q\left|u_{tx_j}\overline{v}_{x_i}\right|dQ\le$$

$$\le c_0\sum_{i,j=1}^{n}\left(\int_Q u_{tx_j}^2dQ\int_Q\overline{v}_{x_j}^2dQ\right)^{\frac{1}{2}}\le c_0n^2\|u\|_{H_0^2}\|v\|_{H_T^0},$$

$$\left|\left(a_iu_{x_i},v\right)_{L_2(Q)}\right|=\left|\left(a_iu_{tx_i},\overline{v}\right)_{L_2(Q)}\right|\le c_0\left(\int_Q u_{tx_j}^2dQ\int_Q\overline{v}^2dQ\right)^{\frac{1}{2}}.$$

Since $\overline{v}(x,T)=0,$ the Friedrichs inequality can be applied to this function. We obtain

$$c_0\left(\int_Q u_{tx_j}^2dQ\int_Q\overline{v}^2dQ\right)^{\frac{1}{2}}\le c_0\|u\|_{H_0^2}c_Q\left(\int_Q v^2dQ\right)^{\frac{1}{2}}\le c_0c_Q\|u\|_{H_0^2}\|v\|_{H_T^0},$$

where c_Q is the Friedrichs constant depending on the domain Q only. Similarly, estimate the last term:

$$\left|\left(au,v\right)_{L_2(Q)}\right|=\left|\left(au_t,\overline{v}\right)_{L_2(Q)}\right|\le c_0\left(\int_Q u_t^2dQ\int_Q\overline{v}^2dQ\right)^{\frac{1}{2}}\le$$

$$\le c_0c_Q\left(\int_Q u_{tt}^2dQ\int_Q v^2dQ\right)^{\frac{1}{2}}\le c_0c_Q\|u\|_{H_0^2}\|v\|_{H_T^0}.$$

Adding the obtained inequalities, we get $|(L_D u, v)_{L_2(Q)}| \leq c\|u\|_{H_0^2}\|v\|_{H_T^0}$, for some constant $c > 0$. Dividing the last inequality by $\|v\|_{H_T^0}$ and taking the supremum over all elements $v \in C_T^\infty(Q)$, we obtain the statement of the lemma.

LEMMA 6.5.– *There exists a positive constant* c *such that for any function* $v \in C_T^\infty(Q)$, *the following inequality holds*:

$$\left\|L_D^* v\right\|_{(H_0^0)^-} \leq c\|v\|_{H_T^2}.$$

The proof is similar to Lemma 6.4.

REMARK 6.1.– *The inequalities proven in Lemmas 6.4 and 6.5 allow us to expand the operators* L_D *and* L_D^* *from their domains* $C_0^\infty(Q)$ *and* $C_T^\infty(Q)$ *to the spaces* H_0^2 *and* H_T^2, *respectively, by continuity. We will use the* same *notations* L_D *and* L_D^* *for the expanded operators. Note that the inequalities claimed in Lemmas 6.4 and 6.5 hold for all functions* $u \in H_0^2$ *and* $v \in H_T^2$.

LEMMA 6.6.– *There exists a positive constant* c *such that for any function* $u \in H_0^2$, *the following inequality holds*:

$$c^{-1}\|u\|_{S_0^1} \leq \|L_D u\|_{(V_T^0)^-}.$$

PROOF.– *Assume that* $u \in C_0^\infty(Q) \subset H_0^2$. *By virtue of Lemma 6.4, we have* $L_D u \in (H_T^0)^-$. *Since there is a dense embedding* $V_T^0 \subset H_T^0$ *(e.g. Anikushyn and Nomirovskyi 2008), it follows that* $(H_T^0)^- \subset (V_T^0)^-$. *Therefore,* $L_D u$ *defines a linear continuous functional* $\langle \cdot, L_D u\rangle_{V_T^0}$ *on the space* V_T^0. *We apply this functional to the element v, which is defined as the solution of the equation*

$$-\int_0^t e^{M\tau}\left(-v(x,\tau) + \int_T^\tau v(x,s)ds\right)d\tau = u(x,t). \qquad [6.19]$$

Recall that $\bar{v}(x,t) = \int_T^t v(x,s)\,ds$. Then,

$$u = -\int_0^t e^{M\tau}(-v+\bar{v})\,d\tau, \qquad [6.20]$$

$$u_t = e^{Mt}(v-\bar{v}), \qquad [6.21]$$

$$u_{tt} = Me^{Mt}(v-\bar{v}) + e^{Mt}(v_t - v). \qquad [6.22]$$

The value of the positive constant M will be chosen later. Note that the function v, defined in this way, belongs to the space V_T^0. Let us transform and estimate $\langle v, L_D u\rangle_{V_T^0}$. Substituting u from equation [6.19], we obtain $\langle v, L_D u\rangle_{V_T^0} = I_1 + I_2 + I_3 + I_4 + I_5$. Consider each term separately.

In the same way as in the proof of Theorem 2 in Anikushyn (2008), we have:

$$I_1 = -(u_t, v_t)_{L_2(Q)} =$$

$$= M\left(e^{Mt}v, v\right)_{L_2(Q)} - M\left(e^{Mt}\bar{v}, v\right)_{L_2(Q)} + \left(e^{Mt}v_t, v\right)_{L_2(Q)} - \left(e^{Mt}v, v\right)_{L_2(Q)}.$$

Here, $M\left(e^{Mt}v, v\right)_{L_2(Q)} = M\int_Q e^{Mt}v^2 dQ$ and $-\left(e^{Mt}v, v\right)_{L_2(Q)} = -\int_Q e^{Mt}v^2 dQ.$

For another two terms, we use the Gauss formula:

$$-M\left(e^{Mt}\bar{v}, v\right)_{L_2(Q)} = -\frac{M}{2}\int_{\partial Q} e^{Mt}\bar{v}^2 \vec{n}\, d\partial Q + \frac{M^2}{2}\int_Q e^{Mt}\bar{v}^2 dQ =$$

$$= \frac{M}{2}\int_\Omega \bar{v}^2\Big|_{t=0} d\Omega + \frac{M^2}{2}\int_Q e^{Mt}\bar{v}^2 dQ,$$

$$\left(e^{Mt}v_t, v\right)_{L_2(Q)} = \int_Q e^{Mt}v_t v\, dQ = \frac{1}{2}\int_{\partial Q} e^{Mt}v^2 \vec{n}\, d\partial Q - \frac{M}{2}\int_Q e^{Mt}v^2 dQ =$$

$$= -\frac{1}{2}\int_\Omega v^2\Big|_{t=0} d\Omega + \frac{1}{2}\int_\Omega e^{MT}v^2\Big|_{t=T} d\Omega - \frac{M}{2}\int_Q e^{Mt}v^2 dQ.$$

Then,

$$I_1 = \frac{M-2}{2}\int_Q e^{Mt} v^2 dQ + \frac{M}{2}\int_\Omega \bar{v}^2\Big|_{t=0} d\Omega + \frac{M^2}{2}\int_Q e^{Mt}\bar{v}^2 dQ +$$

$$+\frac{1}{2}\int_\Omega e^{MT} v^2\Big|_{t=T} d\Omega - \frac{1}{2}\int_\Omega v^2\Big|_{t=0} d\Omega.$$

The function u belongs to the class C_0^∞; hence, $u_t(x,0)=0$. Therefore, from [6.21], it follows that $v(x,0)=\bar{v}(x,0)$.

Since the equality $v(x,T)=e^{-MT}u_t(x,T)$ and the inequality $2e^{2Mt}\left(v^2+\bar{v}^2\right)\geq\left(e^{Mt}\left(v+\bar{v}\right)\right)^2=u_t^2$ hold, for sufficiently large M, we obtain:

$$I_1 \geq \frac{M}{10}\int_Q e^{Mt} v^2 dQ + \frac{M^2}{10}\int_Q e^{Mt}\bar{v}^2 dQ + \frac{M}{10}\int_Q e^{-Mt}u_t^2 dQ + \frac{M}{10}\int_\Omega v^2\Big|_{t=0} d\Omega +$$

$$+\frac{1}{4}\int_\Omega e^{MT} v^2\Big|_{t=T} d\Omega + \frac{1}{4}\int_\Omega e^{-MT} u_t^2\Big|_{t=T} d\Omega.$$

Consider the term I_2. Using the symmetry of the matrix $\{a_{ij}\}_{i,j=1}^n$ and the Gauss formula, we obtain:

$$I_2 = -\frac{1}{2}\sum_{i,j=1}^n \int_Q (a_{ij}u_{x_i})_{x_j} v\, dQ = \frac{1}{2}\sum_{i,j=1}^n \int_Q a_{ij}u_{x_i}v_{x_j} dQ = -\frac{1}{2}$$

$$\sum_{i,j=1}^n \int_Q a_{ij}u_{tx_i}\bar{v}_{x_j} dQ =$$

$$= \frac{1}{2}\sum_{i,j=1}^n \int_Q a_{ij}e^{Mt}\left(-v_{x_i}+\bar{v}_{x_i}\right)\bar{v}_{x_j} dQ \geq \frac{\alpha}{4}$$

$$\sum_{i=1}^n \int_\Omega \bar{v}_{x_i}^2\Big|_{t=0} d\Omega + \frac{\alpha M}{4}\sum_{i=1}^n \int_Q e^{Mt}\bar{v}_{x_i}^2 dQ.$$

Furthermore, we assume that a constant $c_0>0$ is greater to all coefficients $|a_{ij}|, |a_i|, |a|$ in the domain $\bar{\Omega}$. For the term I_3, the following inequalities hold:

$$I_3 = -\frac{1}{2}\sum_{i,j=1}^{n}\int_Q (a_{ij}u_{x_i})_{x_j} v dQ = \frac{1}{2}\sum_{i,j=1}^{n}\int_Q a_{ij}u_{x_i}v_{x_j} dQ =$$

$$= \frac{1}{2}\sum_{i,j=1}^{n}\int_Q a_{ij}e^{-Mt}u_{x_i}u_{tx_j} dQ + \frac{1}{2}\sum_{i,j=1}^{n}\int_Q a_{ij}u_{x_i}\overline{v}_{x_j} dQ.$$

Using the Cauchy inequality, we easily get

$$I_3 \geq \frac{\alpha}{4}\sum_{i=1}^{n}\int_\Omega e^{-MT}u_{x_i}^2\Big|_{t=T} d\Omega + \frac{\alpha M}{4}\sum_{i=1}^{n}\int_Q e^{-Mt}u_{x_i}^2 dQ - \frac{c_0 n}{4}$$

$$\sum_{i=1}^{n}\int_Q e^{-Mt}u_{x_i}^2 + e^{Mt}\overline{v}_{x_i}^2 dQ.$$

For the term I_4, we apply the Cauchy inequality as follows

$$I_4 = \sum_{i=1}^{n}\int_Q a_i u_{x_i} v dQ \geq -\frac{c_0}{2}\sum_{i=1}^{n}\int_Q e^{-Mt}u_{x_i}^2 + e^{Mt}v^2 dQ =$$

$$= -\frac{c_0}{2}\sum_{i=1}^{n}\int_Q e^{-Mt}u_{x_i}^2 dQ - \frac{c_0 n}{2}\int_Q e^{Mt}v^2 dQ.$$

Similarly, using the Cauchy inequality, we estimate I_5 in the following way:

$$I_5 = \int_Q auvdQ = -\int_Q au_t\overline{v}dQ = -\int_Q ae^{Mt}v\overline{v}dQ + \int_Q ae^{Mt}\overline{v}^2 dQ \geq$$

$$\geq -c_0\int_Q e^{Mt}\frac{v^2+\overline{v}^2}{2}dQ - c_0\int_Q e^{Mt}\overline{v}^2 dQ.$$

Recall that $\langle\cdot,\cdot\rangle_V$ is a bilinear form constructed using the spaces $V_T^0, (V_T^0)^-$. Adding the obtained inequalities, we get the general estimate:

$$
\begin{aligned}
\langle v, L_D u\rangle_V &\geq \frac{M}{10}\int_Q e^{Mt}v^2 dQ + \frac{M^2}{10}\int_Q e^{Mt}\overline{v}^2 dQ \\
&+\frac{M}{10}\int_Q e^{-Mt}u_t^2 dQ + \frac{M}{10}\int_\Omega v^2\Big|_{t=0} d\Omega + \\
&+\frac{1}{4}\int_\Omega e^{MT}v^2\Big|_{t=T} d\Omega + \frac{1}{4}\int_\Omega e^{-MT}u_t^2\Big|_{t=T} d\Omega + \frac{\alpha}{4} \\
&\sum_{i=1}^{n}\int_\Omega \overline{v}_{x_i}^2\Big|_{t=0} d\Omega + \frac{\alpha M}{4}\sum_{i=1}^{n}\int_Q e^{Mt}\overline{v}_{x_i}^2 dQ + \\
&+\frac{\alpha}{4}\sum_{i=1}^{n}\int_\Omega e^{-MT}u_{x_i}^2\Big|_{t=T} d\Omega + \frac{\alpha M}{4}\sum_{i=1}^{n}\int_Q e^{-Mt}u_{x_i}^2 dQ - \frac{c_0 n}{4} \\
&\sum_{i=1}^{n}\int_Q e^{-Mt}u_{x_i}^2 + e^{Mt}\overline{v}_{x_i}^2 dQ + \\
&-\frac{c_0}{2}\sum_{i=1}^{n}\int_Q e^{-Mt}u_{x_i}^2 dQ - \frac{c_0 n}{2}\int_Q e^{Mt}v^2 dQ \\
&-c_0\int_Q e^{Mt}\frac{v^2+\overline{v}^2}{2} dQ - c_0\int_Q e^{Mt}\overline{v}^2 dQ.
\end{aligned}
$$

Now, it is easy to conclude that for an arbitrarily large number d, there exists a constant $M > 0$ such that:

$$
\begin{aligned}
\langle v, L_D u\rangle_V &\geq d\int_Q e^{Mt}v^2 dQ + d\int_Q e^{Mt}\overline{v}^2 dQ + d\int_Q e^{-Mt}u_t^2 dQ + \\
&+\frac{1}{4}\int_\Omega e^{-MT}u_t^2\Big|_{t=T} d\Omega + \frac{\alpha}{8}\sum_{i=1}^{n}\int_\Omega \overline{v}_{x_i}^2\Big|_{t=0} d\Omega + d\sum_{i=1}^{n}\int_Q e^{Mt}\overline{v}_{x_i}^2 dQ + \\
&+\frac{\alpha}{4}\sum_{i=1}^{n}\int_\Omega e^{-MT}u_{x_i}^2\Big|_{t=T} d\Omega + d\sum_{i=1}^{n}\int_Q e^{-Mt}u_{x_i}^2 dQ.
\end{aligned}
$$

Using the inequalities $e^{-Mt} \geq e^{-MT}$, $e^{Mt} \geq e^{-MT}$, we obtain:

$$
\langle v, L_D u\rangle_V \geq c^{-1}(\| u \|_{S_0^1}^2 + \| v \|_{V_T^0}^2).
$$

Finally, applying the Schwarz inequality to the left-hand side and the Cauchy inequality to the right-hand side of the last inequality, we get:

$$\| L_D u \|_{(V_T^0)^-} \| v \|_{V_T^0} \geq \langle v, L_D u \rangle_V \geq 2c^{-1} \| u \|_{S_0^1} \| v \|_{V_T^0} .$$

Thus, the statement of the lemma is proven for all $u \in C_0^\infty(Q)$. For other functions in H_0^2, the required statement is established by passing to a limit. Let $u \in H_0^2$ (not necessarily smooth). Due to the density of the embedding $C_0^\infty(Q) \subset H_0^2$, the element u can be approximated by a sequence of elements $u_m \in C_0^\infty(Q)$, i.e. $\| u - u_m \|_{H_0^2} \to 0$ as $m \to \infty$.

As a result of Lemma 1, the inequality $\| L_D u - L_D u_m \|_{(H_T^0)^-} \leq c_1 \| u - u_m \|_{H_0^2}$ holds. Hence, $L_D u_m \to L_D u$ in the space $(H_T^0)^-$, and due to the inclusion of spaces, $L_D u_m \to L_D u$ in the space $(V_T^0)^-$. Moreover, for each element u_m, as proven above, the inequality:

$$c^{-1} \| u_m \|_{S_0^1} \leq \| L_D u_m \|_{(V_T^0)^-}$$

holds.

It remains to note that convergence in the space H_0^2 implies convergence in S_0^1, and therefore, $\| u - u_m \|_{S_0^1} \to 0$. Now, passing to the limit as $m \to \infty$ in the last inequality, we obtain the desired statement.

In the same way, we can prove the following statement.

LEMMA 6.7.– *There exists a positive constant* c *such that for any function* $v \in H_T^0$, *the following inequality holds*:

$$c^{-1} \| v \|_{S_0^1} \leq \| L_D^* v \|_{(V_T^0)^-} .$$

Thus, it follows from Lemmas 6.4–6.7 that the following theorem holds.

THEOREM 6.2.– *For all* $u \in H_0^2$ *and* $v \in H_T^0$, *the following inequalities hold:*

$$\|u\|_{S_0^1} \le c_1 \|L_D u\|_{(V_T^0)^-} \le c_2 \|L_D u\|_{(H_T^0)^-} \le c_3 \|u\|_{H_0^2}, \quad [6.23]$$

$$\|v\|_{S_T^1} \le c_1 \left\|\overset{*}{L_D} v\right\|_{(V_0^0)^-} \le c_2 \left\|\overset{*}{L_D} v\right\|_{(H_0^0)^-} \le c_3 \|v\|_{H_T^2}. \quad [6.24]$$

Note that Theorem 6.2 is a special case of Theorem 4 from the work (Anikushyn and Nomirovskyi 2008). The provided proof is necessary for further investigation. The proof using the "shift" method from the cited work is appropriate for extending in the case of an integral-differential operator.

REMARK 6.2.– *It can be shown that Theorem 1 remains valid, even though the coefficients of the equation* a_{ij}, a_i, a *belong to the space* $L_\infty(\Omega)$.

6.5. A priori inequalities for the integro-differential operator

Now consider the integral part of the operator:

$$Bu = \int_0^t \sum_{i=1}^n K_i(t,\tau) u_{x_i x_i}(x,\tau)\, d\tau.$$

LEMMA 6.8.– *There exists a constant* $c_B > 0$ *such that for any functions* $u \in S_0^1$, $v \in V_T^0$ *and any constant* $M > 0$, *the following inequality holds:*

$$\left|\langle v, Bu\rangle_V\right| \le c_B \sum_{i=1}^n \int_Q e^{Mt} \overline{v}_{x_i}^2(x,t) + e^{-Mt} u_{x_i}^2(x,t)\, dQ.$$

PROOF.– Let us first assume that u and v are the smooth functions. Integrating by parts and considering the boundary conditions [6.3], we have:

$$\left|\langle v, Bu\rangle_V\right| = \left|\int_Q \int_0^t \sum_{i=1}^n K_i(t,\tau) u_{x_i x_i}(x,\tau)\, d\tau\, v(x,t)\, dQ\right| =$$

$$=\left|\int_Q\int_0^t\sum_{i=1}^n K_i(t,\tau)u_{x_i}(x,\tau)d\tau v_{x_i}(x,t)dQ\right|=$$

$$=\left|\sum_{i=1}^n\int_\Omega\int_0^T\int_0^t K_i(t,\tau)u_{x_i}(x,\tau)d\tau v_{x_i}(x,t)dt\,d\Omega\right|\leq$$

$$\leq\sum_{i=1}^n\left|\int_\Omega\int_0^T\int_0^t K_i(t,\tau)u_{x_i}(x,\tau)d\tau v_{x_i}(x,t)dt\,d\Omega\right|.$$

Next, integrating by parts, we obtain

$$\int_0^T\int_0^t K_i(t,\tau)u_{x_i}(x,\tau)d\tau v_{x_i}(x,t)dt=$$

$$=-\int_0^t K_i(t,\tau)u_{x_i}(x,\tau)d\tau\int_t^T v_{x_i}(x,s)ds\Big|_0^T+$$

$$+\int_0^T\frac{\partial}{\partial t}\left(\int_0^t K_i(t,\tau)u_{x_i}(x,\tau)d\tau\right)\int_t^T v_{x_i}(x,s)ds\,dt.$$

Note that

$$\int_0^t K_i(t,\tau)u_{x_i}(x,\tau)d\tau\int_t^T v_{x_i}(x,s)ds\Big|_0^T=$$

$$=\int_0^T K_i(T,\tau)u_{x_i}(x,\tau)d\tau\int_T^T v_{x_i}(x,s)ds-$$

$$-\int_T^T K_i(0,\tau)u_{x_i}(x,\tau)d\tau\int_0^T v_{x_i}(x,s)ds=0.$$

Using the Cauchy inequality, we have:

$$\left|\int_\Omega\int_0^T\int_0^t K_i(t,\tau)u_{x_i}(x,\tau)d\tau v_{x_i}(x,t)dt\,d\Omega\right|=$$

$$=\left|\int_{\Omega}\int_{0}^{T}\frac{\partial}{\partial t}\left(\int_{0}^{t}K_i(t,\tau)u_{x_i}(x,\tau)d\tau\right)\int_{t}^{T}v_{x_i}(x,s)\,ds\,dt\,d\Omega\right|\leq$$

$$\leq\frac{1}{2}\int_{Q}e^{-Mt}\left(\frac{\partial}{\partial t}\left(\int_{0}^{t}K_i(t,\tau)u_{x_i}(x,\tau)d\tau\right)\right)^2 dQ+$$
$$\frac{1}{2}\int_{Q}e^{Mt}\left(\int_{t}^{T}v_{x_i}(x,s)ds\right)^2 dQ.$$

Consider the first term. Since the function $K_i(t,\tau)u_{x_i}(x,\tau)$ is continuous and has a continuous partial derivative with respect to t, the integral $\int_0^t K_i(t,\tau)u_{x_i}(x,\tau)d\tau$ can be differentiated with respect to t.

$$\int_{Q}e^{-Mt}\left(\frac{\partial}{\partial t}\left(\int_{0}^{t}K_i(t,\tau)u_{x_i}(x,\tau)d\tau\right)\right)^2 dQ=$$
$$=\int_{Q}e^{-Mt}\left(\int_{0}^{t}K_i'(t,\tau)u_{x_i}(x,\tau)d\tau+K_i(t,t)u_{x_i}(x,t)\right)^2 dQ\leq$$

$$\leq 2\int_{Q}e^{-Mt}\left(\int_{0}^{t}K_i'(t,\tau)u_{x_i}(x,\tau)d\tau\right)^2+e^{-Mt}\left(K_i(t,t)u_{x_i}(x,t)\right)^2 dQ\leq$$

$$\leq 2\int_{Q}e^{-Mt}\int_{0}^{t}\left(K_i'(t,\tau)\right)^2 d\tau\int_{0}^{t}u_{x_i}^2(x,\tau)d\tau+e^{-Mt}K_i^2(t,t)u_{x_i}^2(x,t)dQ.$$

Here and later, $K_i'(t,\tau)$ denotes the derivative of the kernel K_i with respect to t.

Since $K_i(t,t)$ and $K_i'(t,\tau)$ are the continuous functions on the compact set $[0,T]\times[0,T]$, they are bounded. Let:

$$c_K=\max_{i=1..n}\left\{\max_{t\in[0,T]}|K_i(t,t)|,\max_{t,\tau\in[0,T]}|K_i'(t,\tau)|\right\}.$$

Then

$$2\int_Q e^{-Mt}\int_0^t \left(K_i'(t,\tau)\right)^2 d\tau \int_0^t u_{x_i}^2(x,\tau)d\tau + e^{-Mt}K_i^2(t,t)u_{x_i}^2(x,t)dQ \leq$$

$$\leq 2\int_Q e^{-Mt}c_K^2 T\int_0^t u_{x_i}^2(x,\tau)d\tau + e^{-Mt}c_K^2 u_{x_i}^2(x,t)dQ =$$

$$= 2c_K^2 T\int_Q\int_0^t e^{-Mt}u_{x_i}^2(x,\tau)d\tau dQ + 2c_K^2\int_Q e^{-Mt}u_{x_i}^2(x,t)dQ.$$

Note that:

$$\int_0^T\int_0^t e^{-Mt}u_{x_i}^2(x,\tau)d\tau dt = \int_0^T\int_\tau^T e^{-Mt}u_{x_i}^2(x,\tau)dtd\tau = \int_0^T u_{x_i}^2(x,\tau)$$
$$\int_\tau^T e^{-Mt}dtd\tau$$

$$= \int_0^T u_{x_i}^2(x,\tau)\left.\frac{e^{-Mt}}{-M}\right|_\tau^T d\tau = \frac{1}{M}\int_0^T u_{x_i}^2(x,\tau)(e^{-M\tau} - e^{-MT})d\tau =$$

$$= \frac{1}{M}\int_0^T e^{-M\tau}\left(1 - e^{M(\tau-T)}\right)u_{x_i}^2(x,\tau)d\tau \leq \frac{1}{M}\int_0^T e^{-M\tau}u_{x_i}^2(x,\tau)d\tau.$$

Thus

$$2c_K^2 T\int_Q\int_0^t e^{-Mt}u_{x_i}^2(x,\tau)d\tau dQ + 2c_K^2\int_Q e^{-Mt}u_{x_i}^2(x,t)dQ \leq$$

$$\leq \frac{2c_K^2 T}{M}\int_Q e^{-Mt}u_{x_i}^2(x,t)dQ + 2c_K^2\int_Q e^{-Mt}u_{x_i}^2(x,t)dQ =$$

$$= 2c_K^2\left(\frac{T}{M} + 1\right)\int_Q e^{-Mt}u_{x_i}^2(x,t)dQ.$$

Finally, we arrive at

$$\left|\langle v, Bu\rangle_V\right| \leq \sum_{i=1}^{n} c_K^2\left(\frac{T}{M}+1\right)\int_Q e^{-Mt}u_{x_i}^2(x,t)\,dQ+\frac{1}{2}\int_Q e^{Mt}\overline{v}_{x_i}^2(x,t)\,dQ.$$

For arbitrary functions $u \in S_0^1$ and $v \in V_T^0$, the statement of the lemma follows by passing to the limit.

LEMMA 6.9.– *There exists a constant* $c > 0$ *such that for any function* $u \in C_0^\infty(Q)$, *the inequality holds*:

$$\| Lu \|_{(H_T^0)^-} \leq c \| u \|_{H_0^2}.$$

PROOF.– For smooth functions u,v, we have

$$(Bu,v)_{L_2(Q)} = \sum_{i=1}^{n}\int_Q\int_0^t K_i(t,\tau)u_{x_i x_i}(x,\tau)\,d\tau\, v(x,t)\,dQ$$

$$= -\sum_{i=1}^{n}\int_Q\int_0^t K_i(t,\tau)u_{x_i}(x,\tau)\,d\tau\, v_{x_i}(x,t)\,dQ$$

$$= -\sum_{i=1}^{n}\int_\Omega\int_0^T\int_0^t K_i(t,\tau)u_{x_i}(x,\tau)\,d\tau\, v_{x_i}(x,t)\,dQ.$$

Since

$$\left|\sum_{i=1}^{n}\int_0^T\int_0^t K_i(t,\tau)u_{x_i}(x,\tau)\,d\tau\, v_{x_i}(x,t)\,dQ\right| =$$

$$= \left|\sum_{i=1}^{n}\int_0^t K_i(t,\tau)u_{x_i}(x,\tau)\,d\tau\,\overline{v}_{x_i}(x,t)\Big|_0^T - \int_0^T\sum_{i=1}^{n}\frac{\partial}{\partial t}\left(\int_0^t K_i(t,\tau)u_{x_i}(x,\tau)\,d\tau\right)\overline{v}_{x_i}(x,t)\,dQ\right| =$$

$$= \left|\int_0^T\sum_{i=1}^{n}\left(\int_0^t K_i'(t,\tau)u_{tx_i}(x,\tau)\,d\tau + K_i(t,t)u_{x_i}(x,t)\right)\overline{v}_{x_i}(x,t)\,dQ\right| \leq$$

$$\leq c_K \sum_{i=1}^{n} \int_0^T \left(\int_0^t \left| u_{tx_i}(x,\tau) \right| d\tau \right) \overline{v}_{x_i}(x,t)dt + \int_0^T u_{x_i}(x,t)\overline{v}_{x_i}(x,t)dt \leq$$

$$\leq c_K \sum_{i=1}^{n} \int_0^T \left(\int_0^t \left| u_{tx_i}(x,\tau) \right| d\tau \right)^2 dt \int_0^T \overline{v}_{x_i}^2(x,t)dt + \int_0^T u_{x_i}^2(x,t)dt \int_0^T \overline{v}_{x_i}^2(x,t)dt \leq$$

$$\leq c_K \sum_{i=1}^{n} \int_0^T T \int_0^T u_{tx_i}^2(x,\tau)d\tau dt \int_0^T \overline{v}_{x_i}^2(x,t)dt + \int_0^T u_{x_i}^2(x,t)dt \int_0^T \overline{v}_{x_i}^2(x,t)dt =$$

$$= c_K \sum_{i=1}^{n} T^2 \int_0^T u_{tx_i}^2(x,t)dt \int_0^T \overline{v}_{x_i}^2(x,t)dt + \int_0^T u_{x_i}^2(x,t)dt \int_0^T \overline{v}_{x_i}^2(x,t)dt,$$

it follows that

$$\left| (Bu,v)_{L_2(Q)} \right| \leq c \|u\|_{H_0^2} \|v\|_{H_T^0},$$

for some constant $c > 0$. Dividing by $\| v \|_{H_T^0}$ and taking the supremum over all smooth functions v, and considering the density of the latter in the space H_T^0, we obtain

$$\| Bu \|_{(H_T^0)^-} \leq c \| u \|_{H_0^2}.$$

Now using the result from Lemma 6.4, we have:

$$\| Lu \|_{(H_T^0)^-} \leq \| L_D u \|_{(H_T^0)^-} + \| Bu \|_{(H_T^0)^-} \leq 2c \| u \|_{H_0^2}.$$

The lemma is proved.

REMARK 6.3.– The inequality proven in Lemma 6.9 allows us to extend the operator L from the domain $C_0^\infty(Q)$ to the space H_0^2 by continuity.

We will use the same notation for the extended operator and note that the proven inequality will hold for all functions $u \in H_0^2$.

REMARK 6.4.– Similarly, it can be shown that there exists a constant $c > 0$ such that for any function $v \in H_T^2$, the inequality $\| L^* v \|_{(H_0^0)^-} \leq c \| v \|_{H_T^2}$ holds.

LEMMA 6.10.– *There exists a constant* $c > 0$ *such that for any function* $u \in H_0^2$, *the inequality*

$$c^{-1} \| u \|_{S_0^1} \leq \| Lu \|_{(V_T^0)^-}$$

holds.

PROOF.– *Consider the value of the functional* $Lu \in (H_T^0)^- \subset (V_T^0)^-$ *on the element* $v(x, t)$*, where* $v(x,t)$ *is defined as in Lemma 6.6 according to [6.19]. It has been shown above that for any* $d > 0$*, a number* M *can be determined such that the differential part of the operator* L *satisfies the inequality*:

$$\langle v, L_D u \rangle_V \geq d \int_Q e^{Mt} v^2 \, dQ + d \int_Q e^{Mt} \overline{v}^2 \, dQ + d \int_Q e^{-Mt} u_t^2 \, dQ$$

$$+ \frac{1}{4} \int_\Omega e^{-MT} u_t^2 \Big|_{t=T} d\Omega + \frac{\alpha}{8} \sum_{i=1}^n \int_\Omega \overline{v}_{x_i}^2 \Big|_{t=0} d\Omega$$

$$+ d \sum_{i=1}^n \int_Q e^{Mt} \overline{v}_{x_i}^2 \, dQ + \frac{\alpha}{4} \sum_{i=1}^n \int_\Omega e^{-MT} u_{x_i}^2 \Big|_{t=T} d\Omega + d \sum_{i=1}^n \int_Q e^{-Mt} u_{x_i}^2 \, dQ.$$

Using Lemma [6.8], we have

$$\left| \langle v, Bu \rangle_V \right| \leq c_B \sum_{i=1}^n \int_Q e^{Mt} \overline{v}_{x_i}^2 + e^{-Mt} u_{x_i}^2 \, dQ.$$

Next, choosing $d \geq c_B$ and using the triangle inequality, we obtain that for some constants $c > 0$:

$$\langle v, Lu \rangle_{V_T^0} \geq c^{-1} (\| u \|_{S_0^1}^2 + \| v \|_{V_T^0}^2).$$

Applying the Cauchy–Schwarz inequality on the left and the Schwarz inequality on the right, we get:

$$\| Lu \|_{(V_T^0)^-} \| v \|_{V_T^0} \geq \langle v, Lu \rangle_{V_T^0} \geq 2c^{-1} \| u \|_{S_0^1} \| v \|_{V_T^0} .$$

Dividing by $\| v \|_{V_T^0}$, we obtain the desired inequality. The lemma is proved.

REMARK 6.5.– *Repeating the proof of Lemma 6.10 for the adjoint operator, we obtain that there exists a constant $c > 0$ such that for any function $v \in H_T^2$ the inequality $\| v \|_{S_T^1} \leq c_1 \| L^* v \|_{(V_0^0)^-}$ holds.*

Thus, the following theorem holds.

THEOREM 6.3.– *For some positive constants c_1, c_2, c_3, the inequalities*

$$\|u\|_{S_0^1} \leq c_1 \|Lu\|_{(V_T^0)^-} \leq c_2 \|Lu\|_{(H_T^0)^-} \leq c_3 \|u\|_{H_0^2}, \tag{6.25}$$

$$\|v\|_{S_T^1} \leq c_1 \left\|L^* v\right\|_{(V_0^0)^-} \leq c_2 \left\|L^* v\right\|_{(H_0^0)^-} \leq c_3 \|v\|_{H_T^2} \tag{6.26}$$

are valid for all $u \in H_0^2$ and $v \in H_T^2$.

Now, Lemma 6.1 is straightforward.

6.6. Example of an optimal control problem

For simplicity, assume that the domain Ω is cylindrical with respect to the variable x_1, i.e. $\Omega = [a,b] \times \Omega'$. Let $Q' = \Omega' \times (0,T)$, and consider the auxiliary space $W(Q')$, which is the completion of the set $C^\infty(Q')$ with respect to the norm

$$\| \varphi \|_W = \left(\int_{Q'} \varphi^2 + (\varphi')^2 \, dQ' \right)^{\frac{1}{2}}. \tag{6.27}$$

Here and later, φ' denotes the derivative of the function φ with respect to the variable t.

Consider the optimal control problem for a system whose evolution is described by the equation

$$Lu = f + \mathcal{A}h, \qquad [6.28]$$

with the control operator

$$\mathcal{A}h = \sum_{k=1}^{d} \delta(x_1 - x_{1,k}) \otimes \varphi_k(x', t), \qquad [6.29]$$

where

$$x_1, x_{1,k} \in [a,b], \quad x' = (x_2, x_3, \ldots, x_n), \quad \varphi_k(x', t) \in \mathcal{W}(Q').$$

Here, $h = \{(x_{1,k}, \varphi_k(x',t))\}_{k=1}^{d}$ is the control, belonging to some set of admissible controls $\mathcal{U}$ from the control space $\mathcal{H} = \left([a,b] \times \mathcal{W}(Q') \right)^d$.

By $\mathcal{A}h$, we understand the functional acting on smooth functions in $\bar{Q}$ in the following way

$$(\mathcal{A}h)(v) = \sum_{k-1}^{d} \int_{Q'} v(x_{1,k}, x', t) \varphi_k(x', t) dQ'.$$

Recall that $\bar{v}(x,t) = \int_T^t v(x,\xi) d\xi$. We also introduce the auxiliary notation $\bar{v}(s,x',t) = \int_T^t v(s,x',\xi) d\xi$. Using an approach from Anikushyn and Hranishak (2024), we can use the following properties.

LEMMA 6.11.– *Let* v *be a smooth function and* $\varphi \in \mathcal{W}(Q')$. *For any* $a_1, b_1 \in [a,b]$, *the following inequality holds*

$$\left| \int_{Q'} \int_{a_1}^{b_1} v(s,x',t) ds\, \varphi(x',t) dQ' \right| \le c_\Omega |b_1 - a_1|^{\frac{1}{2}} \|\varphi\|_{\mathcal{W}} \left(\|\bar{v}|_{t=0}\|_{L_2(\Omega)} + \|\bar{v}\|_{L_2(Q)} \right),$$

where c_Ω *is a constant depends only on the domain* Ω.

PROOF.– *Indeed, integrating by parts with respect to t, we get*

$$\int_0^T \int_{a_1}^{b_1} v(s,x',t)ds\varphi(x',t)dt =$$

$$= \int_T^t \int_{a_1}^{b_1} v(s,x',\xi)dsd\xi\varphi(x',t)\Bigg|_{t=0}^{t=T} - \int_0^T \int_T^t \int_{a_1}^{b_1} v(s,x',\xi)dsd\xi\varphi'(x',t)dt =$$

$$= \int_0^T \int_{a_1}^{b_1} v(s,x',\xi)dsd\xi\varphi(x',0) - \int_0^T \int_T^t \int_{a_1}^{b_1} v(s,x',\xi)dsd\xi\varphi'(x',t)dt =$$

$$= \int_{a_1}^{b_1} \int_0^T v(s,x',\xi)d\xi ds\varphi(x',0) - \int_0^T \int_{a_1}^{b_1} \int_T^t v(s,x',\xi)d\xi ds\varphi'(x',t)dt =$$

$$= -\int_{a_1}^{b_1} \overline{v}(s,x',0)ds\varphi(x',0) - \int_0^T \int_{a_1}^{b_1} \overline{v}(s,x',t)ds\varphi'(x',t)dt.$$

By the Cauchy–Schwarz inequality, we get

$$\left|\int_{a_1}^{b_1} \overline{v}(s,x',0)ds\right| \le \left(\int_{a_1}^{b_1} ds \int_{a_1}^{b_1} \overline{v}^2(s,x',0)ds\right)^{\frac{1}{2}} \le \left(|b_1 - a_1| \int_a^b \overline{v}^2(s,x',0)ds\right)^{\frac{1}{2}},$$

$$\left|\int_{a_1}^{b_1} \overline{v}(s,x',t)ds\right| \le \left(\int_{a_1}^{b_1} ds \int_{a_1}^{b_1} \overline{v}^2(s,x',t)ds\right)^{\frac{1}{2}} \le \left(|b_1 - a_1| \int_a^b \overline{v}^2(s,x',t)ds\right)^{\frac{1}{2}}.$$

Therefore,

$$\left|\int_0^T \int_{a_1}^{b_1} v(s,x',t)ds\varphi(x',t)dt\right| \le$$

$$\le |b_1 - a_1|^{\frac{1}{2}} \left(\left(\int_a^b \overline{v}^2(s,x',0)ds\right)^{\frac{1}{2}} |\varphi(x',0)| + \int_0^T \left(\int_a^b \overline{v}^2(s,x',t)ds\right)^{\frac{1}{2}} |\varphi'(x',t)|dt\right) \le$$

$$\leq |b_1 - a_1|^{\frac{1}{2}} \left(\left(\int_a^b \bar{v}^2(s,x',0)ds \right)^{\frac{1}{2}} |\varphi(x',0)| + \left(\int_0^T \int_a^b \bar{v}^2(s,x',t)ds\,dt \int_0^T (\varphi'(x',t))^2 dt \right)^{\frac{1}{2}} \right).$$

Integrating the last inequality over the domain Ω' and applying the Cauchy–Schwarz inequality again to both terms, we obtain

$$\left| \int_{Q'} \int_{a_1}^{b_1} v(s,x',t)ds\,\varphi(x',t)dQ' \right| \leq$$

$$\leq |b_1 - a_1|^{\frac{1}{2}} \int_{\Omega'} \left(\int_a^b \bar{v}^2(s,x',0)ds \right)^{\frac{1}{2}} |\varphi(x',0)|$$

$$+ \left(\int_0^T \int_a^b \bar{v}^2(s,x',t)ds\,dt \int_0^T (\varphi'(x',t))^2 dQ' \right)^{\frac{1}{2}} d\Omega' \leq$$

$$\leq |b_1 - a_1|^{\frac{1}{2}}$$

$$\left(\left(\int_\Omega \bar{v}^2(x,0)d\Omega \int_{\Omega'} \varphi^2(x',0)d\Omega' \right)^{\frac{1}{2}} + \left(\int_Q \bar{v}^2(x,t)dQ \int_{Q'} (\varphi'(x',t))^2 dQ' \right)^{\frac{1}{2}} \right) =$$

$$= |b_1 - a_1|^{\frac{1}{2}} \left(\|\bar{v}|_{t=0}\|_{L_2(\Omega)} \|\varphi|_{t=0}\|_{L_2(\Omega')} + \|\bar{v}\|_{L_2(Q)} \|\varphi\|_W \right). \qquad [6.30]$$

As is known (Berezanskyi 1968), for some constants c_Ω, the following inequality holds for any $x' \in \Omega'$:

$$\varphi^2(x',0) \leq c_\Omega \int_0^T \varphi^2(x',t) + (\varphi'(x',t))^2 dt,$$

hence,

$$\|\varphi|_{t=0}\|_{L_2(\Omega')} = \int_{\Omega'} \varphi^2(x',0)d\Omega' \leq c_\Omega$$
$$\int_{Q'} \varphi^2(x',t) + (\varphi'(x',t))^2 dQ' = c_\Omega \|\varphi\|_W. \qquad [6.31]$$

Combining [6.30] and [6.31], we obtain

$$\left|\int_{Q'}\int_{a_1}^{b_1} v(s,x',t)ds\,\varphi(x',t)dQ'\right| \le c_\Omega |b_1-a_1|^{\frac{1}{2}}\left(\|\bar{v}|_{t=0}\|_{L_2(\Omega)}\|\varphi\|_{\mathcal{W}}+\|\bar{v}\|_{L_2(Q)}\|\varphi\|_{\mathcal{W}}\right).$$

LEMMA 6.12.– *The linear functional* $l_A(v)=(\mathcal{A}h)(v)$ *can be extended to a continuous functional on the space* V_T^0.

PROOF.– *We will show that for smooth functions* $v\in C_T^{\infty}(Q)$, *the inequality* $|l_A(v)|\le c\|v\|_{\mathcal{W}V_T^0}$ *holds. For each* k, *by virtue of Lemma 6.11, we have*

$$\left|\int_{Q'} v(x_{1,k},x',t)\varphi_k(x',t)\,dQ'\right| = \left|\int_{Q'}\left(v(x_{1,k},x',t)-v(a,x',t)\right)\varphi_k(x',t)\,dQ'\right| =$$

$$= \left|\int_{Q'}\int_{a}^{x_{1,k}} v_{x_1}(s,x',t)ds\,\varphi_k(x',t)dQ'\right| \le$$

$$\le c_\Omega |x_{1,k}-a|^{\frac{1}{2}}\|\varphi_k\|_{\mathcal{W}}\left(\|\bar{v}_{x_1}|_{t=0}\|_{L_2(\Omega)}+\|\bar{v}_{x_1}\|_{L_2(Q)}\right)\le$$

$$\le c_\Omega |b-a|^{\frac{1}{2}}\|\varphi_k\|_{\mathcal{W}}\|v\|_{V_T^0}.$$

Therefore, in general,

$$|l_A(v)|\le c_\Omega |b-a|^{\frac{1}{2}}\sum_{k=1}^{d}\|\varphi_k\|_{\mathcal{W}}\|v\|_{V_T^0}.$$

Thus, the boundedness of the functional l_A on the set $C_T^{\infty}(Q)$ is proven, and hence, l_A can be extended to the entire space V_T^0 by continuity. That is, we can assume that $(\mathcal{A}h)(\cdot)\in(V_T^0)^-$.

We outline, that in case $f\in(V_T^0)^-$, then, it follows from Lemma 6.12 that $F(\cdot)=f+(\mathcal{A}h)(\cdot)\in(V_T^0)^-$. Therefore, due to Lemma 6.2, there exists the unique solution to the problem [6.28] in the sense of Definition 6.1.

LEMMA 6.13.– *The operator $\mathcal{A}$ defined by the equation [6.29] is weakly continuous.*

PROOF.– Let $\{h^{(m)}\}_{m=1}^{\infty}$ be a weakly convergent sequence in the space $\mathcal{H}$, i.e. $h^{(m)} \xrightarrow{w} \overset{*}{h}$. Then, $x_{1,i}^{(m)} \to \overset{*}{x}_{1,i}$ in the space $\mathbb{R}$ and $\varphi^{(m)} \xrightarrow{w} \overset{*}{\varphi}$ in the space $\mathcal{W}(Q')$.

First, note that due to Lemma [6.11], the following inequality holds:

$$\left\| \delta(x_1 - x_{1,k}^{(m)}) \otimes \varphi_k^{(m)}(x',t) - \delta(x_1 - \overset{*}{x}_{1,k}) \otimes \varphi_k^{(m)}(x',t) \right\|_{(V_T^0)^-} =$$

$$= \sup_{v \in C_T^\infty} \frac{1}{\|v\|_{V_T^0}} \int_{Q'} \left(v(x_{1,k}^{(m)}, x', t) - v(\overset{*}{x}_{1,k}, x', t) \right) \varphi_k^{(m)}(x',t) dQ' =$$

$$= \sup_{v \in C_T^\infty} \frac{1}{\|v\|_{V_T^0}} \int_{Q'} \int_{\overset{*}{x}_{1,k}}^{x_{1,k}^{(m)}} v_{x_1}(s, x', t) ds \, \varphi_k^{(m)}(x',t) dQ' \le$$

$$\le c_\Omega \, | \overset{*}{x}_{1,k} - x_{1,k}^{(m)} |^{\frac{1}{2}} \sup_{v \in C_T^\infty} \frac{1}{\|v\|_{V_T^0}} \left\| \varphi_k^{(m)} \right\|_{\mathcal{W}} \left(\left\| \overline{v}_{x_1} |_{t=0} \right\|_{L_2(\Omega)} + \left\| \overline{v}_{x_1} \right\|_{L_2(Q)} \right) \le$$

$$\le c_\Omega \, | \overset{*}{x}_{1,k} - x_{1,k}^{(m)} |^{\frac{1}{2}} \left\| \varphi_k^{(m)} \right\|_{\mathcal{W}}.$$

Since the sequence $\varphi^{(m)}$ is weakly convergent in the space $\mathcal{W}(Q')$, it is bounded in norm. Given that $| \overset{*}{x}_{1,k} - x_{1,k}^{(m)} | \to 0$ as $m \to \infty$, we obtain

$$\sum_{k=1}^{d} \left\| \delta(x_1 - x_{1,k}^{(m)}) \otimes \varphi_k^{(m)}(x',t) - \delta(x_1 - \overset{*}{x}_{1,k}) \otimes \varphi_k^{(m)}(x',t) \right\| \to 0, m \to \infty. \tag{6.32}$$

It is also straightforward to show that $\varphi_k^{(m)} \xrightarrow{w} \overset{*}{\varphi}_k$ in the space $L_2(Q')$, therefore

$$\left\langle v, \sum_{k=1}^{d} \delta(x_1 - \overset{*}{x}_{1,k}) \otimes \varphi_k^{(m)}(x',t) \right\rangle_V = \tag{6.33}$$

$$=\sum_{k=1}^{d}\int_{Q'} v(\overset{*}{x}_{1,k},x',t)\varphi_k^{(m)}(x',t)dQ' \to \sum_{k=1}^{d}\int_{Q'} v(\overset{*}{x}_{1,k},x',t)\overset{*}{\varphi}_k(x',t)dQ' =$$

$$=\left\langle v,\sum_{k=1}^{d}\delta(x_1-\overset{*}{x}_{1,k})\otimes\overset{*}{\varphi}_k(x',t)\right\rangle_{\mathrm{V}} = \langle \mathcal{A}h^*,v\rangle_{\mathrm{V}}.$$

Now, for each fixed $v\in C_T^\infty(Q)$, we have

$$\left\langle v,\sum_{k=1}^{d}\delta(x_1-x_{1,k}^{(m)})\otimes\varphi_k^{(m)}(x',t)\right\rangle_{\mathrm{V}} = \left\langle v,\sum_{k=1}^{d}\delta(x_1-\overset{*}{x}_{1,k})\otimes\varphi_k^{(m)}(x',t)\right\rangle_{\mathrm{V}} +$$
$$+\left\langle v,\sum_{k=1}^{d}\delta(x_1-x_{1,k}^{(m)})\otimes\varphi_k^{(m)}(x',t)-\delta(x_1-\overset{*}{x}_{1,k})\otimes\varphi_k^{(m)}(x',t)\right\rangle_{\mathrm{V}}.$$

The first term converges to $\langle v,\mathcal{A}h^*\rangle_{\mathrm{V}}$ due to [6.33], while for the second term, due to [6.32], we have

$$\left|\left\langle v,\sum_{k=1}^{d}\delta(x_1-x_{1,k}^{(m)})\otimes\varphi_k^{(m)}(x',t)-\delta(x_1-\overset{*}{x}_{1,k})\otimes\varphi_k^{(m)}(x',t)\right\rangle_{\mathrm{V}}\right| \le$$

$$\le\sum_{k=1}^{d}\left\|\delta(x_1-x_{1,k}^{(m)})\otimes\varphi_k^{(m)}(x',t)-\delta(x_1-\overset{*}{x}_{1,k})\otimes\varphi_k^{(m)}(x',t)\right\|_{(V_T^0)^-}\|v\|_{V_T^0}\to 0,$$

as $m\to\infty$. Thus, $\langle v,\mathcal{A}h^{(m)}\rangle_{\mathrm{V}}\to\langle v,\mathcal{A}h^*\rangle_{\mathrm{V}}$, which concludes the proof.

Based on Theorem 6.1, considering Lemma 6.13 and inequalities [6.25] and [6.26], we can conclude the following theorem.

THEOREM 6.5.– *Assume that the system is described by the problem [6.28] with the operator* $\mathcal{A}$ *defined in [6.29] and the control space* $\mathcal{H}$. *Furthermore, let the set of allowable controls* $\mathcal{U}$ *be closed, convex, and bounded in* $\mathcal{H}$, *and that the quality criterion* $\Phi(\cdot):S_0^1\to\mathbb{R}$ *is weakly lower semi continuous with respect to the system state* $u(t,x,h)$ *and bounded below. Then, there exists an optimal control for the system [6.28].*

6.7. Conclusion

In this chapter, we addressed the problem of optimal control for hyperbolic integro-differential systems, particularly those influenced by memory effects commonly observed in viscoelastic materials. The study extends existing theoretical frameworks by establishing a priori inequalities for operators in newly defined function spaces. We demonstrated the existence, uniqueness and continuous dependence of solutions on the system's input parameters, confirming the well-posedness of the proposed models. Furthermore, we proved the existence of optimal controls under specific conditions, such as the weak compactness and continuity of the admissible control set. An illustrative example for a cylindrical domain validates the theoretical findings. These results advance the understanding of generalized control for systems governed by integro-differential equations and provide a foundation for future applications in viscoelasticity and other fields.

6.8. References

Anikushyn, A. (2013). Generalized solvability of hyperbolic integro-differential equations. *Bull. Taras Shevchenko Nat. Univ. Kyiv Ser. Phys. Math.*, 4, 60–65 (in Ukrainian).

Anikushyn, A. (2014). Generalized optimal control for systems described by linear integro-differential equations with nonnegative definite integral operators. *J. Automat. Inf. Sci.*, 46(6), 58–67.

Anikushyn, A. (2024). Pulse and point optimal control of linear hyperbolic integro-differential systems with mixed boundary conditions. *J. Optim. Differ. Equ. Appl.*, 32(2), 137–155.

Anikushyn, A. and Andaral, A. (2025). Generalized optimal control for pseudoparabolic integro-differential systems. *J. Math. Sci.* 291, 901–918. doi. 10.1007/s10958-025 07867-1.

Anikushyn, A. and Hranishak, Kh. (2024). Point control of linear hyperbolic integro-differential systems. *Bull. Taras Shevchenko Nat. Univ. Kyiv Ser. Physics Math.*, 79(2), 20–28.

Anikushyn, A. and Hulianytskyi, A. (2014). Generalized solvability of parabolic integro-differential equations. *Diff. Equat.*, 50(1), 98–109.

Anikushyn, A. and Kosteeva, L. (2018). A priori estimates and generalized solvability of the initial boundary value problem for one integro-differential equation. *Journal of Computational and Applied Mathematics*, 3, 2–14 (in Ukrainian).

Anikushyn, A. and Nomirovskyi, D. (2008). Trajectory-financial controllability of hyperbolic systems in different classes of generalized functions. *Bull. Taras Shevchenko Nat. Univ. Kyiv Ser. Phys. Math.*, 3, 119–124 (in Ukrainian).

Anikushyn, A. and Nomirovskyi, D. (2011). Generalized solutions for linear operators with weakened a priori inequalities. *Ukrainian Math. J.*, 62(8), 1175–1186.

Anikushyn, A. and Pokojovy, M. (2017). Multidimensional thermoelasticity for nonsimple materials–well-posedness and long-time behavior. *Appl. Anal.*, 96(9), 1561–1585.

Anikushyn, A. and Pokojovy, M. (2019). Global well-posedness and exponential stability for heterogeneous anisotropic Maxwell's equations under a nonlinear boundary feedback with delay. *J. Math. Anal. Appl.*, 475(1), 278–312.

Anikushyn, A., Lyashko, V., Samosonok, O. (2024). Optimal control of the systems governed by linear hyperbolic integro-differential equations. *Bull. Taras Shevchenko Nat. Univ. Kyiv Ser. Phys. Math.*, 78(1), 111–118.

Anikushyn, A., Hranishak, Kh., Lyashko, V., Samosonok, O. (2025). Optimal control of hyperbolic integro-differential systems. *Cybern. Syst. Anal.*, 61, 66–76.

Berezanskyi, Ju. (1968). *Expansions in Eigenfunctions of Selfadjoint Operators*. American Mathematical Society, Rhode Island.

Cannarsa, P. and Sforza, D. (2011). Integro-differential equations of hyperbolic type with positive definite kernels. *J. Differ. Equ.*, 250(12), 4289–4335.

Conti, M., Gatti, S., Pata, V. (2008). Uniform decay properties of linear Volterra integro-differential equations. *Math. Models Methods Appl. Sci.*, 18(1), 21–45.

Dafermos, C. (1970). An abstract Volterra equation with applications to linear viscoelasticity. *J. Differ. Equ.*, 7(3), 554–569.

Demchenko, H., Anikushyn, A., Pokojovy M. (2019). On a Kelvin-Voigt viscoelastic wave equation with strong delay. *SIAM J. Math. Anal.*, 468(2019), 4382–4412.

Diagana, T. (2014). Existence results for some damped second-order Volterra integro-differential equations. *Appl. Math. Comput.*, 237, 304–317.

Didenko, V.P. (1972). On boundary value problems for multidimensional hyperbolic equations with degeneracy. *Proc. USSR Acad. Sci.*, 205(4), 763–766.

Engler, H. (1991). Weak solutions of a class of quasilinear hyperbolic integro-differential equations describing viscoelastic materials. *Arch. Ration. Mech. Anal.*, 113, 1–38.

Evans, L. (2010). *Partial Differential Equations*. American Mathematical Society, Providence, Rhode Island.

Fabrizio, M. and Morro, A. (1992). *Mathematical Problems in Linear Viscoelasticity*. Society for Industrial and Applied Mathematics, Philadelphia.

Gan, X.-T. and Yin, J.-F. (2015). Symmetric finite volume element approximations of second order linear hyperbolic integro-differential equations. *Computers and Mathematics with Applications*, 70(10), 2589–2600.

Hulianytskyi, A. and Semenov, V. (2012). Integro-differential pseudoparabolic systems: A priori estimates and impulse-point controllability. *Proc. NAS Ukraine*, 4, 43–49 (in Ukrainian).

Karaa, S., Pani, A., Yadav, S. (2015). A priori hp-estimates for discontinuous Galerkin approximations to linear hyperbolic integro-differential equations. *Appl. Numer. Math.*, 96, 1–23.

Kelly, P. (2013). *Solid Mechanics Lecture Notes*. The University of Auckland, Auckland.

Klyushin, D., Lyashko, S., Nomirovskii, D., Petunin, Yu., Semenov, V. (2011). *Generalized Solutions of Operator Equations and Extreme Elements*. Springer, New York.

Lakes, R. (1998). *Viscoelastic Solids*. CRC Press, Boca Raton, FL.

Loreti, P. and Sforza, D. (2010). Reachability problems for a class of integro-differential equations. *J. Differ. Equ.*, 248(7), 1711–1755.

Lyashko, S. (2002). *Generalized Optimal Control of Linear Systems with Distributed Parameters*. Springer, New York.

Lyashko, S. and Nomirovskii, D. (2003a). Generalized solutions and optimal controls in systems describing the dynamics of a viscous stratified fluid. *Diff. Equat.*, 39(1), 90–98.

Lyashko, S. and Nomirovskii, D. (2003b). The generalized solvability and optimization of parabolic systems in domains with thin low-permeable inclusions. *Cybern. Syst. Anal.*, 39, 737–745.

Lyashko, S. and Semenov, V. (2001). Controllability of linear distributed systems in classes of generalized actions. *Cybern. Syst. Anal.*, 37, 13–32.

Lyashko, S., Nomirovskii, D., Sergienko, T. (2001). Trajectory and final controllability in hyperbolic and pseudohyperbolic systems with generalized actions. *Cybern. Syst. Anal.*, 37, 756–763.

Nomirovskii, D. (1999). The control of pseudohyperbolic systems. *J. Math. Sci.*, 97(2), 3945–3951.

Nomirovskii, D. (2004). Generalized solvability of parabolic systems with nonhomogeneous transmission conditions of nonideal contact type. *Diff. Equat.*, 40(10), 1467–1477.

Nomirovskii, D. (2006). Unique solvability of pseudohyperbolic equations with singular right-hand sides. *Math. Notes*, 80, 550–562.

Nomirovskii, D. (2007). Generalized solvability and optimization of a parabolic system with a discontinuous solution. *J. Differ. Equ.*, 233(1), 1–21.

Nomirovskii, D. and Vostrikov, O. (2016). Generalized statements and properties of models of transport processes in domains with cuts. *Cybern. Syst. Anal.*, 52, 931–942.

Renardy, M., Hrusa, W., Nohel, J. (1987). *Mathematical Problems in Viscoelasticity*. Wiley, Hoboken, NJ.

Saedpanah, F. (2014). Well-posedness of an integro-differential equation with positive type kernels modeling fractional order viscoelasticity. *Eur. J. Mech. A Solids*, 44, 201–211.

Semenov, V. (2005). Solvability of a parabolic transmission problem with the condition of a generalized proper lumped source. *Diff. Equat.*, 41, 878–886.

Semenov, V. and Denisov, S. (2005). Optimization and generalized solution in conjugation problems of parabolic systems. *J. Automat. Inf. Scien.*, 37(1), 47–61.

Semenov, V. and Denisov, S. (2023). Impulse trajectory and final controllability of parabolic–hyperbolic systems. *Cybern. Syst. Anal.*, 59(3), 417–427.

Sergienko, I., Khimich, O., Klyushin, D., Lyashko, V., Lyashko, S., Semenov, V. (2023). Formation and development of the scientific school of the mathematical theory of filtration. *Cybern. Syst. Anal.*, 59, 61–70.

Shen, W., Yang, D., Liu, W. (2014). Optimal control problem governed by a linear hyperbolic integro-differential equation and its finite element analysis. *Bound. Value Probl.*, 173.

Tymchyshyn, I. and Nomirovskii, D. (2021). Generalized solvability of a parabolic model describing transfer processes in domains with thin inclusions. *Diff. Equat.*, 57, 1053–1062.

7

Self-adaptive Operator Extrapolation Method for Operator Inclusions in Banach Space

V. SEMENOV[1] **and S. DENYSOV**[2]
[1]*Taras Shevchenko National University of Kyiv, Ukraine*
[2]*Harbour.Space Institute of Technology, Barcelona, Spain*

7.1. Introduction

In this chapter, an iterative splitting algorithm for solving operator inclusion with the sum of the maximal monotone operator and the monotone Lipschitz continuous operator in the Banach space is proposed and studied. The proposed algorithm is an adaptive variant of the "forward-reflected-backward algorithm" developed in Malitsky and Tam (2020), where the rule used to update the step size does not require knowledge of the Lipschitz constant of the operator.

An attractive feature of the algorithm is only one calculation on the iterative step of the resolvent of the maximal monotone operator and the value of the monotone Lipschitz continuous operator. For operator inclusions in two-uniformly convex and uniformly smooth Banach space, the theorem on the weak convergence of the method is proved.

Computational Methods and Mathematical Modeling
in Cyberphysics and Engineering Applications 2,
coordinated by Dmitri KOROLIOUK, Sergiy LYASHKO and Nikolaos LIMNIOS.

This chapter is based on the authors' preliminary articles (Denisov and Semenov 2021; Semenov et al. 2022; Vedel et al. 2027).

Let E be a real Banach space with its dual space E^*. This chapter studies the monotone operator inclusion:

$$\text{find } x \in E \text{ such that } 0 \in (A+B)x, \qquad [7.1]$$

where $A: E \to 2^{E^*}$ is the multivalued maximal monotone operator, and $B: E \to E^*$ is the monotone and Lipschitz continuous operator.

Many actual problems can be written in the form of [7.1]. Among them are variational inequalities and optimization problems arising in the field of optimal control, inverse problem theory, machine learning, image processing, operations research and mathematical physics (see Tibshirani 1996; Alber and Ryazantseva 2006; Lyashko et al. 2013; Raguet et al. 2013; Luke et al. 2020). A prominent example is the saddle problem that plays an important role in mathematical economics:

$$\min_{p \in P} \max_{q \in Q} F(p,q),$$

where $F: P \times Q \to R$ is a smooth convex–concave function, and $P \subseteq R^n$ and $Q \subseteq R^m$ are the closed convex sets, which can be formulated as:

$$\text{find } x \text{ such that } 0 \in (A+B)x,$$

where $x = (p,q) \in R^{n+m}$ and:

$$Ax = \begin{pmatrix} N_P p \\ N_Q q \end{pmatrix}, \quad Bx = \begin{pmatrix} \nabla_1 F(p,q) \\ -\nabla_2 F(p,q) \end{pmatrix},$$

N_P (N_Q) is a normal cone operator for closed convex set P (Q) (Bauschke and Combettes 2017).

Research and development of algorithms for operator inclusions [7.1] and related problems is a rapidly growing area of applied nonlinear analysis (Alber and Ryazantseva 2006; Semenov 2014; Bauschke and Combettes 2017).

The most well-known and popular iterative method for solving monotone operator inclusions [7.1] in Hilbert space is the "forward–backward algorithm" (FBA) (Lions and Mercier 1979; Passty 1979 ; Bauschke and Combettes 2017).

$$x_{n+1} = J_\lambda^A \left(x_n - \lambda B x_n \right),$$

where $J_\lambda^A = \left(I + \lambda A \right)^{-1}$ is the operator resolvent, $A : H \to 2^H$, $\lambda > 0$.

Note that the FBA scheme includes the well-known gradient method and proximal method as special cases (Bauschke and Combettes 2017). For inverse strongly monotone (cocoercive) operators $B : H \to H$, the FBA method is weakly converging (Bauschke and Combettes 2017). However, FBA may diverge for Lipschitz continuous monotone operators *B*.

The condition of the inverse strong monotonicity of the operator B is a rather strong assumption. To weaken it, Tseng (2000) proposed the following modification of the FBA:

$$\begin{cases} y_n = J_\lambda^A \left(x_n - \lambda B x_n \right), \\ x_{n+1} = y_n - \lambda \left(B y_n - B x_n \right), \end{cases}$$

where $B : H \to H$ is a monotone and Lipschitz continuous operator with constant $L > 0$ and $\lambda \in \left(0, L^{-1} \right)$.

Close to the Tseng method are the well-known extragradient method and the method of extrapolation from the past. Recent results regarding these algorithms can be found in Lyashko et al. (2011), Semenov (2018), Chabak et al. (2019), Denisov et al. (2019), Nomirovskii et al. (2019), Semenov et al. (2027).

A main limitation of the Tseng method is the two calls of B per iteration. Further development of this idea led to the so-called "forward-reflected-backward algorithm" (Malitsky and Tam 2020):

$$x_{n+1} = J_\lambda^A \left(x_n - \lambda B x_n - \lambda \left(B x_n - B x_{n-1} \right) \right), \ \lambda \in \left(0, \frac{1}{2L} \right),$$

and related method (Csetnek et al. 2019):

$$x_{n+1} = J_{\lambda}^{A}\left(x_n - \lambda Bx_n\right) - \lambda\left(Bx_n - Bx_{n-1}\right),\ \lambda \in \left(0, \frac{1}{3L}\right).$$

These schemes are naturally called the operator extrapolation schemes. Note that in the three above algorithms, there is a requirement to know operator's Lipschitz constant – which is often unknown or difficult to estimate. To overcome these difficulties, an adaptive strategy for updating parameter $\lambda > 0$ on each step has been proposed in Denisov et al. (2019).

Some progress has been achieved recently in the study of splitting algorithms for inclusions in Banach spaces (e.g. Alber and Ryazantseva 2006; Bonesky et al. 2008; Iiduka and Takahashi 2008; Aoyama and Kohsaka 2014; Takahashi and Takahashi 2016; Cholamjiak and Shehu 2019; Shehu 2019, 2020; Abass et al. 2020; Denisov and Semenov 2021; Vedel et al. 2021; Yang et al. 2021; Semenov et al. 2022. This is largely due to the wide involvement of theoretical results and designs of the geometry of Banach spaces (Diestel 1975; Beauzamy 1985; Cioranescu 1990; Alber 1996). The book by Alber and Ryazantseva (2006) contains extensive material on this topic. In the article by Iiduka and Takahashi (2008) for the solution of inclusions [7.1] in the two-uniformly convex and uniformly smooth Banach space, the next algorithm is proposed:

$$x_{n+1} = J_{\lambda}^{A} \circ J^{-1}\left(Jx_n - \lambda Bx_n\right), \tag{7.2}$$

where $J_{\lambda}^{A} = \left(J + \lambda A\right)^{-1} J$ is the resolvent of operator $A : E \to 2^{E^*}$, $\lambda > 0$, and J is the normalized duality mapping from E to E^*. For inverse strongly monotone (cocoercive) operators $B : E \to E^*$, method [7.2] weakly converges (Iiduka and Takahashi 2008). Recently, Shehu (2019) extended Tseng's result to two-uniformly convex and uniformly smooth Banach spaces. He proposed the following weakly converging process for approximating the solution of inclusion [7.1]:

$$\begin{cases} y_n = J_{\lambda}^{A} \circ J^{-1}\left(x_n - \lambda_n Bx_n\right), \\ x_{n+1} = J^{-1}\left(Jy_n - \lambda_n\left(By_n - Bx_n\right)\right), \end{cases} \tag{7.3}$$

where $\lambda_n > 0$ can be set using knowledge of the Lipschitz constant of the operator B or calculated with linear search. Note that two values of the operator B need to be calculated at the iteration step.

In this chapter, we propose and study a new splitting algorithm for solving operator inclusion [7.1] in Banach space. The algorithm is an adaptive variant of the Malitsky–Tam "forward-reflected-backward algorithm" (Malitsky and Tam 2020), where the step update rule does not require knowledge of the Lipschitz constant for operator B. The algorithm's advantage is only one computation at the iteration step of the resolvent of the maximal monotone operator A and the value of operator B. The method of weak convergence theorem is proved for operator inclusions in two-uniformly convex and uniformly smooth Banach space.

This chapter is structured as follows. Section 7.2 provides the necessary information from the areas of geometry of Banach spaces and theory of monotonous operators. The algorithm is described in section 7.3. Proof of convergence is presented in section 7.4. Theoretical applications to operator equations, convex minimization problems and variational inequalities are given in sections 7.5 and 7.6.

7.2. Preliminaries

Let us recall several concepts and facts of the geometry of Banach spaces (Vainberg 1974; Diestel 1975; Barbu 1976; Beauzamy 1985; Cioranescu 1990; Alber 1996; Agarwal et al. 2009), which are necessary for the formulation and proof of the results.

Let E be a real Banach space with norm $\|\cdot\|$, E^* is the dual space for E, $\langle x^*, x\rangle$ is a value of mapping $x^* \subset E^*$ on element $x \in E$ (i.e. $\langle x^*, x\rangle = x^*(x)$). Let us denote the norm in E^* as $\|\cdot\|_*$.

Let $S_E = \{x \in E : \|x\| = 1\}$. The Banach space is called strictly convex, if for all $x, y \in S_E$ and $x \neq y$, we have $\left\|\frac{x+y}{2}\right\| < 1$. The modulus of convexity of space E is defined as:

$$\delta_E(\varepsilon) = \inf\left\{1 - \left\|\frac{x+y}{2}\right\| : x, y \in S_E, \|x-y\| = \varepsilon\right\} \quad \forall \varepsilon \in (0,2].$$

A Banach space E is called uniformly convex, if $\delta_E(\varepsilon) > 0$ for all $\varepsilon \in (0,2]$. Banach space E is called two-uniformly convex, if such $c > 0$ exist that $\delta_E(\varepsilon) \geq c\varepsilon^2$ for all $\varepsilon \in (0,2]$.

Obviously, a two-uniformly convex space is uniformly convex. It is known that the uniformly convex Banach space is reflexive (Diestel 1975).

A Banach space E is called smooth if the limit:

$$\lim_{t\to 0}\frac{\|x+ty\|-\|x\|}{t} \tag{7.4}$$

exists for all $x, y \in S_E$. A Banach space E is called uniformly smooth if the limit [7.4] exists uniformly over $x, y \in S_E$. There is a duality between the convexity and smoothness of the Banach space E and its dual space E^* (see Beauzamy B. 1985; Diestel 1975): E^* is the strictly convex space $\Rightarrow$ E is the smooth space; E^* is the smooth space $\Rightarrow$ E is the strictly convex space; E is the uniformly convex space $\Rightarrow$ E^* is the uniformly smooth space; E is the uniformly smooth space $\Rightarrow$ E^* is the uniformly convex space. Note that with the reflexivity of space E, the first two implications can be reversed.

It is known that Hilbert spaces and spaces L_p $(1 < p \leq 2)$ are two-uniformly convex and uniformly smooth (spaces L_p are uniformly smooth for $p \in (1,\infty)$) (see Diestel 1975; Beauzamy 1985; Cioranescu 1990).

Also recall, Cioranescu (1990) and Bauschke and Combettes (2017) that a multivalued operator $A: E \to 2^{E^*}$ is called monotone if $\forall x, y \in E$

$$\langle u-v, x-y\rangle \geq 0 \;\; \forall u \in Ax, v \in Ay.$$

A monotone operator $A: E \to 2^{E^*}$ is called maximal monotone if for any monotone operator $B: E \to 2^{E^*}$, we have $\Gamma(A) \subseteq \Gamma(B)$ which implies $\Gamma(A) = \Gamma(B)$, where:

$$\Gamma(A) = \{(x,u) \in E \times E^* : u \in Ax\}$$

is a graph of A (Barbu 1976; Cioranescu 1990).

LEMMA 7.1. (CIORANESCU 1990).– *Let* $A: E \to 2^{E^*}$ *be a maximal monotone operator,* $x \in E$, $u \in E^*$. *Then:*

$$\langle u - v, x - y \rangle \geq 0 \ \ \forall (y, v) \in \Gamma(A) \ \Leftrightarrow \ (x, u) \in \Gamma(A).$$

It is established that if $A: E \to 2^{E^*}$ is the maximal monotone operator, and $B: E \to E^*$ is the Lipschitz continuous monotone operator, then $A+B$ is the maximal monotone operator (Barbu 1976; Cioranescu 1990).

Let us also recall (Bauschke and Combettes 2017) that operator $A: E \to E^*$ is called inverse strongly monotone (cocoercive) if a number $\alpha > 0$ exists (the constant of inverse strong monotonicity) so that:

$$\langle Ax - Ay, x - y \rangle \geq \alpha \|Ax - Ay\|^2.$$

An inverse strongly monotone operator is the Lipschitz continuous, but not every Lipschitz continuous operator is inverse strongly monotone.

Multivalued operator $J: E \to 2^{E^*}$, which acts as:

$$Jx = \left\{ x^* \in E^* : \langle x^*, x \rangle = \|x\|^2 = \|x^*\|_*^2 \right\},$$

is called normalized duality mapping. We know (see Cioranescu 1990; Alber 1996; Agarwal et al. 2009) that: if space E is smooth, then operator J is single-valued; if space E is strongly convex, then operator J is injective and strongly monotone; if space E is reflexive, then operator J is surjective; if space E is uniformly smooth, then operator J is uniformly continuous on bounded subsets of space E.

REMARK 7.1.– For a Hilbert space $J = I$. The explicit form of operator J for Banach spaces ℓ_p, L_p and W_p^m ($p \in (1, \infty)$) is provided in Cioranescu (1990) and Alber (1996).

Let E be a reflexive, strictly convex and smooth Banach space. The maximal monotonicity of operator $A: E \to 2^{E^*}$ is equivalent to equality $R(J + \lambda A) = E^*$ for all $\lambda > 0$ (Barbu 1976).

For the maximal monotone operator $A: E \to 2^{E^*}$ and $\lambda > 0$, the resolvent $J_\lambda^A : E \to E$ is defined as follows (Barbu 1976),

$$J_\lambda^A x = (J + \lambda A)^{-1} Jx, \; x \in E,$$

where J is the normalized duality mapping from E to E^*. We know that

$$A^{-1}0 = F(J_\lambda^A) = \{x \in E : \; J_\lambda^A x = x\} \; \forall \lambda > 0.$$

We also know that the set $A^{-1}0$ is closed and convex (Barbu 1976; Bauschke and Combettes 2017).

Let E be a smooth Banach space. Let us consider the functional introduced by Yakov Alber (1996):

$$D(x, y) = \|x\|^2 - 2\langle Jy, x\rangle + \|y\|^2 \; \forall x, y \in E.$$

The definition of D implies a useful three-point identity:

$$D(x,y) - D(x,z) - D(z,y) = 2\langle Jz - Jy, x - z\rangle \; \forall x, y, z \in E.$$

If the space E is strictly convex, then for $x, y \in E$, we have:

$$D(x,y) = 0 \Leftrightarrow x = y.$$

LEMMA 7.2. (BARBU 1976).– *Let E be a uniformly convex and uniformly smooth Banach space, and (x_n) and (y_n) the bounded sequences of elements from E. Then:*

$$\|x_n - y_n\| \to 0 \;\; \Leftrightarrow \;\; \|Jx_n - Jy_n\|_* \to 0 \;\; \Leftrightarrow \;\; D(x_n, y_n) \to 0.$$

LEMMA 7.3. (Bonesky et al. 2008; Aoyama and Kohsaka 2014).– *Let E be a two-uniformly convex and smooth Banach space. Then, for some $\mu \geq 1$, the next inequality holds:*

$$D(x, y) \geq \frac{1}{\mu}\|x - y\|^2 \; \forall x, y \in E.$$

REMARK 7.2.– For Banach spaces ℓ_p, L_p and W_p^m ($1<p\leq 2$), we have $\mu=\frac{1}{p-1}$ (Xu and Roach 1997). And for a Hilbert space, the inequality for Lemma 7.3 becomes the identity.

7.3. Algorithm

Let E be a two-uniformly convex and uniformly smooth Banach space. Let A be a multivalued operator acting from E into 2^{E^*}, and B be an operator acting from E into dual space E^*.

Consider the next operator inclusion problem:

$$\text{find } x\in E \text{ such that } 0\in (A+B)x, \tag{7.5}$$

Let us denote set of solutions of [7.5] as $(A+B)^{-1}0$.

Assume that the following conditions are satisfied:

– $A:E\to 2^{E^*}$ is a maximal monotone operator;

– $B:E\to E^*$ is a monotone and Lipschitz continuous operator with Lipschitz constant $L>0$;

– set $(A+B)^{-1}0$ is non-empty.

Operator inclusion [7.5] can be formulated as the problem of finding a fixed point:

$$\text{find } x\in E \text{ such that } x=J_\lambda^A\circ J^{-1}(Jx-\lambda Bx), \tag{7.6}$$

where $\lambda>0$. Formulation [7.6] is useful because it contains an obvious algorithmic idea. Calculation scheme:

$$x_{n+1}=J_\lambda^A\circ J^{-1}(Jx_n-\lambda Bx_n)$$

was studied in Iiduka and Takahashi (2008) for inverse strongly monotone operators $B:E\to E^*$ However, the scheme generally does not converge for Lipschitz continuous monotone operators.

Let us use the idea of work by Malitsky and Tam (2020) and consider the modified scheme:

$$x_{n+1} = J_{\lambda}^{A} \circ J^{-1}\left(Jx_n - \lambda Bx_n - \lambda\left(Bx_n - Bx_{n-1}\right)\right)$$

with the extrapolation term $-\lambda\left(Bx_n - Bx_{n-1}\right)$. Let us use the update rule for $\lambda > 0$ similar to the one from Denisov et al. (2019), Denisov and Semenov (2021) and Semenov et al. (2022) to exclude the explicit use of the Lipschitz constant of operator B.

We will assume that we know constant $\mu \geq 1$ from Lemma 7.3.

ALGORITHM 7.1. SELF-ADAPTIVE OPERATOR EXTRAPOLATION ALGORITHM.–

Choose some $x_0 \in E$, $x_1 \in E$, $\tau \in \left(0, \frac{1}{2\mu}\right)$ *and* $\lambda_0, \lambda_1 > 0$. *Set* $n = 1$.

7. Compute

$$x_{n+1} = J_{\lambda_n}^{A} \circ J^{-1}\left(Jx_n - \lambda_n Bx_n - \lambda_{n-1}\left(Bx_n - Bx_{n-1}\right)\right).$$

2. If $x_{n-1} = x_n = x_{n+1}$, *then STOP and* $x_n \in \left(A+B\right)^{-1} 0$, *else go to step 3.*

3. Compute

$$\lambda_{n+1} = \begin{cases} \min\left\{\lambda_n, \tau \frac{\left\|x_{n+1} - x_n\right\|}{\left\|Bx_{n+1} - Bx_n\right\|_*}\right\}, & \text{if } Bx_{n+1} \neq Bx_n, \\ \lambda_n, & \text{otherwise.} \end{cases}$$

Set $n := n+1$ *and go to 7.*

Sequence $\left(\lambda_n\right)$ which is created by the rule in step 3 is non-increasing and bounded from below by $\min\left\{\lambda_1, \tau L^{-1}\right\}$. So, $\lim_{n\to\infty} \lambda_n > 0$ exists.

Let us prove the convergence of Algorithm 7.1.

7.4. Convergence

In the beginning of this section, we state and prove the inequality on which the proof of Algorithm 7.1 weak convergence is based.

LEMMA 7.4.– *For the sequence* (x_n) *generated by Algorithm 7.1, the following inequality holds:*

$$D(z,x_{n+1})+2\lambda_n\langle Bx_n-Bx_{n+1},x_{n+1}-z\rangle+\tau\mu\frac{\lambda_n}{\lambda_{n+1}}D(x_{n+1},x_n)$$

$$\leq D(z,x_n)+2\lambda_{n-1}\langle Bx_{n-1}-Bx_n,x_n-z\rangle+\tau\mu\frac{\lambda_{n-1}}{\lambda_n}D(x_n,x_{n-1})$$

$$-\left(1-\tau\mu\frac{\lambda_{n-1}}{\lambda_n}-\tau\mu\frac{\lambda_n}{\lambda_{n+1}}\right)D(x_{n+1},x_n),$$

where $z\in(A+B)^{-1}0$.

PROOF.– Let $z\in(A+B)^{-1}0$. Then

$$-Bz\in Az.$$

Equality $x_{n+1}=J_{\lambda_n}^A\circ J^{-1}(Jx_n-\lambda_nBx_n-\lambda_{n-1}(Bx_n-Bx_{n-1}))$ can be formulated as:

$$Jx_n-\lambda_nBx_n-\lambda_{n-1}(Bx_n-Bx_{n-1})-Jx_{n+1}\in\lambda_nAx_{n+1}.$$

Use the monotonicity of operator A. We have:

$$\langle Jx_{n+1}+\lambda_n(Bx_n-Bz)+\lambda_{n-1}(Bx_n-Bx_{n-1})-Jx_n,z-x_{n+1}\rangle\geq 0. \qquad [7.7]$$

Let us write a three-point identity:

$$D(z,x_n)-D(z,x_{n+1})-D(x_{n+1},x_n)=2\langle Jx_{n+1}-Jx_n,z-x_{n+1}\rangle. \qquad [7.8]$$

Using [7.8] within [7.7], we get:

$$D(z,x_{n+1}) \le D(z,x_n) - D(x_{n+1},x_n)$$

$$+2\langle \lambda_n (Bx_n - Bz) + \lambda_{n-1}(Bx_n - Bx_{n-1}), z - x_{n+1}\rangle . \qquad [7.9]$$

Using the monotonicity of operator B. We have:

$$\langle \lambda_n (Bx_n - Bz) + \lambda_{n-1}(Bx_n - Bx_{n-1}), z - x_{n+1}\rangle = \lambda_n \langle Bx_n - Bx_{n+1}, z - x_{n+1}\rangle$$

$$+\lambda_{n-1}\langle Bx_n - Bx_{n-1}, z - x_{n+1}\rangle + \underbrace{\lambda_n \langle Bx_{n+1} - Bz, z - x_{n+1}\rangle}_{\le 0}$$

$$\le \lambda_n \langle Bx_n - Bx_{n+1}, z - x_{n+1}\rangle + \lambda_{n-1}\langle Bx_n - Bx_{n-1}, z - x_n\rangle$$

$$+\lambda_{n-1}\langle Bx_n - Bx_{n-1}, x_n - x_{n+1}\rangle . \qquad [7.10]$$

Using [7.10] within [7.9], we get:

$$D(z,x_{n+1}) \le D(z,x_n) - D(x_{n+1},x_n) + 2\lambda_n \langle Bx_n - Bx_{n+1}, z - x_{n+1}\rangle$$

$$+2\lambda_{n-1}\langle Bx_n - Bx_{n-1}, z - x_n\rangle + 2\lambda_{n-1}\langle Bx_n - Bx_{n-1}, x_n - x_{n+1}\rangle . \qquad [7.11]$$

Using the rule for λ_n recalculation, we can estimate the term $2\lambda_{n-1}\langle Bx_n - Bx_{n-1}, x_n - x_{n+1}\rangle$ in [7.11] from above:

$$2\lambda_{n-1}\langle Bx_n - Bx_{n-1}, x_n - x_{n+1}\rangle \le 2\lambda_{n-1}\|Bx_n - Bx_{n-1}\|_* \|x_n - x_{n+1}\|$$

$$\le 2\tau \frac{\lambda_{n-1}}{\lambda_n}\|x_n - x_{n-1}\|\|x_{n+1} - x_n\| \le \tau \frac{\lambda_{n-1}}{\lambda_n}\|x_n - x_{n-1}\|^2 + \tau \frac{\lambda_{n-1}}{\lambda_n}\|x_n - x_{n+1}\|^2$$

$$\le \tau\mu \frac{\lambda_{n-1}}{\lambda_n} D(x_n, x_{n-1}) + \tau\mu \frac{\lambda_{n-1}}{\lambda_n} D(x_{n+1}, x_n) .$$

So, we came to inequality

$$D(z,x_{n+1})+2\lambda_n\langle Bx_n-Bx_{n+1},x_{n+1}-z\rangle+\tau\mu\frac{\lambda_n}{\lambda_{n+1}}D(x_{n+1},x_n)$$

$$\le D(z,x_n)+2\lambda_{n-1}\langle Bx_{n-1}-Bx_n,x_n-z\rangle+\tau\mu\frac{\lambda_{n-1}}{\lambda_n}D(x_n,x_{n-1})$$

$$-\left(1-\tau\mu\frac{\lambda_{n-1}}{\lambda_n}-\tau\mu\frac{\lambda_n}{\lambda_{n+1}}\right)D(x_{n+1},x_n),$$

which completes the proof.

Let us formulate the main result.

THEOREM 7.1.– *Let E be a 2-uniformly convex and uniformly smooth Banach space, $A:E\to 2^{E^*}$ be a maximal monotone operator, $B:E\to E^*$ be a monotone and Lipschitz continuous operator, $(A+B)^{-1}0\neq\varnothing$. Suppose that the normalized duality map J is sequentially weakly continuous. Then, sequence (x_n) generated by Algorithm 7.1 converges weakly to some point $z\in(A+B)^{-1}0$.*

PROOF.– Let $z'\in(A+B)^{-1}0$. Denote:

$$a_n=D(z',x_n)+2\lambda_{n-1}\langle Bx_{n-1}-Bx_n,x_n-z'\rangle+\tau\mu\frac{\lambda_{n-1}}{\lambda_n}D(x_n,x_{n-1}),$$

$$b_n=\left(1-\tau\mu\frac{\lambda_{n-1}}{\lambda_n}-\tau\mu\frac{\lambda_n}{\lambda_{n+1}}\right)D(x_{n+1},x_n).$$

The inequality from Lemma 7.4 becomes $a_{n+1}\le a_n-b_n$. As $\lim_{n\to\infty}\lambda_n>0$ exists, we have:

$$1-\tau\mu\frac{\lambda_{n-1}}{\lambda_n}-\tau\mu\frac{\lambda_n}{\lambda_{n+1}}\to 1-2\tau\mu\in(0,1)\text{ when }n\to\infty.$$

Let us show that $a_n \geq 0$ for all big enough $n \in N$. We have:

$$a_n = D(z', x_n) + 2\lambda_{n-1}\langle Bx_{n-1} - Bx_n, x_n - z'\rangle + \tau\mu\frac{\lambda_{n-1}}{\lambda_n}D(x_n, x_{n-1})$$

$$\geq \frac{1}{\mu}\|x_n - z'\|^2 - 2\lambda_{n-1}\|Bx_{n-1} - Bx_n\|_* \|x_n - z'\| + \tau\frac{\lambda_{n-1}}{\lambda_n}\|x_{n-1} - x_n\|^2$$

$$\geq \frac{1}{\mu}\|x_n - z'\|^2 - 2\tau\frac{\lambda_{n-1}}{\lambda_n}\|x_n - x_{n-1}\|\|x_n - z'\| + \tau\frac{\lambda_{n-1}}{\lambda_n}\|x_{n-1} - x_n\|^2$$

$$\geq \left(\frac{1}{\mu} - \tau\frac{\lambda_{n-1}}{\lambda_n}\right)\|x_n - z'\|^2.$$

As there exists such $n_0 \in N$ that:

$$\frac{1}{\mu} - \tau\frac{\lambda_{n-1}}{\lambda_n} > 0 \text{ for all } n \geq n_0,$$

so $a_n \geq 0$ starting from n_0.

Now, we can conclude that the next limit exists:

$$\lim_{n\to\infty}\left(D(z', x_n) + 2\lambda_{n-1}\langle Bx_{n-1} - Bx_n, x_n - z'\rangle + \tau\mu\frac{\lambda_{n-1}}{\lambda_n}D(x_n, x_{n-1})\right),$$

and:

$$\sum_{n=1}^{\infty}\left(1 - \tau\mu\frac{\lambda_{n-1}}{\lambda_n} - \tau\mu\frac{\lambda_n}{\lambda_{n+1}}\right)D(x_{n+1}, x_n) < +\infty.$$

So, the sequence (x_n) is bounded, and:

$$\lim_{n\to\infty} D(x_{n+1}, x_n) = \lim_{n\to\infty}\|x_{n+1} - x_n\| = 0.$$

From the fact:

$$\lim_{n\to\infty}\left(2\lambda_{n-1}\langle Bx_{n-1}-Bx_n,x_n-z'\rangle+\tau\mu\frac{\lambda_{n-1}}{\lambda_n}D(x_n,x_{n-1})\right)=0,$$

we get the convergence of sequences $(D(z',x_n))$ for all $z'\in(A+B)^{-1}0$.

Let us show that all weak partial limits of sequence (x_n) generated by Algorithm 7.1 belong to the set $(A+B)^{-1}0$.

Consider a subsequence (x_{n_k}) that weakly converges to some point $z\in E$. Let us show that $z\in(A+B)^{-1}0$. If we take any point $(y,v)\in\Gamma(A+B)$, then $v-By\in Ay$. We have:

$$Jx_{n_k}-\lambda_{n_k}Bx_{n_k}-\lambda_{n_k-1}\left(Bx_{n_k}-Bx_{n_k-1}\right)-Jx_{n_k+1}\in\lambda_{n_k}Ax_{n_k+1}.$$

Using the monotonicity of A operator, we get:

$$\left\langle\lambda_{n_k}v+Jx_{n_k+1}+\lambda_{n_k}\left(Bx_{n_k}-By\right)+\lambda_{n_k-1}\left(Bx_{n_k}-Bx_{n_k-1}\right)-Jx_{n_k},y-x_{n_k+1}\right\rangle\geq 0.$$

And then, using the monotonicity of the B operator, we can obtain the estimation:

$$\begin{aligned}
&\left\langle v,y-x_{n_k}\right\rangle+\left\langle v,x_{n_k}-x_{n_k+1}\right\rangle=\left\langle v,y-x_{n_k+1}\right\rangle\\
&\geq\left\langle By+\frac{1}{\lambda_{n_k}}\left(Jx_{n_k}-Jx_{n_k+1}-\lambda_{n_k}Bx_{n_k}-\lambda_{n_k-1}\left(Bx_{n_k}-Bx_{n_k-1}\right)\right),y-x_{n_k+1}\right\rangle\\
&=\frac{1}{\lambda_{n_k}}\left\langle Jx_{n_k}-Jx_{n_k+1},y-x_{n_k+1}\right\rangle-\frac{\lambda_{n_k-1}}{\lambda_{n_k}}\left\langle Bx_{n_k}-Bx_{n_k-1},y-x_{n_k+1}\right\rangle\\
&+\left\langle By-Bx_{n_k+1},y-x_{n_k+1}\right\rangle+\left\langle Bx_{n_k+1}-Bx_{n_k},y-x_{n_k+1}\right\rangle\\
&\geq\frac{1}{\lambda_{n_k}}\left\langle Jx_{n_k}-Jx_{n_k+1},y-x_{n_k+1}\right\rangle-\frac{\lambda_{n_k-1}}{\lambda_{n_k}}\left\langle Bx_{n_k}-Bx_{n_k-1},y-x_{n_k+1}\right\rangle\\
&+\left\langle Bx_{n_k+1}-Bx_{n_k},y-x_{n_k+1}\right\rangle.
\end{aligned}$$

From $\lim_{n\to\infty}\|x_n - x_{n-1}\| = 0$ and the Lipschitz property of the B operator, follows:

$$\lim_{n\to\infty}\|Bx_n - Bx_{n-1}\|_* = 0 .$$

Due to the uniform continuity on bounded sets of the normalized duality mapping J , we obtain:

$$\lim_{n\to\infty}\|Jx_n - Jx_{n+1}\|_* = 0 .$$

Thus:

$$\langle v, y - z\rangle = \lim_{k\to\infty}\langle v, y - x_{n_k}\rangle \geq 0 .$$

The maximal monotonicity of operator $A+B$ and arbitrariness of $(y,v)\in\Gamma(A+B)$ imply $z\in(A+B)^{-1}0$ (Lemma 7.1).

We need to show that the sequence (x_n) converges weakly to z. Let us argue by contradiction. Let a subsequence (x_{m_k}) exist such that $x_{m_k} \to z'$ weakly and $z \neq z'$. It is clear that $z' \in (A+B)^{-1}0$. We have:

$$2\langle Jx_n, z - z'\rangle = D(z', x_n) - D(z, x_n) + \|z\|^2 - \|z'\|^2 .$$

From above, we deduce the existence of the limit:

$$\lim_{n\to\infty}\langle Jx_n, z - z'\rangle .$$

Due to the sequential weak continuity of the normalized duality mapping J, we obtain:

$$\langle Jz, z - z'\rangle = \lim_{k\to\infty}\langle Jx_{n_k}, z - z'\rangle = \lim_{k\to\infty}\langle Jx_{m_k}, z - z'\rangle = \langle Jz', z - z'\rangle ,$$

so $\langle Jz - Jz', z - z'\rangle = 0$. And from this follows the contradiction – $z = z'$.

7.5. Variants

For completeness, let us formulate a variant of the Operator Extrapolation Algorithm with a constant parameter $\lambda > 0$.

ALGORITHM 7.2.–

Choose some $x_0 \in E$, $x_1 \in E$, $\lambda \in \left(0, \frac{1}{2\mu L}\right)$. *Set* $n = 1$.

7. *Compute*

$$x_{n+1} = J_\lambda^A \circ J^{-1}\left(Jx_n - 2\lambda Bx_n + \lambda Bx_{n-1}\right).$$

2. *If* $x_{n-1} = x_n = x_{n+1}$, *then STOP and* $x_n \in (A+B)^{-1} 0$. *Else set* $n := n+1$ *and go to 7.*

Consider the problem of finding the zero of a monotone Lipschitz continuous operator $B : E \to E^*$:

find $x \in E$ such that $Bx = 0$.

For such a case, Algorithm 7.1 becomes Algorithm 7.3.

ALGORITHM 7.3.–

Choose some $x_0 \in E$, $x_1 \in E$, $\tau \in \left(0, \frac{1}{2\mu}\right)$ *and* $\lambda_0, \lambda_1 > 0$. *Set* $n = 1$.

7. *Compute*

$$x_{n+1} = J^{-1}\left(Jx_n - \lambda_n Bx_n - \lambda_{n-1}\left(Bx_n - Bx_{n-1}\right)\right).$$

2. *If* $x_{n-1} = x_n = x_{n+1}$, *then STOP and* $x_n \in B^{-1} 0$. *Else go to step 3.*

3. *Compute*

$$\lambda_{n+1} = \begin{cases} \min\left\{\lambda_n, \tau \dfrac{\|x_{n+1} - x_n\|}{\|Bx_{n+1} - Bx_n\|_*}\right\}, & \text{if } Bx_{n+1} \neq Bx_n, \\ \lambda_n, & \text{otherwise.} \end{cases}$$

Set $n := n+1$ *and go to 7.*

Weak convergence of Algorithm 7.3 follows from Theorem 7.1.

THEOREM 7.2.– *Let* E *be a two-uniformly convex and uniformly smooth Banach space,* $B: E \to E^*$ *– monotone and Lipschitz continuous operator,* $B^{-1}0 \neq \varnothing$. *Assume that normalized duality mapping* J *is sequentially weakly continuous. Then, the sequence* (x_n) *generated by Algorithm 7.3 converges weakly to some point* $z \in B^{-1}0$.

REMARK 7.3.– In Iiduka and Takahashi (2008), the following algorithm was proposed to find the zero of a reverse strongly monotone operator $B: E \to E^*$:

$$x_{n+1} = J^{-1}(Jx_n - \lambda_n Bx_n),\ x_1 \in E,$$

where $\lambda_n \in [a,b] \subseteq \left(0, \frac{\alpha}{\mu}\right)$, $\alpha > 0$ is a constant of inverse strong monotonicity of operator B. This algorithm generally does not converge for Lipschitz continuous monotone operators.

ALGORITHM 7.4.–

Choose some $x_0 \in E$, $x_1 \in E$, $\tau \in \left(0, \frac{1}{2\mu}\right)$ *and* $\lambda_0, \lambda_1 > 0$. *Set* $n = 1$.

7. Compute

$$x_{n+1} = J^{-1}\left(Jx_n - \lambda_n \nabla f(x_n) - \lambda_{n-1}\left(\nabla f(x_n) - \nabla f(x_{n-1})\right)\right).$$

2. If $x_{n-1} = x_n = x_{n+1}$, *then STOP and* $x_n \in \operatorname{arg\,min} f$. *Else go to step 3.*

3. Compute

$$\lambda_{n+1} = \begin{cases} \min\left\{\lambda_n, \tau \frac{\|x_{n+1} - x_n\|}{\|\nabla f(x_{n+1}) - \nabla f(x_n)\|_*}\right\}, & \text{if } \nabla f(x_{n+1}) \neq \nabla f(x_n), \\ \lambda_n, & \text{otherwise.} \end{cases}$$

Set $n := n+1$ *and go to 7.*

Using Theorem 7.2, let us consider the problem of minimizing a convex continuously Frechet differentiable functional:

$$f(x) \to \min_{x \in E}. \qquad [7.12]$$

We can formulate a variant of Algorithm 7.3 for [7.12], which is a convex minimization problem.

For this variant of the algorithm, from Theorem 7.2, we get Theorem 7.3.

THEOREM 7.3.– *Let E be a two-uniformly convex and uniformly smooth Banach space, $f: E \to R$ is a convex continuously Frechet differentiable functional with a Lipschitz continuous derivative, and* $\arg\min f \neq \varnothing$. *Assume that the normalized duality mapping J is sequentially weakly continuous. Then, the sequence (x_n) generated by Algorithm 7.4 converges weakly to a point $z \in \arg\min f$.*

REMARK 7.4.– All results are obtained for the class of two-uniformly convex and uniformly smooth real Banach spaces, which do not contain important for applications spaces L_p and W_p^m ($2 < p < +\infty$). It is highly desirable to get rid of this limitation. Fast and robust algorithms for computing the resolvent for a wide range of maximal monotone operators are needed to effectively apply algorithms for nonlinear problems in Banach spaces.

REMARK 7.5.– An interesting question is the study of the behavior of Algorithm 7.3. Namely, the question of the asymptotic behavior of $\|Bx_n\|_*$. Note that an estimate $\|Bx_n\|_* = O\left(\frac{1}{n}\right)$ is theoretically possible.

7.6. Application to variational inequalities

Let H be a real Hilbert space, C is a non-empty convex and closed subset of space H, and $B: H \to H$ is a monotone and Lipschitz continuous operator. Consider the next variational inequality problem:

$$\text{find } x \in C \text{ such that } (Bx, y - x) \geq 0 \quad \forall y \in C, \qquad [7.13]$$

Let us denote the set of solutions of [7.13] as $VI(B, C)$. Variational inequality [7.13] is equivalent to the operator inclusion (Bauschke and Combettes 2017):

$$\text{find } x \in H \text{ such that } 0 \in (A + B)x,$$

where $A = N_C$ is a normal cone for set C, i.e.:

$$N_C x = \begin{cases} \{w \in H : (w, y - x) \le 0 \ \ \forall y \in C\}, & x \in C, \\ \varnothing, & \text{otherwise.} \end{cases}$$

It is established that:

$$J_\lambda^A = (I + \lambda A)^{-1} = (I + \lambda N_C)^{-1} = P_C,$$

where P_C is the projection operator onto the set C (Bauschke and Combettes 2017).

For problem [7.13], Algorithm 7.1 takes the form of Algorithm 7.5.

ALGORITHM 7.5.–

Choose some $x_0 \in H$, $x_1 \in H$, $\tau \in \left(0, \frac{1}{2}\right)$ *and* $\lambda_0, \lambda_1 > 0$. *Set* $n = 1$.

7. Compute

$$x_{n+1} = P_C\left(x_n - \lambda_n B x_n - \lambda_{n-1}\left(B x_n - B x_{n-1}\right)\right).$$

2. If $x_{n-1} = x_n = x_{n+1}$, *then STOP and* $x_n \in VI(B, C)$. *Else go to step 3.*

3. Compute

$$\lambda_{n+1} = \begin{cases} \min\left\{\lambda_n, \tau \dfrac{\|x_{n+1} - x_n\|}{\|B x_{n+1} - B x_n\|}\right\}, & \text{if } B x_{n+1} \ne B x_n, \\ \lambda_n, & \text{otherwise.} \end{cases}$$

Set $n := n + 1$ *and go to step 7.*

For the variational inequality case, from Theorem 7.1, we get Theorem 7.4.

THEOREM 7.4.– *Let* H *be a real Hilbert space,* C *is a non-empty closed and convex subset of* H, $B : H \to H$ *is a monotone Lipschitz continuous operator, and let* $VI(B, C) \ne \varnothing$. *Then, the sequence* (x_n) *generated by Algorithm 7.5 converges weakly to a point* $z \in VI(B, C)$.

7.7. Conclusion

In this chapter, an iterative splitting algorithm for solving an operator inclusion with the sum of a maximal monotone operator and a monotone Lipschitz continuous operator in a Banach space is proposed and studied.

The algorithm is an adaptive variant of the "forward-reflected-backward algorithm" of Malitsky–Tam (2020) where the used rule for updating the step size does not require knowledge of the Lipschitz constant of operator B. An attractive feature of the algorithm is only one computation of the resolvent of the maximal monotone operator A and the value of the monotone Lipschitz continuous operator B at the iteration step.

The theorem on the weak convergence of the method is proved for operator inclusions in a two-uniformly convex and uniformly smooth Banach space. Theoretical applications to operator equations, convex minimization problems and variational inequalities are presented.

An interesting challenge for the future is the development of strongly convergent modifications of Algorithm 7.7.

7.8. Acknowledgments

This work was partially supported by the Department of Target Training of Taras Shevchenko Kyiv National University of the National Academy of Sciences of Ukraine (project "New subgradient and extragradient methods for non-smooth regression problems", 0124U002162).

7.9. References

Abass, H.A., Izuchukwu, C., Mewomo, O.T., Dong, Q.L. (2020). Strong convergence of an inertial forward-backward splitting method for accretive operators in real Banach space. *Fixed Point Theory*, 21, 397–412.

Agarwal, R.P., O'Regan, D., Sahu, D.R. (2009). *Fixed Point Theory for Lipschitzian-Type Mappings with Applications*. Springer, New York.

Alber, Y.I. (1996). Metric and generalized projection operators in Banach spaces: Properties and applications. In *Theory and Applications of Nonlinear Operators of Accretive and Monotone Type*, Kartsatos, A. (ed.). Dekker, New York.

Alber, Y. and Ryazantseva, I. (2006). *Nonlinear Ill-posed Problems of Monotone Type*. Springer, Dordrecht.

Aoyama, K. and Kohsaka, F. (2014). Strongly relatively nonexpansive sequences generated by firmly nonexpansive-like mappings. *Fixed Point Theory Appl.*, 95. doi: 10.1186/1687-1812-2014-95.

Barbu, V. (1976). *Nonlinear Semigroups and Differential Equations in Banach Spaces.* Springer, Dordrecht.

Bauschke, H.H. and Combettes, P.L. (2017). *Convex Analysis and Monotone Operator Theory in Hilbert Spaces.* Springer, New York.

Beauzamy, B. (1985). *Introduction to Banach Spaces and Their Geometry.* North-Holland, Amsterdam.

Bonesky, T., Kazimierski, K.S., Maass, P., Schopfer, F., Schuster, T. (2008). Minimization of Tikhonov functionals in Banach spaces. *Abstr. Appl. Anal.*, Article ID 192679. doi: 10.1155/2008/192679.

Chabak, L., Semenov, V., Vedel, Y. (2019). A new non-euclidean proximal method for equilibrium problems. In *Recent Developments in Data Science and Intelligent Analysis of Information*, Chertov, O., Mylovanov, T., Kondratenko, Y., Kacprzyk, J., Kreinovich, V., Stefanuk, V. (eds). Springer, Cham. doi: 10.1007/978-3-319-97885-7_6.

Cholamjiak, P. and Shehu, Y. (2019). Inertial forward-backward splitting method in Banach spaces with application to compressed sensing. *Appl. Math.*, 64, 409–435.

Cioranescu, I. (1990). *Geometry of Banach Spaces, Duality Mappings and Nonlinear Problems*. Kluwer Academic, Dordrecht.

Csetnek, E.R., Malitsky, Y., Tam, M.K. (2019). Shadow douglas-rachford splitting for monotone inclusions. *Appl. Math. Optim.*, 80, 665–678.

Denisov, S.V. and Semenov, V.V. (2021). Convergence of adaptive forward-reflected-backward algorithm for solving variational inequalities. In *CEUR Workshop Proceedings*, 3106 [Online]. Available at: https://ceur-ws.org/Vol-3106/Paper_17.pdf.

Denisov, S.V., Nomirovskii, D.A., Rublyov, B.V., Semenov, V.V. (2019). Convergence of extragradient algorithm with monotone step size strategy for variational inequalities and operator equations. *J. Automat. Inf. Sci.*, 51, 12–24. doi: 10.1615/JAutomatInfScien.v57.i6.20.

Diestel, J. (1975). *Geometry of Banach Spaces*. Springer-Verlag, Berlin–Heidelberg.

Iiduka, H. and Takahashi, W. (2008). Weak convergence of a projection algorithm for variational inequalities in a Banach space. *Journal of Mathematical Analysis and Applications*, 339(1), 668–679. doi: 10.1016/j.jmaa.2007.07.019.

Lions, P.L. and Mercier, B. (1979). Splitting algorithms for the sum of two nonlinear operators. *SIAM J. Numer. Anal.*, 16, 964–979.

Luke, D.R., Charitha, C., Shefi, R., Malitsky, Y. (2020). Efficient, quantitative numerical methods for statistical image deconvolution and denoising. In *Nanoscale Photonic Imaging. Topics in Applied Physics*, Salditt, T., Egner, A., Luke, D.R. (eds). Springer, Cham, 313–338.

Lyashko, S.I., Semenov, V.V., Voitova, T.A. (2011). Low-cost modification of Korpelevich's methods for monotone equilibrium problems. *Cybernet. Syst. Analysis*, 47(4), 631–639.

Lyashko, S.I., Klyushin, D.A., Nomirovsky, D.A., Semenov, V.V. (2013). Identification of age-structured contamination sources in ground water. In *Optimal Control of Age-Structured Populations in Economy, Demography, and the Environment*, Boucekkine, R., Hritonenko, N., Yatsenko, Y. (eds). Routledge, London, New York.

Malitsky, Y. and Tam, M.K. (2020). A forward-backward splitting method for monotone inclusions without cocoercivity. *SIAM J. Optim.*, 30, 1451–1472.

Nomirovskii, D.A., Rublyov, B.V., Semenov, V.V. (2019). Convergence of two-stage method with bregman divergence for solving variational inequalities. *Cybernet. Syst. Analysis*, 55, 359–368. doi: 10.1007/s10559-019-00142-7.

Passty, G.B. (1979). Ergodic convergence to a zero of the sum of monotone operators in hilbert spaces. *J. Math. Analysis Appl.*, 72, 383–390.

Raguet, H., Fadili, J., Peyre, G. (2013). A generalized forward-backward splitting. *SIAM J. Imaging Sci.*, 6, 1199–1226.

Semenov, V.V. (2014). A strongly convergent splitting method for systems of operator inclusions with monotone operators. *J. Automat. Inf. Sci.*, 46(5), 45–56.

Semenov, V.V. (2018). Modified extragradient method with bregman divergence for variational inequalities. *J. Automat. Inf. Sci.*, 50(8), 26–37. doi: 10.1615/JAutomatInfScien.v50.i8.30.

Semenov, V.V., Denisov, S.V., Kravets, A.V. (2021). Adaptive two-stage bregman method for variational inequalities. *Cybernet. Syst. Analysis*, 57, 959–967. doi: 10.1007/s10559-021-00421-2.

Semenov, V.V., Denisov, S.V., Sandrakov, G.V., Kharkov, O.S. (2022). Convergence of the operator extrapolation method for variational inequalities in banach spaces. *Cybernet. Syst. Analysis*, 58, 740–753. doi: 10.1007/s10559-022-00507-5.

Shehu, Y. (2019). Convergence results of forward-backward algorithms for sum of monotone operators in banach spaces. *Results Math.*, 74. doi: 10.1007/s00025-019-1061-4.

Shehu, Y. (2020). Single projection algorithm for variational inequalities in Banach spaces with application to contact problem. *Acta Math. Sci.*, 40, 1045–1063.

Takahashi, S. and Takahashi, W. (2016). Split common null point problem and shrinking projection method for generalized resolvents in two Banach spaces. *J. Nonlinear Convex Anal.*, 17, 2171–2182.

Tibshirani, R. (1996). Regression shrinkage and selection via the lasso. *J. Roy. Stat. Soc.*, 58, 267–288.

Tseng, P. (2000). A modified forward-backward splitting method for maximal monotone mappings. *SIAM J. Optim.*, 38, 431–446.

Vainberg, M.M. (1974). *Variational Method and Method of Monotone Operators in the Theory of Nonlinear Equations*. Wiley, New York.

Vedel, Y., Semenov, V., Denisov, S. (2021). A novel algorithm with self-adaptive technique for solving variational inequalities in banach spaces. In *Advances in Optimization and Applications (OPTIMA 2021)*. Springer, Cham. doi: 10.1007/978-3-030-92711-0_4.

Xu, Z.B. and Roach, G.F. (1991). Characteristic inequalities of uniformly convex and uniformly smooth Banach spaces. *J. Math. Analysis Appl.*, 157(1), 189–210.

Yang, J., Cholamjiak, P., Sunthrayuth, P. (2021). Modified Tseng's splitting algorithms for the sum of two monotone operators in Banach spaces. *AIMS Math.*, 6(5), 4873–4900.

8

Forecasting Algorithms Based on Intellectual Analysis of Polynomial Extrapolation and Divided Differences

Y. TURBAL, M. TURBAL and A. BOMBA
National University of Water and Environmental Engineering, Rivne, Ukraine

8.1. Introduction – Problem of the time series forecasting

The task of the data extrapolation is extremely important and relevant for many fields of scientific research nowadays (Lyashko et al. 2003; Bakas 2019; Monroe et al. 2020). The entire apparatus of mathematical modeling and data analysis is aimed in particular at building forecasting of the phenomena and processes. But the problem of forecasting is still relevant. It is obvious that a random process, for example, cannot be predicted in one experiment at all – here, we can only talk about certain averaged characteristics. Deterministic processes often require the construction of an adequate mathematical model for good forecasts. However, in the absence of stationarity and with an insufficient amount of input data, it is practically impossible to build appropriate mathematical models. Under the conditions of small amounts of data, it is difficult to determine what role the stochastic component plays in the certain process, and therefore, whether it is possible to use deterministic approaches for its modeling. After all, it is obvious that for any set of points, there are many curves that pass through them or somehow approximate them, and therefore, it is difficult to assert that any one curve (model) is exactly the law that

Computational Methods and Mathematical Modeling in Cyberphysics and Engineering Applications 2,
coordinated by Dmitri KOROLIOUK, Sergiy LYASHKO and Nikolaos LIMNIOS.

comprehensively describes the phenomenon and will allow us to effectively predict its behavior in the future. A hypothesis about the law to which the point observations correspond is effective if it allows us to predict the relevant process with acceptable accuracy starting from some observation step. But, in practice, it is not always possible to find an effective hypothesis.

If we talk about the short-term forecast of a certain deterministic process, then the problem of extrapolation of the function given in the table arises. Among known methods of extrapolation, methods based on interpolation polynomials can be singled out. Such problems were considered, in particular, by Kostinsky (2014), Bakas (2019) and Turbal et al. (2020). A large number of works are devoted to the approximation of continuous and differentiable functions by polynomials. Well-known theorems of Weierstrass, Dzyadyk and Thiman (Zhan et al. 2012 ; Bakas 2019) are about the approximation of continuous functions by polynomials on a finite interval. In this case, on the basis of the corresponding polynomials, it was proposed to solve the problem of extrapolation according to the rule "interpolate well – extrapolate carefully". At the same time, Newton's interpolation polynomials of the second kind (Kostinsky 2014) and generalized interpolation polynomials according to various Chebyshev function systems (Zhao et al. 2011; Turbal et al. 2012 ; Wang et al. 2015; Klyushin et al. 2020), such as exponents, trigonometric functions, were most often used for extrapolation.

There are known methods based on trend analysis (Wang et al. 2016), radial basis functions, extrapolation methods based on splines (Makridakis et al. 2016) and Bezier-type curves that do not pass directly through the interpolation points. Various statistical approaches (Makridakis et al. 2016) are used; in particular, the geometric method, which consists of taking into account the arithmetic mean ratio of the elements of the time series, Bayesian methods, etc. Machine learning methods and artificial neural networks play an important role in forecasting. At the time when the problem of interpolation is solved, for example, on the basis of the interpolation polynomials, the relevant theorems on the approximation of any smooth function by polynomials with any given accuracy have been proven, but in the case of extrapolation, there are no similar solutions. In practice, predictions based on polynomials often give very poor results for extrapolation. And here the problem arises: is it still possible to use polynomials to construct effective forecasts? And that is why the topic of this study is relevant.

In this chapter, we consider methods of extrapolation, which are based on the principles of searching for the certain regularities in the data and not using a direct mathematical model of the investigated process. These forecasting methods are based on the use of polynomials, but special algorithms are proposed for choosing the optimal degree of the polynomial for extrapolation. It is also considered a pyramidal method. The main idea of this method is to develop a technique that allows us to build the optimal predict point without direct use of any specific classes of extrapolation function. This forecasting uses the maximum of information that focuses on the special kind of finite differences.

This chapter proposes a number of refinements of the algorithm proposed in Turbal et al. (2023). In particular, a criterion for constructing a forecast in the case of convergence of one-step polynomial forecasts is proposed; a detailed numerical study of the extrapolation error for the case of finding a forecast as an arithmetic average of polynomial forecasts is carried out.

8.2. Method of finding the predictive value based on a polynomial of any degree without finding the coefficients of the polynomial

The object of research in the chapter is the problem of extrapolation based on interpolation polynomials. The corresponding problem is formulated classically. Let $f_i = f(i\Delta), i = n, n-1, ..., n-m$ be known values of some function, which are defined in points $i\Delta$ in accordance. The value of this function f_{n+1} must be evaluated, which is defined at some point $x = x_{n+1} > x_n$. Here, we will consider a point $x = (n+1)\Delta$.

To solve such a problem in practice, in particular, Newton's interpolation polynomials of the second kind (Brezinski et al. 2020) of different degrees are used under the condition that the point is not far from the point x_n.

The main hypothesis of our study is that the effective prediction of the function at a point $x = (n+1)\Delta$ can be built as the arithmetic mean of forecasts based on polynomials of different degrees or a polynomial of the maximum-highest degree in the presence of convergence of polynomial forecasts. However, to implement such an approach, it is important to determine the conditions under which such a forecast will be effective. A separate important issue is the error of extrapolation.

In the process of research and development of our prediction algorithm, the method of finding the predictive value for one step based on a polynomial of any degree without finding the coefficients of the polynomial, which was proposed in Turbal et al. (2023), was used. This method makes it much more efficient to find the predictive value per grid step compared to the classical approach. Indeed, in order to build a forecast based on a polynomial of a certain degree, it is necessary to determine the coefficients of the polynomial by solving the SLAE – System of Linear Algebraic Equations. If we consistently consider forecasts based on 2, 3, 4, etc., consecutive points $f_i = f(i\Delta), i = n, n-1, ..., n-m$, then we can see that the result of the projection based on m points cannot be used for m+1 points – it is still necessary to solve a new SLAE. Therefore, in order to optimize the computational procedure, we will abandon the problem of solving the SLAE for constructing a one-step projection and use the result proposed in Turbal et al. (2021).

Due to the simplicity of the corresponding calculation procedure, the predictive value can be easily found. If polynomials of higher degrees are used, it is easy to use an arbitrary high-level programming language to construct a forecast. At the same time, the main element of implementation will be the calculation of binomial coefficients.

STATEMENT 8.1.– Let us have a uniform grid with a step Δ, the values of some function are known $f_i = f(i\Delta), i = n, n-1, ..., n-m$. Then, the predictive value of the function f_{n+1}^{m-1}, built on the basis of Newton's interpolation polynomial of the second kind $P_2^{m-1}(x)$ at the point $(n+1)\Delta$ can be found based on the ratio:

$$f_{n+1}^{m-1} = P_2^{m-1}((n+1)\Delta) = \sum_{k=1}^{m} (-1)^{k-1} C_m^k f_{n-k+1} \quad [8.1]$$

Thus, only the sign-changing sum of products of binomial coefficients and time series values is used to calculate the forecast value for one step for any polynomial.

We recall that Newton's interpolation polynomials of the second kind (Brezinski et al. 2020) are written in the form:

$$P_2^N(x) = f_n + \Delta^1 f_{n-1}(x - x_n) + \Delta^2 f_{n-2}(x - x_{n-1})(x - x_n) + \dots$$

$$+\Delta^N f_{n-N}(x - x_{n-N+1})\dots(x - x_n).$$

where:

$$\Delta^j f_i = \frac{\Delta^{j-1} f_{i+1} - \Delta^{j-1} f_i}{x_{i+j} - x_i}, \quad [8.2]$$

$$\Delta^0 f_i = f_i.$$

The corresponding result can be obtained by analyzing the table of divided differences. Suppose that for some *k*, the condition is fulfilled:

$$\Delta^k f_{n-k+1} = \Delta^k f_{n-k}. \quad [8.3]$$

In this case, the predicted value can be found by the following obvious calculation procedure:

$$\Delta^{i-1} f_{n-i+2} = \Delta^i f_{n-i+1}(x_{n+1} - x_{n+1-i}) + \Delta^i f_{n-i}, \\ i = \overline{k,0}. \quad [8.4]$$

Let us mark $x_{n+1} = x$. Taking [8.2] into account, we have:

$$\Delta^k f_{n-k} = \Delta^k f_{n-k+1} = \frac{\Delta^{k-1} f_{n-k+2} - \Delta^{k-1} f_{n-k+1}}{x - x_{n-k+1}}.$$

From here, it is easy to get the ratio:

$$f_{n+1} = \Delta^k f_{n-k+1}(x - x_{n+k+1})(x - x_{n-k+2})...(x - x_n) + \\ +\Delta^{k-1} f_{n-k+1}(x - x_{n-k+2})(x - x_{n-k+3})...(x - x_n) + ... + f_n.$$

Then, under condition [8.3], the relation holds:

$$f_{n+1} = \Delta^k f_{n-k}(x - x_{n+k+1})(x - x_{n-k+2})...(x - x_n) + \\ +\Delta^{k-1} f_{n-k+1}(x - x_{n-k+2})(x - x_{n-k+3})...(x - x_n) + ... + f_n. \quad [8.5]$$

Thus, when condition [8.3] is fulfilled, the predictive value of the function f_{n+1} at some point $x = x_{n+1}$ is defined as the value of Newton's polynomial of the second form (backward interpolation formula) of the degree *k*, which is equivalent to the calculation procedure [8.4].

Consider an example of a table of divided differences (see Table 8.1). Here, the grid step is equal to 1.

It is easy to see that in the case when the grid step is equal to 1, the relation holds:

$$\Delta^i f_j = \frac{\sum_{k=0}^{i} (-1)^k C_i^k f_{j+i-k}}{i!}. \quad [8.6]$$

6	f_6				
		$\Delta^1 f_5$			
5	f_5		$\Delta^2 f_4$		
		$\frac{f_5 - f_4}{1}$		$\Delta^3 f_3$	
4	f_4		$\frac{f_5 - 2f_4 + f_3}{1*2}$		$\Delta^4 f_2 = \\ = \frac{f_5 - 4f_4 + 6f_3 - 4f_2 + f_1}{1*2*3*4}$
		$\frac{f_4 - f_3}{1}$		$\frac{f_5 - 3f_4 + 3f_3 - f_2}{1*2*3}$	
3	f_3		$\frac{f_4 - 2f_3 + f_2}{1*2}$		$\frac{f_5 - 4f_4 + 6f_3 - 4f_2 + f_1}{1*2*3*4}$
		$\frac{f_3 - f_2}{1}$		$\frac{f_4 - 3f_3 + 3f_2 - f_1}{1*2*3}$	
2	f_2		$\frac{f_3 - 2f_2 + f_1}{1*2}$		
		$\frac{f_2 - f_1}{1}$			
1	f_1				

Table 8.1. *A fragment of the table of divided differences*

In the case of an arbitrary uniform step grid δ:

$$\Delta^i f_j = \frac{\sum_{k=0}^{i} (-1)^k C_i^k f_{j+i-k}}{i!\delta^i} \qquad [8.7]$$

The corresponding ratio is proved by the method of mathematical induction on the index *i*. Indeed, at $i = 1$, relation [8.6] is obvious. Let it be performed for some *n*.

Then:

$$\Delta^{n+1} f_j = \frac{\Delta^n f_{j+1} - \Delta^n f_j}{n+1} \Delta^i f_j = \frac{1}{n+1}\left(\frac{\sum_{k=0}^{n}(-1)^k C_n^k f_{j+1+n-k}}{n!} - \frac{\sum_{k=0}^{n}(-1)^k C_n^k f_{j+n-k}}{n!}\right) =$$

$$= \frac{1}{(n+1)!}\left(\sum_{k=0}^{n}(-1)^k C_n^k f_{j+1+n-k} - \sum_{k=0}^{n}(-1)^k C_n^k f_{j+n-k}\right) =$$

$$= \frac{1}{(n+1)!}\left(\sum_{k=0}^{n}(-1)^k C_n^k f_{j+1+n-k} - \sum_{k=0}^{n}(-1)^k C_n^k f_{j+n-k}\right) =$$

$$= \frac{1}{(n+1)!}\left(f_{j+1+n} + \sum_{k=1}^{n}(-1)^k C_n^k f_{j+1+n-k} - \sum_{k=0}^{n}(-1)^k C_n^k f_{j+n-k}\right) =$$

$$= \frac{1}{(n+1)!}\left(f_{j+1+n} + \sum_{k=0}^{n-1}(-1)^{k+1} C_n^{k+1} f_{j+n-k} - \sum_{k=0}^{n-1}(-1)^k C_n^k f_{j+n-k} - (-1)^n f_j\right) =$$

$$\frac{1}{(n+1)!}\left(f_{j+1+n} - \sum_{k=0}^{n-1}(-1)^k C_n^{k+1} f_{j+n-k} - \sum_{k=0}^{n-1}(-1)^k C_n^k f_{j+n-k} - (-1)^n f_j\right) =$$

$$= \frac{1}{(n+1)!}\left(f_{j+1+n} - \sum_{k=0}^{n-1}(-1)^k C_{n+1}^{k+1} f_{j+n-k} - (-1)^n f_j\right) =$$

$$= \frac{1}{(n+1)!}\left(f_{j+1+n} - \sum_{k=1}^{n}(-1)^{k-1} C_{n+1}^k f_{j+n-k+1} - (-1)^n f_j\right) = \sum_{k=0}^{n+1}(-1)^k C_{n+1}^k f_{j+n+1-k}.$$

If we use assumption [8.3] and construct the predicted value according to procedure [8.4], it is easy to see that the predicted value is simply the sum of the corresponding finite differences from the previous diagonal of the table, multiplied by the corresponding values. In particular, for the corresponding values from Table 8.2, we have:

$$\Delta^3 f_3 = 4\frac{f_5 - 4f_4 + 6f_3 - 4f_2 + f_1}{2*3*4} + \frac{f_5 - 3f_4 + 3f_3 - f_2}{2*3} = \frac{2f_5 - 7f_4 + 9f_3 - 5f_2 + f_1}{2*3},$$

$$\Delta^2 f_4 = 3\frac{2f_5 - 7f_4 + 9f_3 - 5f_2 + f_1}{2*3} + \frac{f_5 - 2f_4 + f_3}{2} = \frac{3f_5 - 9f_4 + 10f_3 - 5f_2 + f_1}{2},$$

$$\Delta^1 f_5 = 2\frac{3f_5 - 9f_4 + 10f_3 - 5f_2 + f_1}{2} + f_5 - f_4 = 4f_5 - 10f_4 + 10f_3 - 5f_2 + f_1,$$

$$f_6 = f_5 + \Delta^1 f_5 = f_5 + 4f_5 - 10f_4 + 10f_3 - 5f_2 + f_1 = 5f_5 - 10f_4 + 10f_3 - 5f_2 + f_1.$$

Or, on another note,

$$f_6 = 1(2(3(4\Delta^4 f_1 + \Delta^3 f_2) + \Delta^2 f_3) + \Delta^1 f_4) + f_5 = 5f_5 - 10f_4 + 10f_3 - 5f_2 + f_1.$$

We see that the predicted value is the sum of all rows of the "significant variable" Pascal triangle (see Figure 8.1), the elements of which are multiplied by the corresponding function values. In the general case, we can write the following recurrence relation:

$$\phi_1 = \Delta^{n-1} f_1,$$
$$\phi_i = (n-i+1)\phi_{i-1} + \Delta^{n-i} f_i, i = 2,3,...n,$$
$$f_{n+1} = \phi_n.$$

It is easy to see that all the considerations above are valid when the grid step is arbitrary and uniform δ. Then, taking into account the known combinatorial relation $\sum_{k=i}^{n} C_k^i = C_{n+1}^{i+1}, i = 0,1,2,...$, we get Statement 8.1.

Note that to calculate the predictive value for large degrees of polynomials, the corresponding binomial coefficients should be calculated exactly according to Pascal's triangle in order to avoid the calculation of factorials of large numbers in the classical definition of binomial coefficients.

So, we will build predictive values on the basis of Newton's interpolation polynomials of the second kind. Based on Statement 8.1, any prediction using an interpolation polynomial on a uniform grid can be easily constructed and of arbitrary degree.

8.3. The optimal polynomial extrapolation problem

It is easy to see that it is not always possible to build an effective forecast based on interpolation polynomials. Consider, for example, the following simple sequence: *-a,a,-a,a,-a,a,...,-a*. We can write in more detail $f_1, f_2, ..., f_n$, where:

$$f_i = \begin{cases} -a, i = 2m+1 \\ a, i = 2m \end{cases}, \quad m = 0,1,2,....$$

Let $n = 2k$. Then, according to relation [8.1], the predictive value based on the m-1 degree polynomial is determined by the formulas:

$\{-a\}$:

$$f_{n+1}^{0} = \sum_{k=1}^{1} (-1)^{k-1} C_1^k f_{n-k+1} = -a \,,$$

$\{a,-a\}$:

$$f_{n+1}^{1} = \sum_{k=1}^{2} (-1)^{k-1} C_2^k f_{n-k+1} = -2a - a = -3a \,,$$

$\{-a,a,-a\}$

$$f_{n+1}^{2} = \sum_{k=1}^{3} (-1)^{k-1} C_3^k f_{n-k+1} = -3a - 3a - a = -7a = -(2^3 - 1)a \,,$$

...

$$f_{n+1}^{m-1} = \sum_{k=1}^{m} (-1)^{k-1} C_m^k f_{n-k+1} = -(2^m - 1)a \,.$$

But obviously, for our sequence, the next predictive value is *a*.

Let us consider the same sequence; the last element is *a*: *-a,a,-a,a,-a,a, ...,a*. The predictive value based on the m-1 degree polynomial is determined by the formulas:

$\{a\}$:

$$f_{n+1}^{0} = \sum_{k=1}^{1} (-1)^{k-1} C_1^k f_{n-k+1} = a \,,$$

$\{-a,a\}$:

$$f_{n+1}^{1} = \sum_{k=1}^{2} (-1)^{k-1} C_2^k f_{n-k+1} = 2a + a = 3a \,,$$

$\{a,-a,a\}$

$$f_{n+1}^{2}=\sum_{k=1}^{3}(-1)^{k-1}C_{3}^{k}f_{n-k+1}=3a+3a+a=-7a=(2^{3}-1)a\,,$$

...

$$f_{n+1}^{m-1}=\sum_{k=1}^{m}(-1)^{k-1}C_{m}^{k}f_{n-k+1}=(2^{m}-1)a\,.$$

But obviously, for our sequence, the next predictive value is *-a*. We see how much the predictive value deteriorates with increasing degree of the polynomial.

The task arises: how to choose the degree of the interpolation polynomial so that the extrapolation error is minimal. Obviously, in such a statement, the problem looks incorrect, because we do not know anything about the next value of the function, except for the corresponding values specified in the previous points. Here, it is impossible to introduce the concept of error (because we know nothing about the next value of the function, and therefore, there is nothing to compare the predicted value constructed in some way) without considering certain additional requirements for the extrapolated function or assumptions about its future behavior. Let us consider several examples.

Let us have a function $f(x)=x^{6}\sin(x)$, defined in points 1, 2, 3, 4, ...,10. In Table 4.2, the first row shows the values of the function at points 13, 14, 15, 16, 17, 18, 19 and 20. In the following rows, the values of interpolation polynomials of the corresponding degree are constructed from the previous points at the same point. Then, we can determine which degree of the polynomial is best to calculate the predicted value and find the error. In Table 8.1, the best predictive value is highlighted in color. As a result, we get that for the value of the function at point 13, the best prediction gives a polynomial of degree 8, 14–6, 15–9, 16–3, 17–10, 18–4, 19–6 and 20–10. Thus, we see that looking at the values of an arbitrary function at different consecutive points, we have different degrees of polynomials that give a minimum error.

Table 8.3 shows the values of the arithmetic mean of 10 interpolation forecasts, the minimum and maximum values that determine the interval of polynomial uncertainty, the length of the corresponding interval and its middle. As we can see from this example, the best degree of interpolation polynomial for forecasting is the one for which the forecast value deviates minimally from the average value of all forecasts. In other words, the average of all polynomial predictions gives the best prediction.

2,028,066	7,458,814	7,407,185	–4,830,216	–23,205,798	–25,542,746	7,051,105
–1,432,853	5,658,330	12,889,561	7,355,556	–17,067,617	–41,581,380	–27,879,695
–35,985	9,119,249	14,690,045	1,873,180	–29,253,389	–47,719,562	–11,841,061
1,825,368	11,183,300	13,029,610	–5,409,680	–35,956,785	–41,671,970	10,335,754
3,428,506	11,385,997	9,305,124	–11,032,105	–35,377,321	–28,920,984	26,464,978
4,273,126	9,985,558	5,377,940	–12,930,044	–29,175,432	–16,749,461	29,843,216
4,173,055	7,740,498	2,851,196	–10,900,800	–21,075,603	–10,779,828	21,049,930
3,259,596	5,595,508	2,569,512	–6,344,812	–15,005,019	–12,910,023	6,287,012
1,910,319	4,363,978	4,432,818	–1,507,139	–13,490,424	–21,110,802	–6,345,712
625,813	4,481,726	7,527,653	1,467,228	–16,813,501	–30,826,176	–10,777,657
–117,658	5,883,979	10,504,741	1,346,760	–23,110,946	–37,218,473	–5,494,227

Table 8.2. *Illustration of choosing the best degree of the interpolating polynomial for extrapolation*

Consider the following example – the value of the same function at points 1, 1.5, 2, 2.5, 3, 3.5, ...,. Table 4.4 shows similar values for the function at points 7, 7.5, 8, 8.5, 9, 9.5, 10 and 10.5.

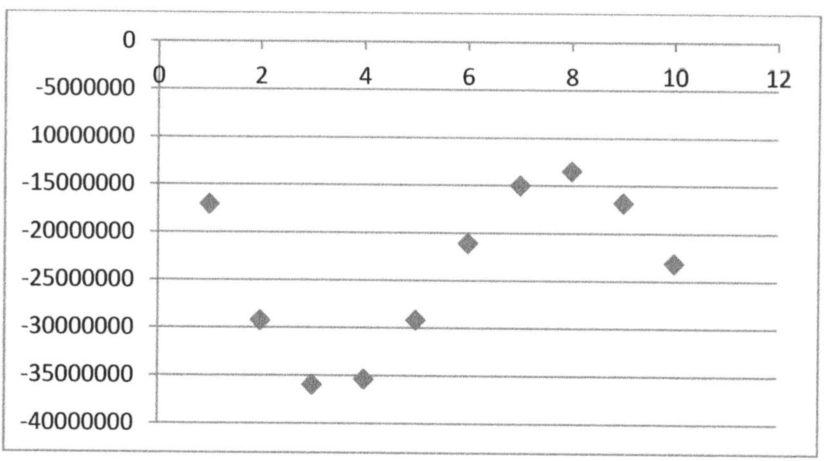

Figure 8.1. *Extrapolation points based on interpolation polynomials of degrees 1, 2, ... respectively (value of column 5 of Table 4.2).*

Average	1,790,929	7,539,812	8,317,820	-3,608,186	-23,632,603	-28,948,865
Min	-1,432,853	4,363,978	2,569,512	-12,930,044	-35,956,785	-47,719,562
Max	4,273,126	11,385,997	14,690,045	7,355,556	-13,490,424	-10,779,828
Middle	1,420,137	7,874,988	8,629,779	-2,787,244	-24,723,605	-29,249,695
Interval	5,705,979	7,022,019	12,120,533	20,285,601	22,466,361	36,939,734

Table 8.3. *Additional parameters in the analysis of forecast values*

We see that there is a significant shift to polynomials of higher degrees. At the same time, convergence is observed in a certain sense starting from the seventh-degree polynomial; the corresponding forecast values differ relatively little. Obviously, in such a situation, a polynomial of the maximum degree should be chosen. At the same time, average values will also give a good forecast with a large number of observations.

The graph for the data of the first column of Table 4.2 is shown in Figure 8.2. The figure shows a clear tendency towards convergence. Therefore, the optimal value of the exponent is chosen as the maximum for the available data.

Let us consider another example. Let us have a function $f(x) = \exp(x)$, defined in points 1, 2, 3, 4, ...,10. 1, 1.5, 2, 2.5, 3, 3.5,…,. Table 4.4 shows similar values for the function at points 7, 7.5, 8, 8.5, 9, 9.5, 10, 10.5. It is easy to see that there is a tendency to convergence; the sequence of forecast values is monotonous and limited, and the distance between neighboring points tends to 0. In such a situation, we choose a polynomial of the maximum degree.

77,294	166,944	259,354	301,149	219,017	-55,243	-544,021
45,485	138,364	256,594	351,765	342,944	136,884	-329,503
68,252	170,173	285,174	354,525	292,328	12,957	-521,630
79,979	179,215	281,945	328,705	238,952	-60,354	-589,830
82,985	176,530	269,674	306,114	211,395	-80,289	-584,719
81,439	170,838	260,089	295,794	206,430	-72,668	-559,673
78,795	166,693	256,194	295,060	211,785	-60,081	-542,248
76,939	165,191	256,445	298,220	217,874	-52,849	-537,410
76,336	165,546	258,198	301,129	220,802	-51,706	-539,804
76,608	166,504	259,596	302,286	220,822	-53,491	-543,341
77,140	167,190	260,035	302,045	219,685	-55,297	-545,093

Table 8.4. *Illustration of optimal degree selection for the interpolation polynomial*

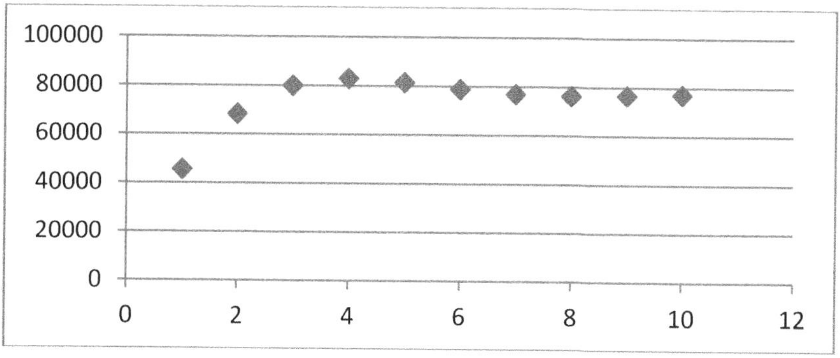

Figure 8.2. *Extrapolation points based on interpolation polynomials of degrees 1, 2, ..., respectively (values of the first column of Table 4.4)*

Art. many	1,097	1,808	2,981	4,915	8,103	13,360	22,026	36,316
1	927	1,528	2,519	4,154	6,849	11,291	18,616	30,693
2	1,030	1,698	2,799	4,615	7,609	12,546	20,685	34,103
3	1,070	1,765	2,910	4,797	7,909	13,040	21,499	35,445
4	1,086	1,791	2,953	4,868	8,027	13,234	21,819	35,973
5	1,093	1,801	2,970	4,897	8,073	13,310	21,945	36,181
6	1,095	1,805	2,977	4,908	8,091	13,340	21,994	36,262
7	1,096	1,807	2,979	4,912	8,098	13,352	22,014	36,295
8	1,096	1,808	2,980	4,914	8,101	13,357	22,021	36,307
9	1,097	1,808	2,981	4,914	8,102	13,359	22,025	36,312
10	1,097	1,808	2,981	4,915	8,103	13,359	22,026	36,314

Table 8.5. *Illustration of optimal degree selection for the interpolation polynomial*

The above examples allow us to build an algorithm for choosing the best polynomial for extrapolation among interpolation polynomials:

1) Building forecasts for all possible interpolation polynomials based on previous data based on ratio [8.1].

2) Analyzing the obtained sequence for the presence of a tendency towards convergence (the distance between adjacent points decreases monotonically).

3) If there is a tendency to convergence, the forecast of the polynomial of the highest degree is selected as the predictive value.

4) Under conditions of non-monotonicity of the sequence of forecasts, the forecast is constructed as an arithmetic average of polynomial forecasts.

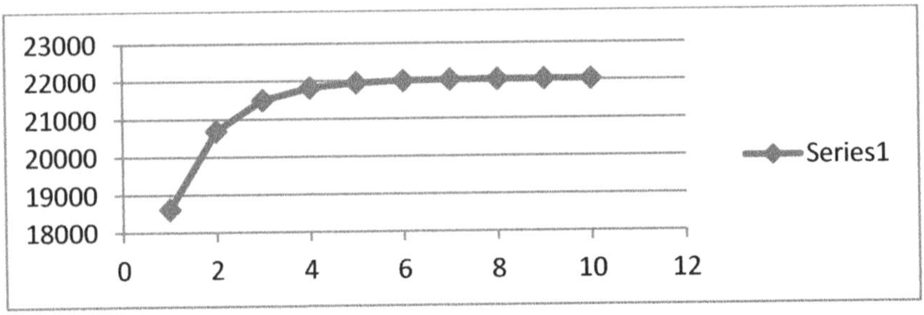

Figure 8.3. *Extrapolation points based on interpolation polynomials of degrees 1, 2, ..., respectively (values of the seventh column of Table 4.5)*

8.4. Condition of forecast efficiency based on the arithmetic mean of polynomial forecasts

Let us find the condition when the forecast defined as the arithmetic mean of forecasts based on interpolation polynomials can be effective. We have:

$$f_{n+1}^{m-1} = \sum_{k=1}^{m} (-1)^{k-1} C_m^k f_{n-k+1},$$

$$f_{n+1}^{m-1} = \sum_{k=1}^{m} (-1)^{k-1} C_m^k f_{n-k+1} = C_m^1 f_n + \sum_{k=2}^{m} (-1)^{k-1} C_m^k f_{n-k+1} =$$

$$= C_m^1 f_n + \sum_{k=2}^{m} (-1)^{k-1} C_m^k f_{n-k+1} - \sum_{k=2}^{m} (-1)^{k-1} C_m^k f_{n-k+2} + \sum_{k=2}^{m} (-1)^{k-1} C_m^k f_{n-k+2} =$$

$$= C_m^1 f_n + \sum_{k=2}^{m} (-1)^{k-1} C_m^k (f_{n-k+1} - f_{n-k+2}) - \sum_{i=0}^{m-2} f_{n+1}^i.$$

$$\sum_{i=0}^{m-2} f_{n+1}^i = \sum_{i=0}^{m-2} \sum_{k=1}^{i+1} (-1)^{k-1} C_{i+1}^k f_{n-k+1} = \sum_{k=1}^{m-1} \sum_{i=k-1}^{m-2} (-1)^{k-1} C_{i+1}^k f_{n-k+1} =$$

$$= \sum_{k=1}^{m-1} (-1)^{k-1} f_{n-k+1} \sum_{i=k-1}^{m-2} C_{i+1}^k.$$

Let us use the well-known combinatorial ratio:

$$C_{k+m}^{k} = \sum_{i=0}^{m} C_{k-1+i}^{k-1} .$$

Then:

$$\sum_{i=k-1}^{m-2} C_{i+1}^{k} = [i1 = i - k + 1] = \sum_{i=0}^{m-k-1} C_{i+k}^{k} = [k1 = k+1] = \sum_{i=0}^{m-k1} C_{i+k1-1}^{k1-1} =$$
$$= C_{k1+m-k1}^{k1} = C_{m}^{k+1}.$$

From the last expression:

$$\sum_{k=1}^{m-1} (-1)^{k-1} f_{n-k+1} \sum_{i=k-1}^{m-2} C_{i+1}^{k} = \sum_{k=1}^{m-1} (-1)^{k-1} f_{n-k+1} C_{m}^{k+1} = [k1 = k+1] = \sum_{k=2}^{m} (-1)^{k-2} f_{n-k+2} C_{m}^{k}$$

$$0 = C_{m}^{1} f_{n} + \sum_{k=2}^{m} (-1)^{k-1} C_{m}^{k} (f_{n-k+1} - f_{n-k+2}) - \sum_{i=1}^{m-1} f_{n+1}^{i} ,$$

$$f_{n} = \frac{1}{m} \sum_{i=1}^{m} f_{n+1}^{i} - \frac{1}{m} \sum_{k=2}^{m} (-1)^{k-1} C_{m}^{k} (f_{n-k+1} - f_{n-k+2}) ,$$

$$f_{n+1} = \frac{1}{m} \sum_{i=1}^{m} f_{n+1}^{i} - \frac{1}{m} \sum_{k=1}^{m} (-1)^{k-1} C_{m}^{k} (f_{n-k+1} - f_{n-k+2}) =$$
$$\frac{1}{m} \sum_{i=1}^{m} f_{n+1}^{i} - \frac{1}{m} \sum_{k=2}^{m} (-1)^{k-1} C_{m}^{k} (f_{n-k+1} - f_{n-k+2}) - (f_{n} - f_{n+1}) =$$
$$= \frac{1}{m} \sum_{i=1}^{m} f_{n+1}^{i} - \frac{1}{m} \sum_{k=2}^{m} (-1)^{k-1} C_{m}^{k} (f_{n-k+1} - f_{n-k+2}) - (f_{n} - f_{n+1}).$$

The following extrapolation method directly follows from the last relation. Let $f_{n+1}^{i} = P_{2}^{i}((n+1)\Delta)$. Then, the predictive value $f_{n+1} = f((n+1)\Delta)$ found in the form:

$$f_{n+1} = \frac{1}{\hat{m}} \sum_{i=1}^{\hat{m}} f_{n+1}^{i} , \qquad [8.8]$$

where $\hat{m} = \arg\min_{m} \left| \frac{1}{m} \left(\sum_{k=2}^{m} (-1)^{k-1} C_{m}^{k} (f_{n-k+1} - f_{n-k+2}) + \sum_{i=1}^{m} f_{n+1}^{i} \right) - f_{n} \right| .$ [8.9]

At the same time, the estimate of the absolute error of extrapolation is determined by the expression:

$$error = \left| \frac{1}{m} \left(\sum_{k=2}^{m} (-1)^{k-1} C_m^k (f_{n-k+1} - f_{n-k+2}) + \sum_{i=1}^{m} f_{n+1}^i \right) - f_n \right|. \qquad [8.10]$$

In practice, the relative error can be represented in the form:

$$verror = \left| \frac{1}{m} \left(\sum_{k=2}^{m} (-1)^{k-1} C_m^k (f_{n-k+1} - f_{n-k+2}) + \sum_{i=1}^{m} f_{n+1}^i \right) - f_n \right| / \frac{1}{m} \sum_{i=1}^{m} f_{n+1}^i. \qquad [8.11]$$

8.5. Improved algorithm for optimal polynomial forecasting

Despite the fact that polynomials are well suited for solving function interpolation problems, large errors often occur when solving extrapolation problems based on polynomials.

Let us investigate whether it is possible to reduce the extrapolation error by using certain combinations of polynomial forecast values, for example, the arithmetic mean of polynomial forecasts. Turbal et al. (2023) proposed an algorithm (Algorithm 8.1) for identifying the optimal degree of the extrapolation polynomial. The essence of the algorithm is to perform the following steps:

1) Data mining. Forming the input array.

2) Polynomial predictive set construction using the ratio [8.1] (Algorithm 8.2).

3) The polynomial predictive set is scrutinized to detect any inclination towards convergence, where the gap between adjacent forecast points either consistently diminishes or increases.

4) If a tendency towards convergence is observed, the prediction from the polynomial with the highest feasible degree is chosen as the predictive value, and the algorithm concludes.

5) In cases where the sequence of forecasts lacks monotonicity, expressions [8.10] and [8.11] are computed. If either the error [8.10] or the relative error [8.11] proves to be insignificant (with the threshold determined through expert judgment), a forecast is formulated as the arithmetic mean of polynomial predictions [8.9], and the algorithm concludes. Otherwise, refining the polynomial forecast becomes unfeasible.

Let us consider the steps of this algorithm in more detail. The first important point is the analysis of a set of polynomial forecasts for convergence to a certain value. We will introduce some explanations related to the construction of the forecast in the presence of a tendency to convergence. Let us have some time series $\{(1, x_1), (2, x_2), ..., (n + 1, x_{n+1})\}$, and we study it using some generalized polynomials that are built on the system of Chebyshev functions $\{u_i(t), i = 1.n\}$.

We will build all possible forecasts based on generalized polynomials based on the time series: $(1, x_1)$, $(2, x_2)$, ..., (n , x_n). Suppose that there is a generalized polynomial of some degree m that models the specified time series, i.e. the relation holds: $x_i = \sum_{k=1}^{m} a_k u_k(i)$.

ALGORITHM 8.1.–

Algorithm 1: Constructing the predictive value of the function for one step forward using m-th degree polinomials according (2)

Initialization:
set: vector (time series) of values of observed process $f_1, f_2, ..., f_{n-1}, f_n$;
n-vector length;
$m - 1$ -degree of the extrapolation polinomial;
$C(i, j)$-matrix of binomial coeffocients;
$C(1,0) = 1; C(1,1) = 1$;
iteration counter $i = 2$;
while $i \leq round(m)$ **do**
  $j = 0$;
  while $j \leq round(i)$ **do**
    C(i,j)=C(i-1,j-1)+C(i-1,j);
  end
end
$S = 0$;
$k = 1$;
$sign = 1$;
while $k \leq round(m)$ **do**
  $S = S + C(m, k) * f(n - k + 1) * sign$;
  $sign = sign * (-1)$
end
Return S

It is obvious that the prediction built on the basis of all generalized polynomials, starting with degree m, will converge. Otherwise, the unity property of the generalized interpolation polynomial passing through the specified n points will be violated.

This is the main idea that explains the situation of convergence of the sequence of polynomial forecasts. In the case when there is convergence, the predicted values of all polynomials, starting from a certain degree, differ by small amounts. This implies the existence of a value around which the predicted values of all generalized polynomials, starting from some fixed degree, are clustered. Such a value can be considered predictive. However, here it is necessary to specify the concept of the tendency to convergence. Because the criterion of monotonicity and monotonous reduction of distances between prognostic values clearly does not exhaust all possible cases.

ALGORITHM 8.2.–

Algorithm 2: The polinomial predictive set formation

Initialization:
set: vector (time series) of values of observed process $f_1, f_2, \ldots, f_n$;
n-vector length;
iteration counter $i = 1$;
while $i \leq round(n-1)$ **do**
| Using Algorithm 1 ; m=i; calculation $SP(i)$;
end
Return SP

Assume that we do not have a case where the distances between adjacent forecast points are monotonically decreasing or the sequence is monotonically increasing. In this case, the sequence can behave arbitrarily; it can make some oscillations near the convergence point, but the amplitude of these oscillations must decrease. In this case, we can supplement Algorithm 8.4. Let the sequence monotonically increase or decrease, while the distance between adjacent points monotonically decreases. Then, the optimal degree of the polynomial for forecasting is maximal. Let the sequence be non-monotonic, i.e. it has points of local extrema. We will fix the points of local extrema. If the intervals between adjacent maxima-minima form a nested sequence – each subsequent one is nested in the previous one, then we can claim that there is convergence to some constant.

We find the constant as the middle of the interval of change of convexity of the last points.

To find the interval of the convexity change, we look at all the points from right to left from the end and find the first pair (x_k, x_{k+1}) for which the relation is fulfilled: $(x_{k+1} - (x_{k+2} - x_k)/2 - x_k)(x_k - (x_{k+1} - x_{k-1})/2 - x_{k-1}) < 0$.

Then, the predicted value is equal to $(x_k+x_{k+1})/2$. These considerations are implemented in Algorithm 8.4.

ALGORITHM 8.3.–

Algorithm 3: General stepwise procedure of building the predictive value in monotonic case

```
Initialization:
creating set of all possible polinomial prediction {SP[i], i = 1,n-1;}
 using Algorithm 2;
set: iteration counter k = 1; l = 1;;
monotonicity analysis:
while k ≤ round(n − 2) AND SP[k] <= SP[k + 1] do
 | k=k+1;
end
while l ≤ round(n − 2) AND SP[l] >= SP[l + 1] do
 | l=l+1;
end
if k = n − 2 OR l = n − 2 then
 |  analysis of the reduction of average distances;
 |  s=1; k1=1;
 |  while s <= round((n − 2)/5) do
 |   |  while k1 ≤ round(n − 2 − s) do
 |   |   |  k2=k1; dist[k1]=0;
 |   |   |  while k2 ≤ round(k1 + s − 1) do
 |   |   |   | dist[k1]=dist[k1]+ABS(SP[k2+1]-SP[k2]);
 |   |   |  end
 |   |   |  dist[k1]=dist[k1]/s;
 |   |  end
 |   |  while k1 ≤ round(n − 2 − s) AND dist[k1] >= dist[k1 + 1]] do
 |   |   | k1=k1+1;
 |   |  end
 |   |  if k1 = n − 2 − s then
 |   |   | return SP[n-1];
 |   |  end
 |   |  s=s+1;
 |  end
end
Return -1; nonmonotonic case;
```

ALGORITHM 8.4.–

Algorithm 4: Stepwise procedure of building the predictive value in the case of nonmonotonic convergence

```
Initialization:
creating set of all possible polinomial prediction {SP[i], i = 1,n-1;}
  using Algorithm 2;
set: iteration counter k = 2; iextr=1;n=0;
if Algorithm 3 returns -1 then
    formation of arrays of points of local extrema: while
      k ≤ round(n − 2) do
        if SP[k] > SP[k − 1]ANDSP[k + 1] < SP[k]] then
          | SPextr[iextr] = SP[k]; iextr = iextr + 1;
        end
        if SP[k] < SP[k − 1]ANDSP[k + 1] > SP[k] then
          | SPextr[iextr] = SP[k]; iextr = iextr + 1;
        end
    end
    k=1; while k ≤ round(iextr − 1 − 2) do
        if ABS(SPextr[k] − SPextr[k + 1]) >
          ABS(SPextr[k + 1] − SPextr[k + 2]) then
          | n=n+1;
        end
    end
    if n = iextr − 3 then
        k=n-1;
        while k >= 1 do
            if (SP[k − 1] − (SP[k] − SP[k − 1])/2 − SP[k − 1])(SP[k −
              2] − (SP[k − 1] − SP[k − 2])/2 − SP[k − 2]) < 0 then
              | Return (SP[k − 2] + SP[k − 1])/2;
            end
            k=k-1;
        end
    end
end
```

8.6. Numerical results of the polynomial forecast

Let us consider the test function $f(x) = x^6 sin(x)$ at the points 1, 1.5, 2, 2.5, ..., 9.5, 10, 10.5. The value at the point 10.5 is -1178876.453. As discussed above, we will construct polynomial predictive values using (2) and the function values at the points 10, 9.5, 9, 8.5, …, .

The predictive set is {-1032799.243;-47317.735;-1269708.956;-1223899.974; -1183202.069;-1167550.283;-1169323.276;-1175934.332;-1180151.429;1180831.432;

- 1179759.926;-1178742.174;-1178446.598;-1178645.963;-1178898.415;-1178982.539;-1178932.321}.

D/T	77293	166943	259354	301149	219016	-55242	-544021	-1178876
2	45484	138363	256593	351764	342943	136884	-329502	-1032799
3	68251	170172	285174	354525	292327	712957	-521629	-1247317
4	79978	179214	281945	328705	238951	-60354	-589830	-1269708
5	82985	176529	269674	306114	211395	-80289	-584719	-1223899
6	81439	170838	260088	295794	206429	-72667	-559672	-1183202
7	78795	166692	256194	295060	211784	-60081	-542248	-1167550
8	76939	165191	256445	298220	217873	-52849	-537410	-1169323
9	76335	165546	258197	301129	220802	-51705	-539804	-1175934
10	76607	166503	259595	302285	220821	-53491	-543341	-1180151
11	77139	167189	260035	302044	219685	-55296	-545092	-1180831
12	77488	167343	259789	301363	218789	-55965	-545038	-1179759
13	77529	167149	259389	300928	218574	-55737	-544316	-1178742
14	77384	166913	259183	300893	218795	-55296	-543821	-1178446
15	77243	166822	259213	301064	219050	-55074	-543768	-1178645
16	77207	166872	259334	301204	219135	-55108	-543936	-1178898
17	77242	166959	259405	301224	219079	-55227	-544071	-1178982
18	77274	167010	259390	301172	219004	-55290	-544086	-1178932

Table 8.6. *Constructing the polynomial predictions for the test function $f(x) = x^6 \sin(x)$ at the point 13, 14, 15,...,19. T – true value, D – degree of the polynomial*

Let us consider in more detail the last column of Table 8.6. A graphic illustration of convergence is presented in Figure 8.3. The average prediction is -1,181,222.

When we use Algorithm 8.4, we get -1,178,772. True value is -1,178,876. As we can see, Algorithm 8.4 gives significant advantages in comparison, for example, with the calculation of the forecast value as the arithmetic mean of all polynomial forecasts.

Similar results are obtained by examining sets of polynomial forecasts from other columns of Table 8.2 (see Figure 8.2). As we can see in Figure 8.2, in all cases, we have convergence and can use Algorithm 8.3 to build forecasts.

Let us consider the function f(x) = xsin(x)cos(x) at the points -2, -1.5, -1, -0,5, ..., 9.5 . The function values and polynomial predictive set can be seen in Figure 8.3.

In this case, we have no convergence. However, all points of the set of polynomial forecasts are within some fixed interval, and the conditions of small error are fulfilled.

Therefore, we can find the predictive value as the arithmetic mean of all points of the set of polynomial predictions. The average of polynomial predictions is 4,568,633,838. The true value is 4,392,442,102. A good prediction result was obtained.

In this subsection, we study polynomial predictions in the case of stochastic input data.

Let us consider the bitcoin exchange rate in US dollars from October 22, 2023 until April 7, 2024 with a weekly interval. We can see the relevant data in Figure 8.4: 34,082, 35,065, 37,150, 36,568, 37,787, 39,458, 43,718, 42,271, 43,710, 42,136, 43,967, 42,851, 41,695, 42,120, 43,005, 47,758, 51,646, 51,571, 61,994, 68,366, 65,314, 64,037, 69,611, 68,890 and 70,937.

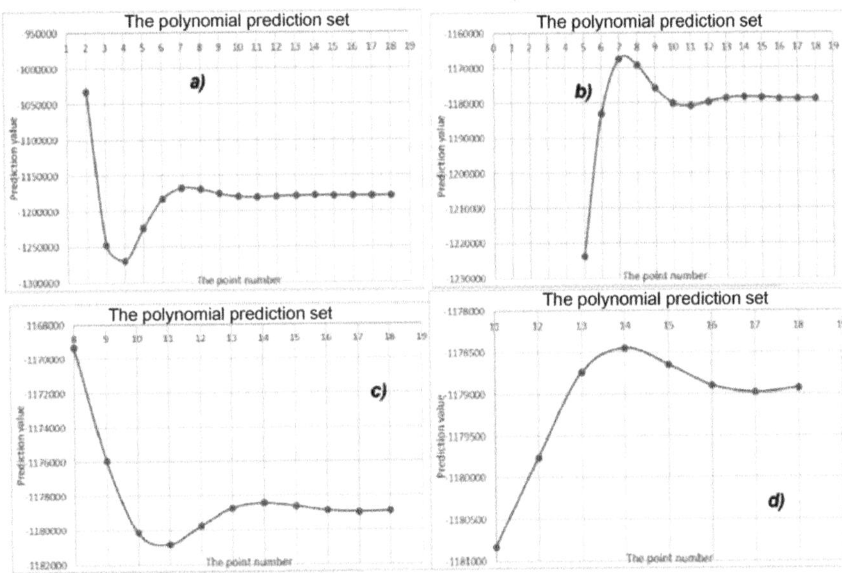

Figure 8.4. *The polynomial prediction set: a) graph for 17 points; b) graph for the point numbers 5–17; c) graph for the points 8–17; d) graph for the points 10–17*

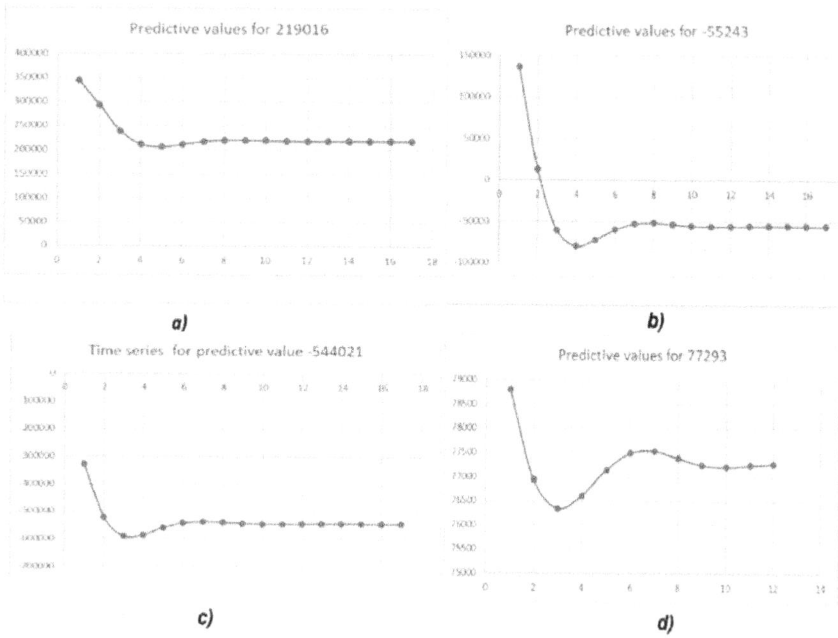

Figure 8.5. *The polynomial prediction set for the different columns in Table 4.2: a) column 219,016; b) column -55,243; c) column -544,021; d) column 77,293*

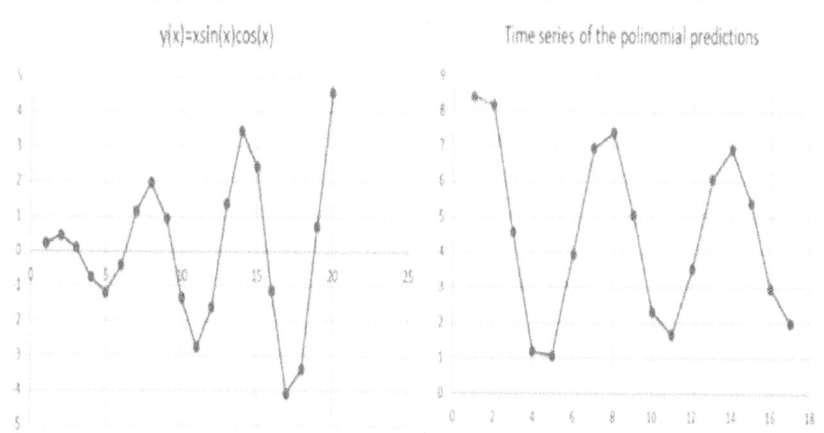

Figure 8.6. *For the function xsin(x)cos(x): a) the function values at the points -2, -1.5, -1,...,9.5; b) polynomial predictive set*

If we construct a series of polynomial forecasts in the order of increasing degrees of polynomials, the series diverges (see Figure 8.5). In such a case, based on the algorithm described in Turbal et al. (2023), the forecast is based on the polynomial of the minimum degree of 1 or equal to the value of the studied process at the last known point. But here we can offer some clarification. Let us pay attention to how a series of polynomial predictions behaves with small degrees of polynomials. In this case, there is some fluctuation at the beginning of the forecast series (see Figure 8.5b) and d)). This means that the process is inside some limited interval at the corresponding segment of the series, and we can build a forecast as an arithmetic average of the corresponding values. We will propose the following heuristic criterion: we will consider as many consecutive values from a series of polynomial predictions so that the corresponding arithmetic mean differs minimally from the last known value of the function.

We can consider value 68,890 and build all possible polynomial predictions starting from the value before 68,890: 69,611, 64,037, 65,314, ... Results are shown in Figure 8.5 c)–d). Then, we find the arithmetic mean of consecutive values of polynomial forecasts, which deviates minimally from the last known value of the function 69,611: we use the first size predictive points and find the average value 68,636. This is our predictive value; the true value is 68,890. Thus, in the case of a divergence of the series of forecast values, we can build a forecast.

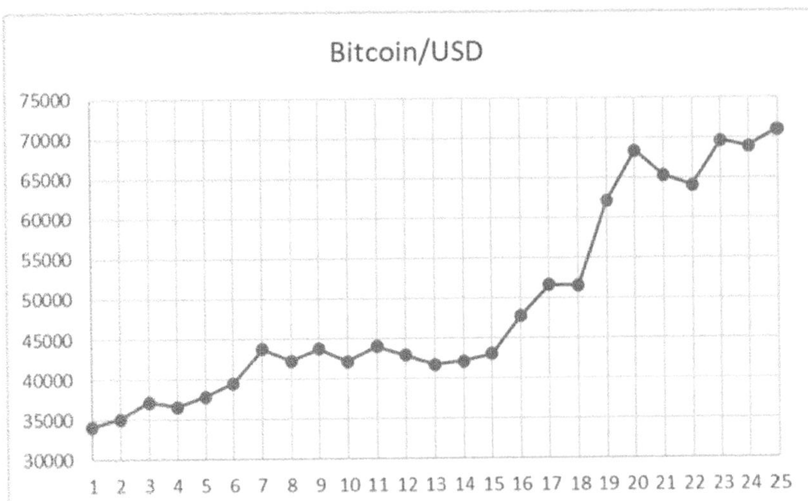

Figure 8.7. *The bitcoin exchange rate in US dollars from October 22, 2023 to April 7, 2024 a one week interval*

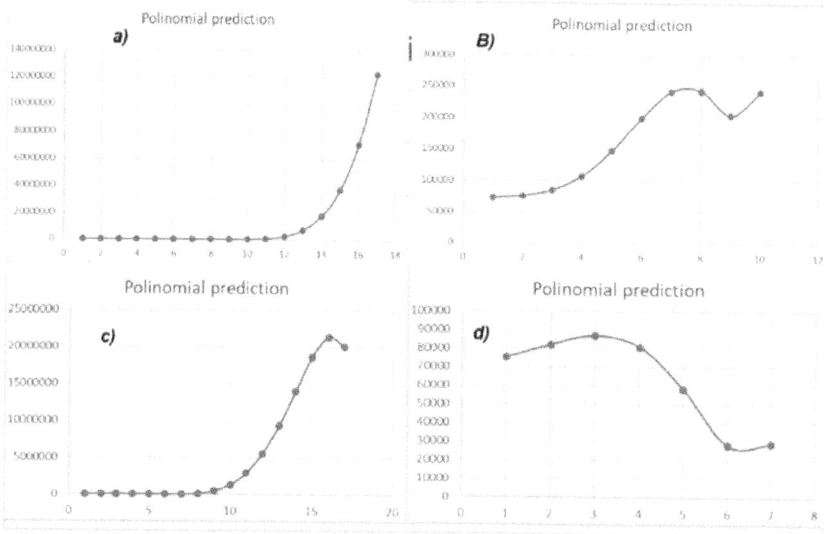

Figure 8.8. *Polynomial predictive set for the bitcoin exchange rate time series: a) predictive set using last value 70,937; b) first 10 points of the predictive series; c) predictive set using last value 68890; d) first 7 points of the predictive series*

8.7. Pyramidal method of extrapolation

Let *f(x)* be a continuous function of one independent variables defined over a closed interval [a,b] having continuous derivatives, and let values f_1, f_2 ...,f_n of this function be known at the points x_1,x_2,...,x_n respectively. We need to assess the value of this function at the point $x > x_n$.

Define the respective middle points $x_i^c = (x_i + x_{i+1})/2$. Consider the finite differences modified as follows:

$$\Delta^1 f_i = \frac{f_{i+1} - f_i}{x_{i+1} - x_i}, i = \overline{1, n-1},$$

$$\Delta^2 f_i = \frac{\Delta^1 f_{i+1} - \Delta^1 f_i}{x^c{}_{i+1} - x^c{}_i}, i = \overline{1, n-2},$$

$$\Delta^3 f_i = \frac{\Delta^2 f_{i+1} - \Delta^2 f_i}{x_{\ i+2} - x_{\ i+1}}, i = \overline{1, n-3},$$

...

In general, we have:

$$\Delta^j f_i = \begin{cases} \frac{\Delta^{j-1} f_{i+1} - \Delta^{j-1} f_i}{x_{i+\left[\frac{j}{2}\right]+1} - x_{i+\left[\frac{j}{2}\right]}}, j = 2k+1, \\ \frac{\Delta^{j-1} f_{i+1} - \Delta^{j-1} f_i}{x^c_{i+\frac{j}{2}} - x^c_{i+\frac{j}{2}-1}}, j = 2k, \end{cases} \quad [8.12]$$

$j = \overline{1, n-1}, i = \overline{1, n-j}.$

Let:

$$r_i^j = \begin{cases} x^c_{i+\frac{j}{2}}, j = 2k, \\ x_{i+\left[\frac{j}{2}\right]+1}, j = 2k+1, \end{cases}$$

$$l_i^j = \begin{cases} x^c_{i+\frac{j}{2}-1}, j = 2k, \\ x_{i+\left[\frac{j}{2}\right]}, j = 2k+1. \end{cases}$$

Then:

$$\Delta^j f_i = \frac{\Delta^{j-1} f_{i+1} - \Delta^{j-1} f_i}{r_i^j - l_i^j}. \quad [8.13]$$

Obviously, the differences [8.13] better approximate the function derivative and differ from the classic differences used in the Newton–Gregory backward difference formula and calculated using the formula:

$$\Delta^j f_i = \frac{\Delta^{j-1} f_{i+1} - \Delta^{j-1} f_i}{x_{i+j+1} - x_i} \quad [8.14]$$

If for any number k the finite difference $\Delta^k f_{n-k+1}$ is known, we can easily build the predictive value of the function at the x_{n+1} (see Figure 4.9) using the following procedure:

$$\Delta^{j-1} f_{n-j+2} = \Delta^{j-1} f_{n-j+1} + \Delta^j f_{n-j+1} (r^j_{n-j+1} - l^j_{n-j+1}), \quad [8.15]$$

$j = \overline{k, 1}$

Note that when the classical finite difference [8.14] is considered and $\Delta^k f_{n-k+1} = \Delta^k f_{n-k}$ from the procedure [8.15], we obtain the Newton–Gregory backward difference formula. The main idea of our extrapolation method is to find such a number of tables of divided differences [8.13] for which we can find the approximately predictive value of the corresponding finite differences using an approach that is discussed below.

					$\Delta^5 f_1$	…	$\Delta^5 f_{n-5}$							
				$\Delta^4 f_1$		…		$\Delta^4 f_{n-4}$		$\Delta^4 f_{n-3}$				
			$\Delta^3 f_1$		$\Delta^3 f_2$	…	$\Delta^3 f_{n-4}$		$\Delta^3 f_{n-3}$		$\Delta^3 f_{n-2}$			
		$\Delta^2 f_1$		$\Delta^2 f_2$		…		$\Delta^2 f_{n-3}$		$\Delta^2 f_{n-2}$		$\Delta^2 f_{n-1}$		
	$\Delta^1 f_1$		$\Delta^1 f_2$		$\Delta^1 f_3$	…	$\Delta^1 f_{n-3}$		$\Delta^1 f_{n-2}$		$\Delta^1 f_{n-1}$		$\Delta^1 f_n$	
f_1		f_2		f_3		…		f_{n-2}		f_{n-1}		f_n		f_{n+1}
x_1	x_1^c	x_2	x_2^c	x_3	x_3^c	…	x_{n-3}^c	x_{n-2}	x_{n-2}^c	x_{n-1}	x_{n-1}^c	x_n	x_n^c	x_{n+1}

Figure 8.9. *The table of divided differences*

Let $f_i^c, i = \overline{1, n-1}$ be the functions values in intermediate points $x_i^c, i = \overline{1, n-1}$.

Define the finite differences in the intermediate points as follows:

$$\Delta^j f^c{}_i = \begin{cases} \dfrac{\Delta^{j-1} f^c{}_{i+1} - \Delta^{j-1} f^c{}_i}{x^c{}_{i+\left[\frac{j}{2}\right]+1} - x^c{}_{i+\left[\frac{j}{2}\right]}}, j = 2k+1, \\ \dfrac{\Delta^{j-1} f^c{}_{i+1} - \Delta^{j-1} f^c{}_i}{x_{i+\frac{j}{2}+1} - x_{i+\frac{j}{2}}}, j = 2k, \end{cases} \qquad [8.16]$$

$j = \overline{1, n-2}, i = \overline{1, n-j}$.

Let $\hat{r}_i^j, \hat{l}_i^j$ be defined as:

$$\hat{r}_i^j = \begin{cases} x_{i+\frac{j}{2}+1}, j = 2k, \\ x^c{}_{i+\left[\frac{j}{2}\right]+1}, j = 2k+1, \end{cases}$$

$$\hat{l}_i^j = \begin{cases} x_{i+\frac{j}{2}}, j = 2k, \\ x^c{}_{i+\left[\frac{j}{2}\right]}, j = 2k+1, \end{cases}$$

From [8.16], we get:

$$\Delta^j f_i^c = \frac{\Delta^{j-1} f^c{}_{i+1} - \Delta^{j-1} f^c{}_i}{\hat{r}_i^j - \hat{l}_i^j}.$$

Consider the values of the form:

$$\tilde{\Delta}^2 f_{n-2}^c = \frac{\left(\frac{f_n - f_{n-1}^c}{x_n - x_{n-1}^c} - \frac{f_{n-1}^c - f_{n-1}}{x_{n-1}^c - x_{n-1}}\right)}{(x_n - x_{n-1})/2},$$

...

$$\tilde{\Delta}^i f_{n-i}^c = \left(\frac{\left(\Delta^{i-2} f_{n-i+2} - \Delta^{i-2} f_{n-i+1}^c\right)}{(r_i - c_i)(r_i - l_i)} - \frac{\left(\Delta^{i-2} f_{n-i+1}^c - \Delta^{i-2} f_{n-i+1}\right)}{(c_i - l_i)(r_i - l_i)}\right) 2$$

where:

$$i = \overline{2, n-1},$$

$$r_i = \begin{cases} x_{n-\frac{i}{2}+1}, i = 2k, \\ x^c_{n-\left[\frac{i}{2}\right]}, i = 2k+1, \end{cases}$$

$$c_i = \begin{cases} x^c_{n-\frac{i}{2}}, i = 2k, \\ x_{n-\left[\frac{i}{2}\right]}, i = 2k+1, \end{cases}$$

$$l_i = \begin{cases} x_{n-\frac{i}{2}}, i = 2k, \\ x^c_{n-\left[\frac{i}{2}\right]-1}, i = 2k+1, \end{cases}$$

$$ll_i = \begin{cases} x^c_{n-\frac{i}{2}-1}, i = 2k, \\ x_{n-\left[\frac{i}{2}\right]-1}, i = 2k+1. \end{cases}$$

Suppose that we have the equality:

$$\tilde{\Delta}^i f_{n-i}^c = \Delta^i f_{n-i}^c. \qquad [8.17]$$

Using [8.15], we can build a procedure for finding the function of the unknown value at the point ${x^c}_n$ as follows:

$$\Delta^{j-1}{f^c}_{n-j+1} = \Delta^{j-1}{f^c}_{n-j} + \Delta^j {f^c}_{n-j}(\hat{r}^j_{n-j} - \hat{l}^j_{n-j}), \quad [8.18]$$

$$j = \overline{\iota, 1}$$

Note that method of calculating the "midpoint" x_i^c may be different. After all, finite difference approximates the derivative or slope of the tangent. However, if we assume, for example, a convexity function, then the difference will determine the true slope of the tangent at some point within the interval. The corresponding displacement depends on the nature of the curvature changes – if the curvature is reduced, the bias will have to start growing at the end of the interval. So, we can enter correction coefficients to calculate values $x_i^c = \frac{x_i + x_{i+1}}{2} + \alpha_i$.

Let us find the necessary and sufficient conditions of relation [8.17]. Suppose that [8.17] is satisfied.

We have:

$$\Delta^j {f^c}_i = \begin{cases} \dfrac{\Delta^{j-1}{f^c}_{i+1} - \Delta^{j-1}{f^c}_i}{{x^c}_{i+\left[\frac{j}{2}\right]+1} - {x^c}_{i+\left[\frac{j}{2}\right]}}, j = 2k+1, \\ \dfrac{\Delta^{j-1}{f^c}_{i+1} - \Delta^{j-1}{f^c}_i}{x_{i+\frac{j}{2}+1} - x_{i+\frac{j}{2}}}, j = 2k, \end{cases}$$

$$\Delta^i {f^c}_{n-i} = \begin{cases} \dfrac{\Delta^{i-1}{f^c}_{n-i+1} - \Delta^{i-1}{f^c}_{n-i}}{{x^c}_{n-i+\left[\frac{i}{2}\right]+1} - {x^c}_{n-i+\left[\frac{i}{2}\right]}}, i = 2k+1, \\ \dfrac{\Delta^{i-1}{f^c}_{n-i+1} - \Delta^{i-1}{f^c}_{n-i}}{x_{n-\frac{i}{2}+1} - x_{n-\frac{i}{2}}}, i = 2k, \end{cases} =$$

$$= \begin{cases} \dfrac{\Delta^{i-1}{f^c}_{n-i+1} - \Delta^{i-1}{f^c}_{n-i}}{{x^c}_{n-\left[\frac{i}{2}\right]} - {x^c}_{n-\left[\frac{i}{2}\right]-1}}, i = 2k+1, \\ \dfrac{\Delta^{i-1}{f^c}_{n-i+1} - \Delta^{i-1}{f^c}_{n-i}}{x_{n-\frac{i}{2}+1} - x_{n-\frac{i}{2}}}, i = 2k, \end{cases} = \frac{\Delta^{i-1}{f^c}_{n-i+1} - \Delta^{i-1}{f^c}_{n-i}}{r_i - l_i}$$

Thus, the condition [8.17] can be written as:

$$\frac{\left(\frac{\left(\Delta^{i-2}f_{n-i+2}-\Delta^{i-2}f^{c}_{n-i+1}\right)}{r_i-c_i}-\frac{\left(\Delta^{i-2}f^{c}_{n-i+1}-\Delta^{i-2}f_{n-i+1}\right)}{c_i-l_i}\right)2}{r_i-l_i}=\frac{\Delta^{i-1}f^{c}{}_{n-i+1}-\Delta^{i-1}f^{c}{}_{n-i}}{r_i-l_i}$$

or:

$$\left(\frac{\left(\Delta^{i-2}f_{n-i+2}-\Delta^{i-2}f^{c}_{n-i+1}\right)}{r_i-c_i}-\frac{\left(\Delta^{i-2}f^{c}_{n-i+1}-\Delta^{i-2}f_{n-i+1}\right)}{c_i-l_i}\right)2=\Delta^{i-1}f^{c}{}_{n-i+1}-\Delta^{i-1}f^{c}{}_{n-i} \quad [8.19]$$

Let $i = 2k$. Then:

$$\Delta^{i-1}f^{c}{}_{n-i+1}=\frac{\Delta^{i-2}f^{c}{}_{n-i+2}-\Delta^{i-2}f^{c}{}_{n-i+1}}{x^{c}{}_{n-i+\left[\frac{i-1}{2}\right]+2}-x^{c}{}_{n-i+\left[\frac{i-1}{2}\right]+1}}=\frac{\Delta^{i-2}f^{c}{}_{n-i+2}-\Delta^{i-2}f^{c}{}_{n-i+1}}{x^{c}{}_{n-\frac{i}{2}+1}-x^{c}{}_{n-\frac{i}{2}}},$$

$$\Delta^{i-1}f^{c}{}_{n-i}=\frac{\Delta^{i-2}f^{c}{}_{n-i+1}-\Delta^{i-2}f^{c}{}_{n-i}}{x^{c}{}_{n-i+\left[\frac{i-1}{2}\right]+1}-x^{c}{}_{n-i+\left[\frac{i-1}{2}\right]}}=$$

$$=\frac{\Delta^{i-2}f^{c}{}_{n-i+1}-\Delta^{i-2}f^{c}{}_{n-i}}{x^{c}{}_{n-\frac{i}{2}}-x^{c}{}_{n-\frac{i}{2}-1}}.$$

Then, equation [8.19] can be written as:

$$2\left(\frac{\left(\Delta^{i-2}f_{n-i+2}-\Delta^{i-2}f^{c}_{n-i+1}\right)}{r_i-c_i}-\frac{\left(\Delta^{i-2}f^{c}_{n-i+1}-\Delta^{i-2}f_{n-i+1}\right)}{c_i-l_i}\right)=$$

$$=\Delta^{i-1}f^{c}{}_{n-i+1}-\Delta^{i-1}f^{c}{}_{n-i}=$$

$$=\frac{\Delta^{i-2}f^{c}{}_{n-i+2}-\Delta^{i-2}f^{c}{}_{n-i+1}}{x^{c}{}_{n-\frac{i}{2}+1}-x^{c}{}_{n-\frac{i}{2}}}-\frac{\Delta^{i-2}f^{c}{}_{n-i+1}-\Delta^{i-2}f^{c}{}_{n-i}}{x^{c}{}_{n-\frac{i}{2}}-x^{c}{}_{n-\frac{i}{2}-1}}.$$

Consider the right side of the last equality. We have an obvious sequence of ratios:

$$\frac{\Delta^{i-2}f^c_{n-i+2}-\Delta^{i-2}f^c_{n-i+1}}{x^c_{n-\frac{i}{2}+1}-x^c_{n-\frac{i}{2}}}-\frac{\Delta^{i-2}f^c_{n-i+1}-\Delta^{i-2}f^c_{n-i}}{x^c_{n-\frac{i}{2}}-x^c_{n-\frac{i}{2}-1}}=$$

$$\frac{\Delta^{i-2}f^c_{n-i+2}-\Delta^{i-2}f_{n-i+2}+\Delta^{i-2}f_{n-i+2}-\Delta^{i-2}f^c_{n-i+1}}{x^c_{n-\frac{i}{2}+1}-x^c_{n-\frac{i}{2}}}-$$

$$\frac{\Delta^{i-2}f^c_{n-i+1}-\Delta^{i-2}f_{n-i+1}+\Delta^{i-2}f_{n-i+1}-\Delta^{i-2}f^c_{n-i}}{x^c_{n-\frac{i}{2}}-x^c_{n-\frac{i}{2}-1}}=$$

$$\frac{\Delta^{i-2}f^c_{n-i+2}-\Delta^{i-2}f_{n-i+2}}{x^c_{n-\frac{i}{2}+1}-x_{n-\frac{i}{2}+1}}\frac{x^c_{n-\frac{i}{2}+1}-x_{n-\frac{i}{2}+1}}{x^c_{n-\frac{i}{2}+1}-x^c_{n-\frac{i}{2}}}+$$

$$+\frac{\Delta^{i-2}f_{n-i+2}-\Delta^{i-2}f^c_{n-i+1}}{x_{n-\frac{i}{2}+1}-x^c_{n-\frac{i}{2}}}\frac{x_{n-\frac{i}{2}+1}-x^c_{n-\frac{i}{2}}}{x^c_{n-\frac{i}{2}+1}-x^c_{n-\frac{i}{2}}}-$$

$$\frac{\Delta^{i-2}f^c_{n-i+1}-\Delta^{i-2}f_{n-i+1}}{x^c_{n-\frac{i}{2}}-x_{n-\frac{i}{2}}}\frac{x^c_{n-\frac{i}{2}}-x_{n-\frac{i}{2}}}{x^c_{n-\frac{i}{2}}-x^c_{n-\frac{i}{2}-1}}-$$

$$-\frac{\Delta^{i-2}f_{n-i+1}-\Delta^{i-2}f^c_{n-i}}{x_{n-\frac{i}{2}}-x^c_{n-\frac{i}{2}-1}}\frac{x_{n-\frac{i}{2}}-x^c_{n-\frac{i}{2}-1}}{x^c_{n-\frac{i}{2}}-x^c_{n-\frac{i}{2}-1}}=$$

$$=\frac{\Delta^{i-2}f^c_{n-i+2}-\Delta^{i-2}f_{n-i+2}}{x^c_{n-\frac{i}{2}+1}-x_{n-\frac{i}{2}+1}}\frac{x^c_{n-\frac{i}{2}+1}-x_{n-\frac{i}{2}+1}}{x^c_{n-\frac{i}{2}+1}-x^c_{n-\frac{i}{2}}}-$$

$$-\frac{\Delta^{i-2}f_{n-i+2}-\Delta^{i-2}f^c_{n-i+1}}{x_{n-\frac{i}{2}+1}-x^c_{n-\frac{i}{2}}}\frac{x^c_{n-\frac{i}{2}+1}-x_{n-\frac{i}{2}+1}}{x^c_{n-\frac{i}{2}+1}-x^c_{n-\frac{i}{2}}}+$$

$$+\frac{\Delta^{i-2}f_{n-i+2}-\Delta^{i-2}f^c_{n-i+1}}{x_{n-\frac{i}{2}+1}-x^c_{n-\frac{i}{2}}}\frac{x^c_{n-\frac{i}{2}+1}-x_{n-\frac{i}{2}+1}}{x^c_{n-\frac{i}{2}+1}-x^c_{n-\frac{i}{2}}}+$$

$$+\frac{\Delta^{i-2}f_{n-i+2}-\Delta^{i-2}f^c_{n-i+1}}{x_{n-\frac{i}{2}+1}-x^c_{n-\frac{i}{2}}}\frac{x_{n-\frac{i}{2}+1}-x^c_{n-\frac{i}{2}}}{x^c_{n-\frac{i}{2}+1}-x^c_{n-\frac{i}{2}}}-$$

$$-\frac{\Delta^{i-2}f^c{}_{n-i+1}-\Delta^{i-2}f_{\ n-i+1}}{x^c{}_{n-\frac{i}{2}}-x_{\ n-\frac{i}{2}}}\frac{x^c{}_{n-\frac{i}{2}}-x_{\ n-\frac{i}{2}}}{x^c{}_{n-\frac{i}{2}}-x^c{}_{n-\frac{i}{2}-1}}+$$

$$+\frac{\Delta^{i-2}f_{\ n-i+1}-\Delta^{i-2}f^c{}_{n-i}}{x_{\ n-\frac{i}{2}}-x^c{}_{n-\frac{i}{2}-1}}\frac{x^c{}_{n-\frac{i}{2}}-x_{\ n-\frac{i}{2}}}{x^c{}_{n-\frac{i}{2}}-x^c{}_{n-\frac{i}{2}-1}}-$$

$$-\frac{\Delta^{i-2}f_{\ n-i+1}-\Delta^{i-2}f^c{}_{n-i}}{x_{\ n-\frac{i}{2}}-x^c{}_{n-\frac{i}{2}-1}}\frac{x^c{}_{n-\frac{i}{2}}-x_{\ n-\frac{i}{2}}}{x^c{}_{n-\frac{i}{2}}-x^c{}_{n-\frac{i}{2}-1}}-$$

$$-\frac{\Delta^{i-2}f_{\ n-i+1}-\Delta^{i-2}f^c{}_{n-i}}{x_{\ n-\frac{i}{2}}-x^c{}_{n-\frac{i}{2}-1}}\frac{x_{\ n-\frac{i}{2}}-x^c{}_{n-\frac{i}{2}-1}}{x^c{}_{n-\frac{i}{2}}-x^c{}_{n-\frac{i}{2}-1}}=$$

$$\left(\frac{\Delta^{i-2}f^c{}_{n-i+2}-\Delta^{i-2}f_{\ n-i+2}}{x^c{}_{n-\frac{i}{2}+1}-x_{\ n-\frac{i}{2}+1}}-\frac{\Delta^{i-2}f_{\ n-i+2}-\Delta^{i-2}f^c{}_{n-i+1}}{x_{\ n-\frac{i}{2}+1}-x^c{}_{n-\frac{i}{2}}}\right)\frac{x^c{}_{n-\frac{i}{2}+1}-x_{\ n-\frac{i}{2}+1}}{x^c{}_{n-\frac{i}{2}+1}-x^c{}_{n-\frac{i}{2}}}+$$

$$+\frac{\Delta^{i-2}f_{\ n-i+2}-\Delta^{i-2}f^c{}_{n-i+1}}{x_{\ n-\frac{i}{2}+1}-x^c{}_{n-\frac{i}{2}}}-$$

$$-\left(\frac{\Delta^{i-2}f^c{}_{n-i+1}-\Delta^{i-2}f_{\ n-i+1}}{x^c{}_{n-\frac{i}{2}}-x_{\ n-\frac{i}{2}}}-\frac{\Delta^{i-2}f_{\ n-i+1}-\Delta^{i-2}f^c{}_{n-i}}{x_{\ n-\frac{i}{2}}-x^c{}_{n-\frac{i}{2}-1}}\right)\frac{x^c{}_{n-\frac{i}{2}}-x_{\ n-\frac{i}{2}}}{x^c{}_{n-\frac{i}{2}}-x^c{}_{n-\frac{i}{2}-1}}$$

$$-\frac{\Delta^{i-2}f_{\ n-i+1}-\Delta^{i-2}f^c{}_{n-i}}{x_{\ n-\frac{i}{2}}-x^c{}_{n-\frac{i}{2}-1}}=$$

$$=2\left(\frac{\left(\Delta^{i-2}f_{n-i+2}-\Delta^{i-2}f^c_{n-i+1}\right)}{x_{n-\frac{i}{2}+1}-x^c_{n-\frac{i}{2}}}-\frac{\left(\Delta^{i-2}f^c_{n-i+1}-\Delta^{i-2}f_{n-i+1}\right)}{x^c_{n-\frac{i}{2}}-x_{n-\frac{i}{2}}}\right)$$

Hence:

$$\Bigg(\frac{\Delta^{i-2}f^{c}{}_{n-i+2}-\Delta^{i-2}f_{\;n-i+2}}{x^{c}{}_{n-\frac{i}{2}+1}-x_{\;n-\frac{i}{2}+1}}-$$

$$-\frac{\Delta^{i-2}f_{\;n-i+2}-\Delta^{i-2}f^{c}{}_{n-i+1}}{x_{\;n-\frac{i}{2}+1}-x^{c}{}_{n-\frac{i}{2}}}\Bigg)\frac{x^{c}{}_{n-\frac{i}{2}+1}-x_{\;n-\frac{i}{2}+1}}{x^{c}{}_{n-\frac{i}{2}+1}-x^{c}{}_{n-\frac{i}{2}}}+$$

$$+\frac{\Delta^{i-2}f_{\;n-i+2}-\Delta^{i-2}f^{c}{}_{n-i+1}}{x_{\;n-\frac{i}{2}+1}-x^{c}{}_{n-\frac{i}{2}}}-$$

$$-\left(\frac{\Delta^{i-2}f^{c}{}_{n-i+1}-\Delta^{i-2}f_{\;n-i+1}}{x^{c}{}_{n-\frac{i}{2}}-x_{\;n-\frac{i}{2}}}\right.$$

$$\left.-\frac{\Delta^{i-2}f_{\;n-i+1}-\Delta^{i-2}f^{c}{}_{n-i}}{x_{\;n-\frac{i}{2}}-x^{c}{}_{n-\frac{i}{2}-1}}\right)\frac{x^{c}{}_{n-\frac{i}{2}}-x_{\;n-\frac{i}{2}}}{x^{c}{}_{n-\frac{i}{2}}-x^{c}{}_{n-\frac{i}{2}-1}}-$$

$$-\frac{\Delta^{i-2}f_{\;n-i+1}-\Delta^{i-2}f^{c}{}_{n-i}}{x_{\;n-\frac{i}{2}}-x^{c}{}_{n-\frac{i}{2}-1}}-$$

$$-\frac{\left(\Delta^{i-2}f^{c}_{n-i+1}-\Delta^{i-2}f_{n-i+1}\right)}{x^{c}_{n-\frac{i}{2}}-x_{n-\frac{i}{2}}}+\frac{\left(\Delta^{i-2}f^{c}_{n-i+1}-\Delta^{i-2}f_{n-i+1}\right)}{x^{c}_{n-\frac{i}{2}}-x_{n-\frac{i}{2}}}=$$

$$=2\left(\frac{\left(\Delta^{i-2}f_{n-i+2}-\Delta^{i-2}f^{c}_{n-i+1}\right)}{x_{n-\frac{i}{2}+1}-x^{c}_{n-\frac{i}{2}}}-\frac{\left(\Delta^{i-2}f^{c}_{n-i+1}-\Delta^{i-2}f_{n-i+1}\right)}{x^{c}_{n-\frac{i}{2}}-x_{n-\frac{i}{2}}}\right),$$

$$\Bigg(\frac{\Delta^{i-2}f^{c}{}_{n-i+2}-\Delta^{i-2}f_{\;n-i+2}}{x^{c}{}_{n-\frac{i}{2}+1}-x_{\;n-\frac{i}{2}+1}}-$$

$$-\frac{\Delta^{i-2}f_{\;n-i+2}-\Delta^{i-2}f^{c}{}_{n-i+1}}{x_{\;n-\frac{i}{2}+1}-x^{c}{}_{n-\frac{i}{2}}}\Bigg)\frac{x^{c}{}_{n-\frac{i}{2}+1}-x_{\;n-\frac{i}{2}+1}}{x^{c}{}_{n-\frac{i}{2}+1}-x^{c}{}_{n-\frac{i}{2}}}+$$

$$+\left(\frac{\Delta^{i-2}f^{c}{}_{n-i+1}-\Delta^{i-2}f{}_{n-i+1}}{x^{c}{}_{n-\frac{i}{2}}-x{}_{n-\frac{i}{2}}}\right.$$
$$\left.-\frac{\Delta^{i-2}f{}_{n-i+1}-\Delta^{i-2}f^{c}{}_{n-i}}{x{}_{n-\frac{i}{2}}-x^{c}{}_{n-\frac{i}{2}-1}}\right)\frac{x{}_{n-\frac{i}{2}}-x^{c}{}_{n-\frac{i}{2}-1}}{x^{c}{}_{n-\frac{i}{2}}-x^{c}{}_{n-\frac{i}{2}-1}}=$$

$$=\left(\frac{\left(\Delta^{i-2}f_{n-i+2}-\Delta^{i-2}f^{c}_{n-i+1}\right)}{x{}_{n-\frac{i}{2}+1}-x^{c}{}_{n-\frac{i}{2}}}-\frac{\left(\Delta^{i-2}f^{c}_{n-i+1}-\Delta^{i-2}f_{n-i+1}\right)}{x^{c}{}_{n-\frac{i}{2}}-x{}_{n-\frac{i}{2}}}\right).$$

We obtain a linear combination that proves that the points $\left(x^{c}_{n-\frac{i}{2}},\tilde{\Delta}^{i}f^{c}_{n-i}\right),\left(x{}_{n-\frac{i}{2}},\tilde{\Delta}^{i}f_{n-i}\right),\left(x{}_{n-\frac{i}{2}+1},\tilde{\Delta}^{i}f_{n-i+1}\right)$ are collinear if the following conditions are satisfied:

$$\frac{x^{c}{}_{n-\frac{i}{2}+1}-x{}_{n-\frac{i}{2}+1}}{x^{c}{}_{n-\frac{i}{2}+1}-x^{c}{}_{n-\frac{i}{2}}}=\frac{x^{c}{}_{n-\frac{i}{2}}-x{}_{n-\frac{i}{2}}}{x^{c}{}_{n-\frac{i}{2}}-x^{c}{}_{n-\frac{i}{2}-1}}$$

or:

$$\frac{x{}_{n-\frac{i}{2}}-x^{c}{}_{n-\frac{i}{2}-1}}{x^{c}{}_{n-\frac{i}{2}}-x^{c}{}_{n-\frac{i}{2}-1}}=\frac{x{}_{n-\frac{i}{2}+1}-x^{c}{}_{n-\frac{i}{2}}}{x^{c}{}_{n-\frac{i}{2}+1}-x^{c}{}_{n-\frac{i}{2}}}.$$

Thus, if equation [8.17] is satisfied, the corresponding points lie on a straight line.

Let us obtain a necessary condition for satisfying [8.17]. It is easy to show that the resulting condition is also sufficient proof, considered above in reverse order. Similarly, consider the *case* $i = 2k + 1$. In general, we can formulate Statement 8.2:

STATEMENT 8.2.– The statement of condition [8.17] is necessary and sufficient to points $\left(c_i,\tilde{\Delta}^{i}f^{c}_{n-i}\right),\left(l_i,\tilde{\Delta}^{i}f^{c}_{n-i-1}\right),\left(r_i,\tilde{\Delta}^{i}f^{c}_{n-i+1}\right)$ lie on a straight line.

It is clear that the latter condition is satisfied (which is consistent with the second-derivative finite-difference equation [8.16]) if, on the interval $[l_i,r_i]$ the curve passing through the corresponding points is a cubic polynomial.

Suppose that some i satisfy the condition:

$$\frac{\left(\Delta^{i-2} f_{n-i+1}^{c}-\Delta^{i-2} f_{n-i+1}\right)}{c_i-l_i}=\frac{\left(\Delta^{i-2} f_{n-i+1}^{c}-\Delta^{i-2} f_{n-i}^{c}\right)}{c_i-ll_i} \qquad [8.20]$$

Let us show that the points$(c_i, \Delta^{i-2} f_{n-i+1}^{c})$, $(l_i, \Delta^{i-2} f_{n-i+1})$ $(ll_i, \Delta^{i-2} f_{n-i}^{c})$ lie on a straight line.

From the condition [8.20], we have:

$$\Delta^{i-2} f_{n-i+1}=\Delta^{i-2} f_{n-i+1}^{c}-\frac{(c_i-l_i)(\Delta^{i-2} f_{n-i+1}^{c}-\Delta^{i-2} f_{n-i}^{c})}{c_i-ll_i}=$$

$$\frac{\Delta^{i-2} f_{n-i+1}^{c}(c_i-ll_i)-(c_i-l_i)(\Delta^{i-2} f_{n-i+1}^{c}-\Delta^{i-2} f_{n-i}^{c})}{c_i-ll_i}=$$

$$\frac{\Delta^{i-2} f_{n-i+1}^{c}(l_i-ll_i)+(c_i-l_i)\Delta^{i-2} f_{n-i}^{c}}{c_i-ll_i}=$$

$$=\frac{\Delta^{i-2} f_{n-i+1}^{c}(l_i-ll_i)}{c_i-ll_i}+\frac{(c_i-l_i)\Delta^{i-2} f_{n-i}^{c}}{c_i-ll_i};$$

$$\frac{c_i(l_i-ll_i)}{c_i-ll_i}+\frac{ll_i(c_i-l_i)}{c_i-ll_i}=l_i.$$

For example, in the case $i=2, n=7$, equation [8.20] yields:

$$\frac{f_6^c-f_6}{x_6^c-x_6}=\Delta^1 f_5^c=\frac{f_6^c-f_5^c}{x_6^c-x_5^c}.$$

We see that the point (x_6, f_6) lies on the line that passes through the points $(x_6^c, f_6^c), (x_5^c, f_5^c)$.

Note that from [8.20], we get:

$$\frac{\Delta^{i-2} f_{n-i+1}^{c}-\Delta^{i-2} f_{n-i+1}}{c_i-l_i}-\frac{\Delta^{i-2} f_{n-i+1}-\Delta^{i-2} f^{c}{}_{n-i}}{l_i-ll_i}=0$$

The last equation follows from the equation:

$$\frac{\Delta^{i-2} f^{c}{}_{n-i+1}-\Delta^{i-2} f{}_{n-i+1}}{c_i-l_i}-\frac{\Delta^{i-2} f{}_{n-i+1}-\Delta^{i-2} f^{c}{}_{n-i}}{l_i-ll_i}=$$

$$= \frac{(\Delta^{i-2} f_{n-i+1}^{c} - \Delta^{i-2} f_{n-i}^{c})}{c_i - ll_i} - \frac{\Delta^{i-2} f_{\ n-i+1} - \Delta^{i-2} f^{c}_{\ n-i}}{l_i - ll_i} =$$

$$= \frac{(\Delta^{i-2} f_{n-i+1}^{c} - \Delta^{i-2} f_{n-i+1} + \Delta^{i-2} f_{n-i+1} - \Delta^{i-2} f_{n-i}^{c})}{c_i - ll_i} -$$

$$- \frac{\Delta^{i-2} f_{\ n-i+1} - \Delta^{i-2} f^{c}_{\ n-i}}{l_i - ll_i} =$$

$$= \frac{(\Delta^{i-2} f_{n-i+1}^{c} - \Delta^{i-2} f_{n-i+1})}{c_i - l_i} \frac{c_i - l_i}{c_i - ll_i} + \frac{(\Delta^{i-2} f_{n-i+1} - \Delta^{i-2} f_{n-i}^{c})}{l_i - ll_i} \frac{l_i - ll_i}{c_i - ll_i} -$$

$$- \frac{\Delta^{i-2} f_{\ n-i+1} - \Delta^{i-2} f^{c}_{\ n-i}}{l_i - ll_i} = \frac{(\Delta^{i-2} f_{n-i+1}^{c} - \Delta^{i-2} f_{n-i+1})}{c_i - l_i} \frac{c_i - l_i}{c_i - ll_i} +$$

$$+ \frac{(\Delta^{i-2} f_{n-i+1} - \Delta^{i-2} f_{n-i}^{c})}{l_i - ll_i} \frac{l_i - c_i}{c_i - ll_i} = \Big(\frac{(\Delta^{i-2} f_{n-i+1}^{c} - \Delta^{i-2} f_{n-i+1})}{c_i - l_i} -$$

$$- \frac{(\Delta^{i-2} f_{n-i+1} - \Delta^{i-2} f_{n-i}^{c})}{l_i - ll_i}\Big) \frac{c_i - l_i}{c_i - ll_i};$$

Note if a continuous function passes through the corresponding points $(c_i, \Delta^{i-2} f_{n-i+1}^{c})$, $(l_i, \Delta^{i-2} f_{n-i+1})$, and $(ll_i, \Delta^{i-2} f_{n-i}^{c})$ which lie on the line, this function changes the convexity in the points l_i. or is linear.

Consequently, we can offer the following algorithm of extrapolation:

1) Construct the table of finite difference according to (1).

2) The function is interpolated at the intermediate and is found to have finite differences (6).

3) We get i^* according to the formula

$$\mathrm{i}^* = \mathrm{argmin_i} \left| \frac{(\Delta^{i-2} f_{n-i+1}^{c} - \Delta^{i-2} f_{n-i+1})}{c_i - l_i} - \frac{(\Delta^{i-2} f_{n-i+1}^{c} - \Delta^{i-2} f_{n-i}^{c})}{c_i - ll_i} \right|$$

4) Calculate the value $\tilde{\Delta}^{\mathrm{i}^*} \mathrm{f}_{\mathrm{n-i}^*}^{\mathrm{c}}$

5) Located predictive value function formulas:

$$\Delta^{j-1}f^{c}{}_{n-j+1} = \Delta^{j-1}f^{c}{}_{n-j} + \Delta^{j}f^{c}{}_{n-j}(\hat{r}^{j}_{n-j} - \hat{l}^{j}_{n-j}), \qquad [8.21]$$

$$j = \overline{i^{*},1}$$

8.8. Numerical results for the pyramidal methods

Consider the examples of applications of the above-described method. Let the function $f(x) = x^{6}\sin(x)$ be defined in points 1, 1.5, 2, 2.5, 3, ..., 11. The fragment tablc of finite differences is shown in Figure 8.10.

7	7.5	8	8.5	9	9.5	10	10.5	11	
77293.8164	166943.844	259354.3	301149.028	219017	-55242.97922	-544021.1109	-1178876.45	-1771543.65	-1941908.279
150719.733	182060.512	134205.2	-40337.668	-356392	-763037.7708	-1123633.474	-1227522.54	-763031.8255	
91730.286	-16514.548	-222398.2	-490597.19	-722700	-767241.4668	-464484.7686	360601.648		
-131480.36	-314128.47	-474082.6	-500301.92	-276644	258215.3339	1127843.115			
-317475.31	-342602.28	-186173.5	197438.367	758517	1404487.391				
-143322.65	131301.852	540040.6	944690.713	1207049					
377570.671	683363.299	813388.9	667008.376						
610713.879	435818.189	-16354.92							
-64772.179	-703068.27								
-1140076.8									

Figure 8.10. *The table of divided differences and optimal row finding*

The first line of the table is the value argument, and then the function and the corresponding finite difference. The optimal value i^{*}, which satisfied the condition [8.17] is - $i^{*} = 9$, $i^{*} - 2 = 7$ –index of corresponding finite difference, which corresponds to the ninth row of the table. $\frac{(\Delta^{i-2}f^{c}_{n-i+1} - \Delta^{i-2}f^{c}_{n-i})}{c_i - ll_i}$ =610,713,879, $\frac{(\Delta^{i-2}f^{c}_{n-i+1} - \Delta^{i-2}f_{n-i+1})}{c_i - l_i}$ =611,585,256.

The corresponding value of finite difference $\tilde{\Delta}^{i^{*}}f^{c}_{n-i^{*}}$.= -703,068,269.

In this case, using procedures [8.21], we find the predictive value of the function at point 11.5; it is equal to -2017907.745, and the exact value is -2024974.077. Thus, the relative error is 0.0035.

Consider the same function and find its predictive value at point 10.5. Details are in the finite difference table shown in Figure 8.11. The optimal value $i^{*} = 7$ and $\tilde{\Delta}^{i^{*}}f^{c}_{n-i^{*}}$=534939.6. The predictive value is -1183977.5; the corresponding exact value is -1178876.5; the relative error is 0.0043. The corresponding polynomial extrapolation error is given in Table 8.3. Analysis of the relative errors shows that the proposed method gives significantly better results compared to the Newton's extrapolation of the best degree.

6	6.5	7	7.5	8	8.5	9	9.5	10	10.5
13036.4	16224.111	77293.8164	166943.844	259354.3	301149.028	219017	-55242.97922	-544021.1109	-1183977.51
35753.9	90330.226	150719.733	182060.512	134205.2	-40337.668	-356392	-763037.7708	-1128734.534	
88383.4	114965.81	91730.286	-16514.548	-222398.2	-490597.19	-722700	-772342.5266		
67799.3	3346.8425	-131480.36	-314128.47	-474082.6	-500301.92	-281745			
71206.5	-199279.63	-317475.31	-342602.28	-186173.5	192337.307				
-215972	-246268.82	-143322.65	131301.852	534939.6					
-123020	72649.42	377570.671	678262.239						
256948	500590.37	605612.819							
476262	348664.59								
21040.3									

Figure 8.11. *Illustration of divided differences construction*

Let us consider, in which cases the method will be most effective. As reviewed above, the optimal replacement of the finite difference into the corresponding modified finite difference is when the finite difference of two smaller orders at the end of the interval of definition behaves just as in Figure 8.6 – the last four points are on the range of convexity.

Obviously, in the case of irregular grids, condition [8.20] is not always satisfied. However, the underlying principle is that the region where the function changes convexity is best approximated by a cubic polynomial, and this region occurs near the boundary of the finite-difference domain. In this case, step 3 of the algorithm can be changed, for example, there is a value i^* for which point $(ll_i + l_i)/2$ minimum deviates from the point of changing convexity of the functions defined by the finite difference which is order i-2 (the feature by finite differences can be determined, for example, a cubic spline).

Polynomial degree	Modified finite differences		Newton's forward interpolation	
	Projected value	Relative error	Projected value	Relative error
9	-852308.1	0.2770	-827646.1	0.2979
8	-1056370.1	0.1039	-812423.4	0.3108
7	-1289000.1	0.0934	-845657.5	0.2827

6	-1484668.7	0.2593	-866348.2	0.2651
5	-1587615.1	0.3467	-874351.3	0.2583
4	-1562487.8	0.3254	-861504.3	0.2692
3	-1402534.1	0.1897	-823699.5	0.3013
2	-1134334.9	0.0378	-772984.9	0.3443
1	-818280.9	0.3059	-818280.9	0.3059

Table 8.7. *Comparison of the modified finite differences methods within Newton's forward interpolation*

Note that the range of convexity at the end of determining interval arises when the function defined by the finite differences, one less order, has the appropriate range of local extreme. If in the table of finite differences there is no such situation, we obtain the accuracy that matches the Newton's extrapolation. Note also that the grid spacing should be such that at least four points describe the area where the function has a local extreme or is changing its prominence.

8.9. Conclusions

The conducted research concerning the further detailing of the extrapolation algorithms based on the analysis of the polynomial forecasts set confirmed the effectiveness of such an approach. Indeed, polynomials are very sensitive to changes in the values of the coefficients and especially the degree. When using it directly for forecasting, good results can be obtained only in special cases. However, such properties make it possible to clearly define the situations when the series of polynomial forecasts converge and when there is no convergence. This enables us to choose the optimal methodology for building forecasts. At the same time, this forecasting technique demonstrates its effectiveness in the case of small time series (it is enough to have about 10–20 observation points). An important refinement of the basic forecasting algorithm in the stochastic case is proposed. Using the polynomials for forecasting in the case of stochastic time series in our research is carried out in the context of finding the average of polynomial forecasts series. A procedure for determining the presence of the fact of convergence in the case of monotonic and non-monotonic series of polynomial forecasts is proposed.

Using the test functions, we show that, in the case of polynomial predictive set convergence, it is possible to construct better forecasts than the forecast based on the arithmetic mean of the polynomial predictive set. With this approach, in fact, all available information in the time series is used. The corresponding random variable is predicted not as its mathematical expectation in the usual sense but as the mathematical expectation of polynomial predictive values.

Algorithms 8.1–8.4 are effective from the point of view of their computational complexity, as they are not based on classical approaches to the construction of the forecast value based on polynomials, but use a fast algorithm for calculating the forecast one step ahead. In this work, the algorithms are presented in a form that is easy for their software implementation.

The further perspectives of the authors' research are the analysis of hybrid approaches, for example, a combination of polynomial and pyramidal predictions. In the future, we can use our approach in the study of processes such as drug transport processes in tumors or the dynamics of a viscous stratified fluid (Lyashko and Nomirovskii 2003) to predict individual characteristics based on experimental data.

8.10. References

Bakas, N.P. (2019). Numerical solution for the extrapolation problem of analytic functions research. *AAAS Research*, 3, 1–27.

Brezinski, C. and Redivo-Zaglia, M. (2020). *Extrapolation and Rational Approximation: The Works of the Main Contributors*. Springer, Cham.

Demiris, N., Lunn, D., Sharples, L.D. (2011). Survival extrapolation using the polyweibull model. *Statistical Methods in Medical Research*, 24(2), 287–301.

Guliyev, V.S., Ghorbanalizadeh, A., Sawano, Y. (2018). Approximation by trigonomepolynomials in variable exponent Morrey spaces. *Analysis and Mathematical Physics*, 9(3), 1265–1285.

Hyndman, R.J. and Kostenko, A.V. (2007). Minimum sample size requirements for seasonal forecasting model. *Foresight*, 7, 175–184.

Israfilov, D.M. and Testici, A. (2018). Some inverse and simultaneous approximation theorems in weighted variable exponent Lebesgue spaces. *Analysis Mathematica*, 44(4), 475–492.

Klyushin, D.A., Lyashko, S.I., Lyashko, N.I., Bondar, O.S., Tymoshenko, A.A. (2020). Generalized optimization of processes of drug transport in tumors. *Cybernetics and Systems Analysis*, 56(5), 758–765.

Kostinsky, A.S. (2014). On the principles of a spline extrapolation concerning geophysical data. *Reports of the National Academy of Sciences of Ukraine*, 2, 111–117.

Lyashko, S.I. and Nomirovskii, D.A. (2003). Generalized solutions and optimal controls in systems describing the dynamics of a viscous stratified fluid. *Differential Equations*, 39(1), 90–98.

Makridakis, S. and Bakas, N. (2016). Forecasting and uncertainty: A survey. *Risk and Decision Analysis*, 6(1), 37–64.

Monroe, J.I., Hatch, H.W., Mahynski, N.A., Shell, M.S. (2020). Extrapolation and interpolation strategies for efficiently estimating structural observables as a function of temperature and density. *The Journal of Chemical Physics*, 153(14), 144101.

Turbal, Y., Bomba, A., Turbal, M., Sokh, A., Radoveniuk, O. (2019). Pyramidal method of extrapolation for short time series. *International Journal of Computing Science and Mathematics*, 10(6), 525.

Turbal, Y., Bomba, A.T.M., Alkaleg Hsen Drivi, A. (2021). Some aspects of extrapolation based on interpolation polynomials. *Physico-Mathematical Modelling and Informational Technologies*, 33, 175–180.

Turbal, Y., Shlikhta, G., Turbal, M., Turbal, B. (2023). The polynomial forecasts improvement based on the algorithm of optimal polynomial degree selecting. *Eastern-European Journal of Enterprise Technologies*, 5(4(125)), 34–42.

Wang, L.Y. and Lee, W.C. (2015). One-step extrapolation of the prediction performance of a gene signature derived from a small study. *BMJ Open*, 5(4), e007170.

Wang, J., Chen, H., Li, Y., Wu, Y., Zhang, Y. (2016). A review of the extrapolation method in load spectrum compiling. *Journal of Mechanical Engineering*, 62(1), 60–75.

Zhan, Z., Fu, Y., Yang, R.J., Xi, Z., Shi, L. (2012). A Bayesian inference based model interpolation and extrapolation. *SAE International Journal of Materials and Manufacturing*, 5(2), 357–364.

Zhao, S., Yang, K., Xintie, X. (2011). Reconstruction from truncated projections using mixed extrapolations of exponential and quadratic functions. *Journal of X-Ray Science and Technology*, 19(2), 155–172.

9

Transformer with BPE Tokenization for Analysis of Interactions of Chemical Substances and Proteins

M. ZOZIUK[1], P. KRYSENKO[2], S. DOVGIY[2], V. MAKAROV[2], Y. YAKIMENKO[3] and D. KOROLIOUK[1,2,3]

[1]*CIMETA, Interdepartmental Center for Comparative Medicine, Alternative Techniques and Aquaculture, University of Rome Tor Vergata, Italy*

[2]*Institute of Telecommunications and Global Information Space of the National Academy of Sciences of Ukraine, Kyiv, Ukraine*

[3]*National Technical University of Ukraine "Igor Sikorsky Kyiv Polytechnic Institute", Kyiv, Ukraine*

In modern natural language processing (NLP), tokenization is a key step that significantly affects the performance of transformer-based models. Byte Pair Encoding (BPE) has become a standard approach due to its ability to effectively balance between vocabulary size, model quality and data processing volume. This chapter explores the application of BPE in bioinformatics, in particular for the analysis of protein sequences and chemical compounds. The proposed universal transformer-based model, which uses BPE to represent chemical compounds (in SMILES format) and protein sequences (amino acid chains), demonstrates its ability to predict biomolecular interactions. The application model opens up new opportunities for the application of NLP methods in the development of therapeutics and the analysis of biological information, due to the ability of the model to adapt to a wide range of tasks without the need for deep modification.

Computational Methods and Mathematical Modeling in Cyberphysics and Engineering Applications 2, coordinated by Dmitri KOROLIOUK, Sergiy LYASHKO and Nikolaos LIMNIOS.

9.1. Introduction

In modern natural language processing (NLP), the choice of an efficient tokenization is an important factor for building high-performance transformer models. Among the many tokenization methods, Byte Pair Encoding (BPE) has become the standard due to its ability to provide a balance between text processing efficiency, dictionary size and model performance. This method, proposed by Sennrich et al. (2016), was originally developed for machine translation tasks but quickly found application in many other aspects of NLP, in particular in training transformers such as GPT or BERT.

The main advantage of BPE is that it combines the advantages of tokenization at the symbol and word levels. This allows the model to reduce the dictionary size without losing the ability to handle rare words, by dividing them into subsequences that occur frequently in the text (Vaswani et al. 2017). Kudo (2018) shows that the use of subdictionary methods, in particular BPE, improves the quality of neural machine translation due to improved generalization and processing of various lexical forms.

BPE in bioinformatics research lies in its ability to work efficiently with high-level hierarchical structures, such as nucleotide or amino acid sequences. Similar to NLP, BPE allows us to reduce the size of the dictionary while preserving key information about rare subsequences. This is especially important for tasks where some parts of genomic or protein data have low frequency but high biological interest (e.g. mutations or rare motifs).

Predicting the interaction of chemical compounds with proteins is one of the key tasks in bioinformatics and chemical biology, as it underlies the development of new drugs and the understanding of molecular mechanisms. Traditionally, molecular docking methods or classical machine learning algorithms have been used for such tasks, but in recent years, deep learning, in particular transformer models, has shown significant progress in modeling the interactions of biomolecules (Rao et al. 2020).

One of the main problems in applying transformers to biological tasks is effective tokenization, which allows the model to work with sequences of chemical compounds and proteins. The representation of chemical compounds as linear text records, such as SMILES (Simplified Molecular Input Line Entry System), and protein amino acid sequences provide unique opportunities for integrating NLP methods into this field (Ofer et al. 2021). In this context, BPE has proven to be an extremely effective tool for reducing the size of the vocabulary and preserving important chemical and biological structures (Chithrananda et al. 2020).

The article investigates how different tokenization methods affect the performance of transformers when analyzing biological sequences such as DNA, RNA or proteins. Approaches ranging from character tokenization (each character is a separate token) to more complex ones such as k-mers or BPE are considered. It is shown that the choice of tokenization method affects the performance of the model: detailed approaches store more information but increase the size of the sequence, while generalized ones reduce the amount of data but may lose important features. The article presents experimental results and recommendations for the optimal choice of tokenization depending on the task and available resources.

We propose a universal model that can be used without detailed specification of the model for a dataset or task, i.e. using the model for training and prediction without deep modification of the model or system. For universal tokenization, the best option is BPE, since its use maintains a balance between the quality of encoding and the amount of memory that the information will occupy. For the transformer model, the GPT model was used, since it is better suited for a wide range of tasks and data.

This chapter presents a transformer-based model in combination with BPE tokenization for predicting the interactions of chemical compounds and proteins. It is shown that the representation of chemical data in the form of SMILES and protein sequences in the form of amino acid chains in combination with BPE and the transformer architecture allows for the analysis of various types of interactions between components of biological systems of any complexity.

9.2. Methods and data

The STITCH database was used for training and testing (Kuhn et al. 2008). The use of the STITCH database has disadvantages in the context of training an artificial intelligence model and obtaining the ability to predict the influence of a chemical component on genetic expression. This is because the interaction values range from 100 to 1,000, which means the strength of the influence. This database summarizes all experimental information collected from various sources. The number of unique chemical elements in the STITCH database is 372,820 and the number of unique genes is 13,858. The number of their combinations is 6,311,815, when 5,166,539,560 are possible, i.e. only ~0.0012% of the possible combinations have been measured experimentally. This is not taking into account how many possible combinations of all known chemical compounds and genes are possible. When creating a dataset for training, it is necessary to give combinations of marks – labels, but if with the designation of known combinations (~0.0012%) everything is simple (we designate, for example, as "1"), then with the designation of unknown ones everything is more complicated. Here, there is no possibility of designating them as "0," because we do not know their interaction. If we train only on known combinations and see the distribution of influence values, it will turn out that the deviation from the average value is quite low – Figure 9.1, which indicates that there are no statistically significant differences in the influence forces, depending on the chemical compound and protein sequence. Despite this, we have the opportunity to establish the presence of correlations, since the deviations are present.

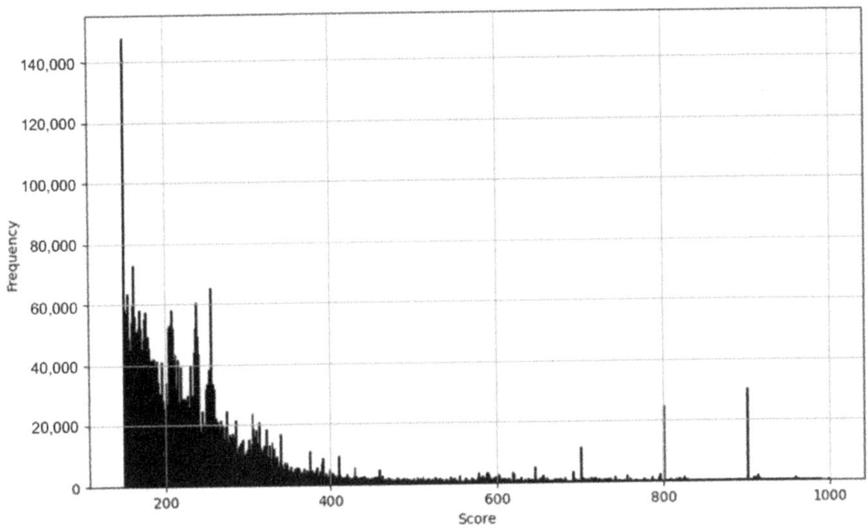

Figure 9.1. *Distribution of the values of the interaction force of a chemical component and a protein. The value of influence is on the X axis. The number of experiments is on the Y axis*

To represent information about protein sequences and chemicals, the BPE algorithm was proposed. This algorithm transforms a sequence of size N into a matrix of size MxV, where V is the dictionary size of the BPE model, and M is the length of the encoded sequence. In this work, a dictionary of size 384 was used. Two different models with the same dictionary size were trained and used for chemical and protein sequences. All sequences with a size greater than 5,000 elements were skipped during training due to the quadratic complexity of the transformer architecture.

9.3. The model's architecture

The architecture of the original transformer (Figure 9.2) is based on a decoder and encoder for amino acid and chemical sequences. The encoder block encodes the chemical sequence – encoded by the BPE algorithm into its own matrix representation. This representation is passed to the cross-attention layer of the mechanism in the Decoder layer, which also encodes the protein sequence, which is also represented by encoding by the BPE algorithm. After obtaining the representation of the protein sequence in the form of a sequence of vectors taking into account the chemical substance, this sequence is processed by the gated recurrent unit (GRU) cell to hold the chemical interaction representation vector. This vector is passed through the layers of a fully connected network with rectified linear unit (ReLU) activation functions and the last layer with one neuron to predict the scalar value combined_score (the presence and strength of the interaction).

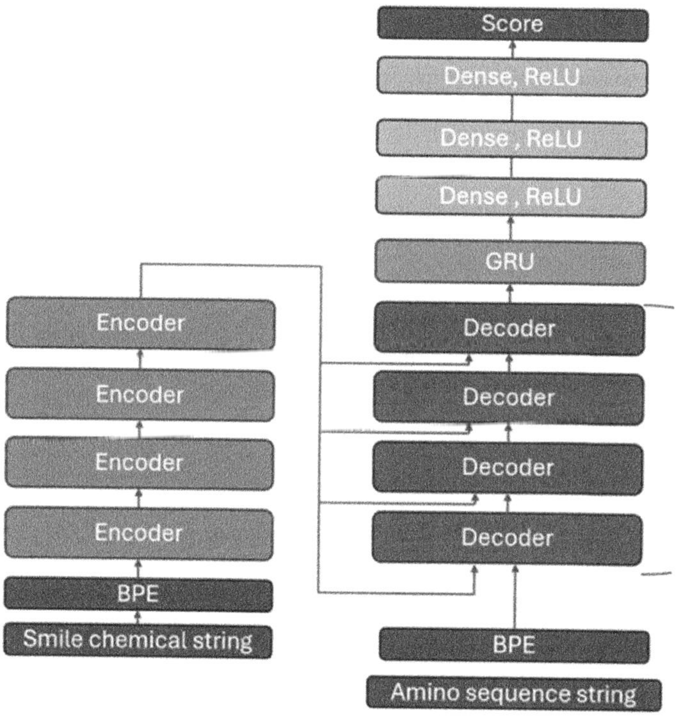

Figure 9.2. *Architecture of an artificial intelligence model for predicting the presence and strength of a chemical compound–protein interaction. BPE – byte pair encoding method. GRU – gated recurrent units. For a color version of this figure, see www.iste.co.uk/koroliouk/computational2.zip*

The architecture of both encoder and decoder layers is presented in Figure 9.3. It contains successive layers of multi-head attention, a fully connected network, normalization and residual connections. The difference between the encoder and decoder is that for the decoder, the representation from the encoder is used for the query and key values.

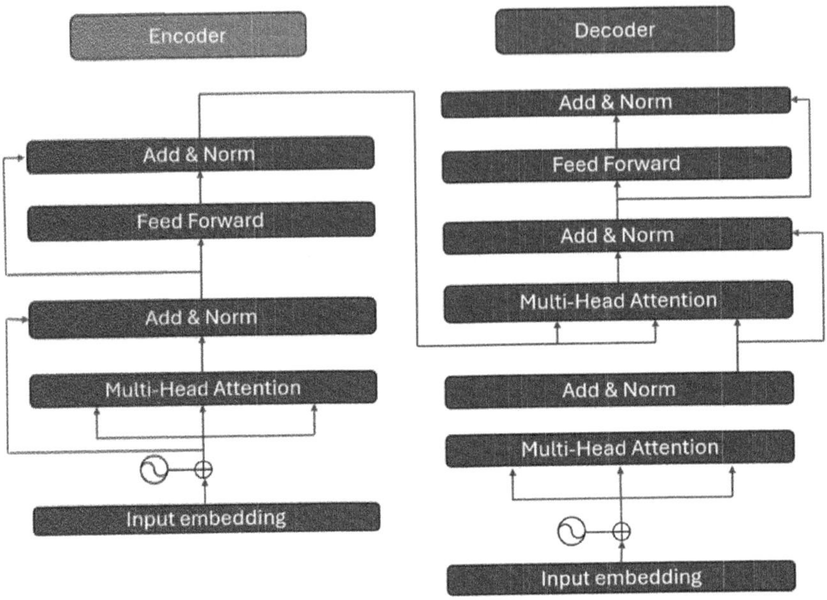

Figure 9.3. *Schematic of the encoder and decoder architecture for processing tokenized data on smiles sequences and amino acid sequences. A similar scheme was used in Zozyuk et al. (2023). For a color version of this figure, see www.iste.co.uk/koroliouk/computational2.zip*

It is worth noting that the architecture is universal, because any symbolic values (such as biological sequences) can be studied in interaction with numerical ones. The system itself performs tokenization and training. The researcher only has to choose the data and combine the dependent parameters. The architecture is also original, as it is designed for a wide range of tasks and data without the need for deep modification of the model parameters or the architecture of the entire system for specific conditions.

9.4. The model's training

The model parameters are determined in Table 9.1.

Layer (type)	*Param #*
Encoder	88,576
Decoder	156,032
GRU	18,816
Dense_16	1,664
Dense_17	32,896
Dense_18	129
dropout_10	0
layer_normalization_20	128

Table 9.1. *The model parameters represented in tabular format. The number of parameters is sufficient and balanced to obtain high-quality training and relatively short training and generation time*

A dataset with 300,000 unique rows indicating protein, chemical bond and combined_score was used for training. The dataset is reduced, as the training was conducted on a regular server without using large supercomputers, and this was enough to test the capabilities of the model. The dataset is balanced and was tested for a large number of combinations to ensure that there were no discrepancies between the samples. For training and validation, the dataset was split into two parts with a sample ratio of 80:20. The Adam algorithm was chosen as the optimizer. The dropout system was used to combat overfitting. The dimension of multi-head attention is 32, the fully connected network inside the encoder and decoder is 64, the number of layers of the encoder and decoder is 4, and the number of heads in multi-head attention is 4. The dimension of the GRU cell is 64; the dimensions of the layers in the fully connected network are 256, 128, 1 according to the scheme.

The training result was a mean square error of 0.017 on the validation dataset. The training graph is shown in Figure 9.4.

The training graph shows that the network is learning and is able to generalize the knowledge gained during training. The retraining process begins at epoch 35. Let us note the parameters of the training process. The time required for training is 10 hours on an RTuzX 4060, testing on one sequence is 1 second, but the testing time strongly depends on the length of the input sequences, both chemical and protein, due to the quadratic complexity of the Transformer architecture itself.

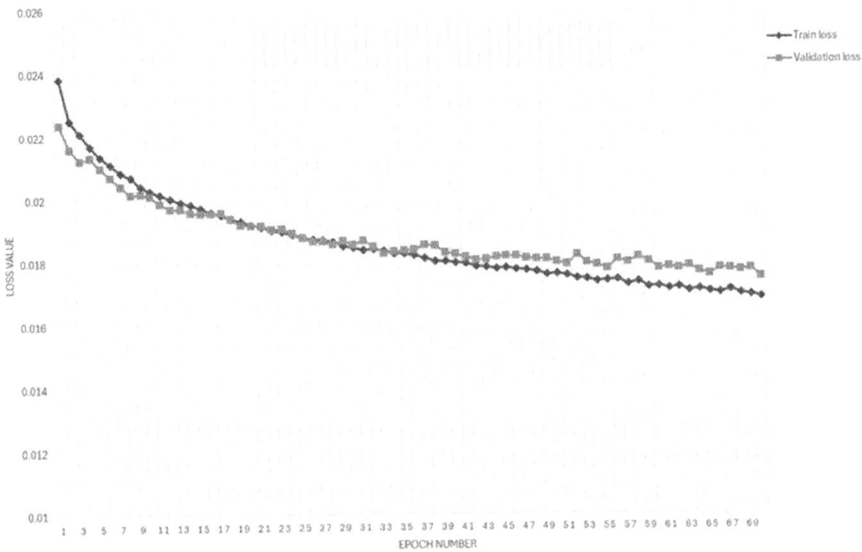

Figure 9.4. *Model training graph for 70 epochs. Red indicates the loss in predicting the presence and strength of interactions for test combinations of sequences, and blue indicates training data. For a color version of this figure, see www.iste.co.uk/ koroliouk/computational2.zip*

We emphasize that using other tokenization methods, the time would be disproportionately longer (Dotan et al. 2024), but the accuracy would not change significantly. Since this dataset was used as an example, we emphasize that the main advantage is the ability to use this system for all types of tasks and data (especially biological due to the presence of a large amount of symbolic information and the need to study correlations), where one of the system outputs is numerical and the other is symbolic. Depending on the need, the ability to change input and output data is significantly simplified by adding an automatic system for determining the data type and setting the optimal encoding method (most often this is BPE tokenization, since it is universal for the symbolic data type). The application model is presented in Figure 9.5. The system is simplified for the convenient adjustment of parameters to specific research needs.

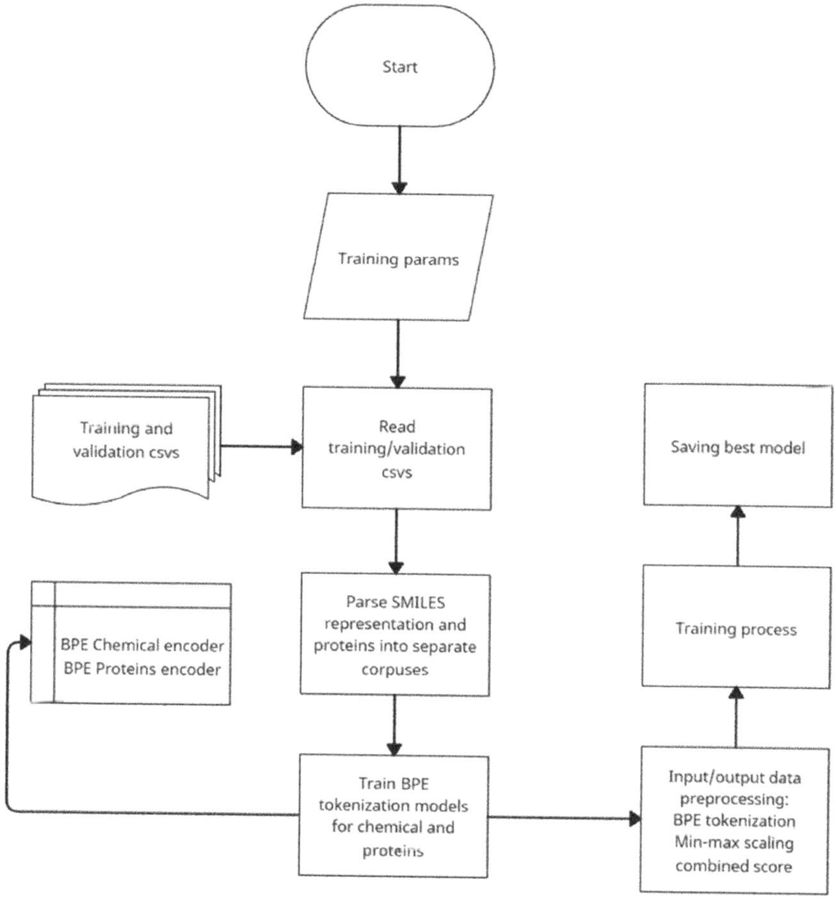

Figure 9.5. *Application model that was used to analyze the possibility of predicting the interaction of a chemical compound and proteins, or the effect of a chemical compound on genetic expression. The model covers all stages from data loading and encoding to storing the neural network model for future use in prediction*

The same application model has been used to analyze short non-coding miRNAs in cancer organisms and their effects on the genetic expression of a wide range of genes (Koroliouk et al. 2024; Zoziuk et al. 2025).

9.5. Conclusion

The effectiveness of using transformers with BPE tokenization for analyzing the interaction of chemical compounds and proteins has been studied. It is shown that the representation of chemical compounds in the SMILES format and protein sequences in the form of amino acid chains in combination with BPE provides an optimal balance between preserving key information and the size of the dictionary. The use of transformer architecture, in particular the GPT model, allows us to effectively solve a wide range of problems due to its versatility and high performance.

Experiments have shown that such a combination of methods contributes to the generalization and accuracy of the analysis, with a mean square error of 0.017 on the validation dataset. However, from the 35th epoch, an overtraining process was observed, which emphasizes the importance of hyperparameter optimization. The results confirm the promising nature of BPE for bioinformatics tasks, ensuring efficient data processing even in complex systems. The universality of the model is the main advantage of the system, especially for biological/medical data, where there is a large amount of related information, and a need to establish correlations for further use of this knowledge.

9.6. References

Ashish, V., Noam, S., Niki, P., Jakob, U., Llion, J., Aidan, N.G., Łukasz, K., Illia, P. (2017). Attention is all you need. In *Proceedings of the 31st International Conference on Neural Information Processing Systems (NIPS'17)*. Curran Associates Inc., Red Hook, NY, 6000–6010.

Chithrananda, S., Grand, G., & Ramsundar, B. (2020). ChemBERTa: Large-scale self-supervised pretraining for molecular property prediction. *arXiv*, abs/2010.09885.

Dan, O., Nadav, B., Michal, L. (2021). The language of proteins: NLP, machine learning & protein sequences. *Computational and Structural Biotechnology Journal*, 19, 1750–1758. doi: 10.1016/j.csbj.2021.03.022.

Dotan, E., Jaschek, G., Pupko, T., Belinkov, Y. (2024). Effect of tokenization on transformers for biological sequences. *Bioinformatics*, 40(4), btae196. doi: 10.1093/bioinformatics/btae196.

Koroliouk, D., Mattei, M., Zoziuk, M., Montesano, C., Bernardini, R., Potestà, M., Wondeu, L.D., Pirrò, S., Galgani, A., Colizzi, V. (2024). Artificial Intelligence and MicroRNA: Role in cancer evolution. In *Digital Ecosystems: Interconnecting Advanced Networks with AI Applications*, Luntovskyy, A., Klymash, M., Melnyk, I., Beshley, M., Schill, A. (eds). Springer, Cham, 229–254. doi: 10.1007/978-3-031-61221-3_11.

Kuhn, M., von Mering, C., Campillos, M., Jensen, L.J., Bork, P. (2008). STITCH: Interaction networks of chemicals and proteins. *Nucleic Acids Res.*, 36(Database issue), D684–D688. doi: 10.1093/nar/gkm795.

Rico, S., Barry, H., Alexandra, B. (2016). Neural machine translation of rare words with subword units. In *Proceedings of the 54th Annual Meeting of the Association for Computational Linguistics (Volume 1: Long Papers)*, 1715–1725. Association for Computational Linguistics, Berlin.

Roshan, R., Joshua, M., Tom, S., Sergey, O., Alexander, R. (2012). Transformer protein language models are unsupervised structure learners. *bioRxiv*, 12(15), 422761. doi: 10.1101/2020.12.15.422761.

Taku, K. (2018). Subword regularization: Improving neural network translation models with multiple subword candidates. In *Proceedings of the 56th Annual Meeting of the Association for Computational Linguistics (Volume 1: Long Papers)*, 66–75. Association for Computational Linguistics, Melbourne.

Zoziuk, M., Colizzi, V., Krysenko, P., Mattei, M., Bernardini, R., Zanzotto, F.M., Marini, S., Koroliouk, D. (2025). Plant miRNAs for improved gene regulation in a wide range of human cancers. *Curr. Issues Mol. Biol.*, 47, 42. doi: 10.3390/cimb47010042.

Zozyuk, M., Koroliouk, D., Krysenko, P., Yurikov, A., Yakymenko, Y. (2023). Prediction of characteristics using a convolutional neural network based on experimental data on the structure and composition of metamaterials. *Stat., Optim. In. Comput.*, 11(3), 730–739. doi: 10.19139/soic-2310-5070-1707.

List of Authors

A.V. ANIKUSHYN
Taras Shevchenko National University of Kyiv
Ukraine

V. BOHAIENKO
V.M. Glushkov Institute of Cybernetics of the NAS of Ukraine
Kyiv
Ukraine

A. BOMBA
National University of Water and Environmental Engineering
Rivne
Ukraine

O.S. BONDAR
Taras Shevchenko National University of Kyiv
Ukraine

V.M. BULAVATSKY
V.M. Glushkov Institute of Cybernetics of the NAS of Ukraine
Kyiv
Ukraine

S. DENYSOV
Harbour.Space Institute of Technology
Barcelona
Spain

S. DOVGIY
Institute of Telecommunications and Global Information Space of the National Academy of Sciences of Ukraine
Kyiv
Ukraine

Kh.M. HRANISHAK
Taras Shevchenko National University of Kyiv
Ukraine

O.F. KASHPUR
Taras Shevchenko National University of Kyiv
Ukraine

D.A. KLYUSHIN
Taras Shevchenko National University of Kyiv
Ukraine

D. KOROLIOUK
CIMETA
Interdepartmental Center for Comparative Medicine, Alternative Techniques and Aquaculture
University of Rome Tor Vergata
Italy
and
Institute of Telecommunications and Global Information Space of the National Academy of Sciences of Ukraine
Kyiv
and
National Technical University of Ukraine "Igor Sikorsky Kyiv Polytechnic Institute"
Kyiv
Ukraine

P. KRYSENKO
Institute of Telecommunications and Global Information Space of the National Academy of Sciences of Ukraine
Kyiv
Ukraine

N. LIMNIOS
Université de Technologie de Compiègne
Sorbonne University Alliance
Compiègne Laboratory of Applied Mathematics (LMAC)
France

S. LYASHKO
Department of Computational Mathematics
Taras Shevchenko National University of Kyiv
Ukraine

V.S. LYASHKO
V.M. Glushkov Institute of Cybernetics of the NAS of Ukraine
Kyiv
Ukraine

V. MAKAROV
Institute of Telecommunications and Global Information Space of the National Academy of Sciences of Ukraine
Kyiv
Ukraine

V.L. MAKAROV
The Institute of Mathematics of the National Academy of Sciences of Ukraine
Kyiv
Ukraine

O. MARCHENKO
V.M. Glushkov Institute of Cybernetics of the NAS of Ukraine
Kyiv
Ukraine

V.V. ONOTSKYI
Taras Shevchenko National University of Kyiv
Ukraine

T. SAMOILENKO
National Technical University of Ukraine "Igor Sikorsky Kyiv Polytechnic Institute"
Kyiv
Ukraine

G. V. SANDRAKOV
Taras Shevchenko National
University of Kyiv
Ukraine

V. SEMENOV
Taras Shevchenko National
University of Kyiv
Ukraine

M. TURBAL
National University of Water and
Environmental Engineering
Rivne
Ukraine

Y. TURBAL
National University of Water and
Environmental Engineering
Rivne
Ukraine

Y. YAKIMENKO
National Technical University of Ukraine
"Igor Sikorsky Kyiv Polytechnic
Institute"
Kyiv
Ukraine

M. ZOZIUK
CIMETA
Interdepartmental Center for Comparative
Medicine, Alternative Techniques
and Aquaculture
University of Rome Tor Vergata
Italy

Index

1D, 72, 80, 81, 83–86
2D, 72, 80, 82–86
3D, 72, 80, 82, 83, 86, 92

A, B

AMLI preconditioner, 46, 47, 54, 67
amino acid
  chains, 289, 291, 298
  sequences, 290, 294
Banach spaces, 226, 227, 229, 231, 241, 243–246
bioinformatics, 289, 290, 298
biopolymers, 182
black box problem, 133
Byte Pair Encoding (BPE), 289

C, D

Caputo
  –Fabrizio derivatives, 159, 174, 175
  derivative, 169, 170, 174, 175
    fractional, 157, 159, 170
Cauchy problem, 4, 10, 38, 71–73, 75–78, 86
  –Buniakovsky inequality, 7
Chebyshev functions, 248, 263
chemical compounds, 289–292, 298
cone-shaped microneedle(s), 72, 86
convex–concave, 224
Coulomb–Mohr criterion, 42
Crank–Nicolson scheme, 10, 18, 32, 185
differential model, 15, 70
DS-algorithm, 78, 81
dynamic consolidation of water-saturated soils, 18, 24, 25, 29, 167

F, G

FEM polynomials, 9, 14
Fourier heat conduction, 156
fractional-differential, 155, 157–159, 170, 174, 175, 178, 179
Frechet differentiable functional, 240, 241
Galerkin
  method, 17, 37, 185
  solution, 3, 15
Galileo group, 156
geoporous medium, 157
GPT model, 291, 298

H, I, K

Hilbert space(s), 136, 139, 153, 186, 191
homogenization
  methods, 96, 97, 129
  theory, 96
inequality
  Cauchy–Schwarz, 198, 212, 214, 215
  Schwarz, 193, 194, 196, 204, 212

initial-boundary value problem, 1, 3, 5, 10, 15, 21, 22, 26, 27, 31, 34, 78, 97–100, 108, 112, 125, 129, 182–184, 189
  axisymmetric, 14
  homogenized, 104, 114
interpolant, 134, 136, 139, 140, 143, 145–147, 150
interpolation polynomials, 134, 138, 143, 145
Kergin interpolation, 143, 151

L, M

Lagrange
  interpolation problem, 139
  polynomial, 138, 147
Laplace transform(s), 98, 101, 104–107, 109, 112–115, 120, 121, 125, 126
Lipschitz
  constant, 223, 226, 227, 231, 232, 243
  continuous operator, 223–225, 229, 231, 235, 239–243
  boundary, 98, 107, 109
Machine learning (ML), 248
memory effect, 97, 132
meromorphic Fredholm family, 122
Mittag–Leffler function, 161–163, 170
monotone
  maximal, 223, 224, 227–231, 235, 241, 243, 246
  operator(s), 224, 225, 228, 229, 231, 240, 245, 246
monotonicity, 260, 262, 264
Moore-Penrose matrix, 136

N, P

natural language processing (NLP), 289, 290
Neumann problems, 107, 110, 185
Newton–Gregory, 272
non-isothermal moisture transfer, 1, 2
parabolic–hyperbolic system(s), 15, 26
Pascal triangle, 254
periodic photonic crystals, 95
periodicity cell, 98
photonic crystals, 95
polymeric soluble shell, 71
polymers, 181, 182
polynomial(s)
  cubic, 280, 284
  generalized, 263, 264
  interpolation, 248–250, 254, 256, 257, 259, 260, 287
protein sequences, 289, 291, 292, 298
pyramidal
  method, 249, 271, 283, 286
  microneedle(s), 72, 73, 80, 81

R, S

Runge–Kutta method, 78, 81
safety factor, 42, 46
Schrödinger equation, 95
SMILES format, 289, 298
soluble microneedles, 71
space(s)
  –time, 178
  Euclidean, 133, 143, 153
  Hilbert, 136, 138, 139, 144, 150, 153
STITCH database, 291
stress–strain state of soil massif, 15, 46
Sturm–Liouville problem, 161

T, V, W

Terzaghi effective stress principle, 19, 24, 167
transdermal drug transfer, 71
viscoelastic
  materials, 181, 182, 219, 220
  media, 181
Volterra, 181, 183, 185–188, 220
  -type, 183, 187, 188
Weierstrass
  M-test, 164, 168, 173, 177
  theorem, 134, 152, 153

14961010-7-Text.pdf TPS: 235x156 Plates: 0 Spine: 17.00

and bound by CPI Group (UK) Ltd, Croydon, CR0 4YY

13/09/2026

14961010-0007